THE STRUCTURE OF AMERICAN INDUSTRY

THE STRUCTURE OF AMERICAN INDUSTRY

NINTH EDITION

WALTER ADAMS
Trinity University (Texas) and Michigan State University

JAMES W. BROCK
Miami University (Ohio)

EDITORS

PRENTICE HALL
Englewood Cliffs New Jersey 07632

Library of Congress Cataloging-in-Publication Data

The structure of American Industry / Walter Adams, James W. Brock
  editors.—9th ed.
    p. cm.
  Includes index.
  ISBN 0-02-300833-4
  1. United States—Industries.   I. Adams, Walter, 1922 Aug. 27–
II. Brock, James W.
  HC106.S85   1995
  338.0973—dc20   94-6773
                                              CIP

Editor-in-Chief: Valerie Ashton
Assistant Editor: Teresa Cohan
Production Supervisor: Dora Rizzuto
Production Managers: Alexandra Odulak and Kurt Scherwatzky
Text Designer: Eileen Burke
Cover Designer: Eileen Burke
Cover Art (Background): Westlight

© 1995 by Prentice Hall, Inc.
A Simon & Schuster Company
Englewood Cliffs, New Jersey 07632

Previous edition's copyright © 1990

Printed in the United States of America

10 9 8 7 6 5 4 3 2 1

0-02-300833-4

Prentice-Hall International (UK) Limited, *London*
Prentice-Hall of Australia Pty. Limited, *Sydney*
Prentice-Hall Canada Inc., *Toronto*
Prentice-Hall Hispanoamericana, S.A., *Mexico*
Prentice-Hall of India Private Limited, *New Delhi*
Prentice-Hall of Japan, Inc., *Tokyo*
Simon & Schuster Asia Pte. Ltd., *Singapore*
Editora Prentice-Hall do Brasil, Ltda., *Rio de Janeiro*

For
Kenneth E. Boulding
(1910–1993)
A pioneer who chose not to tinker
with celestial mechanics for a
nonexistent universe

CONTENTS

PREFACE

The 1980's provided an ideal laboratory for testing rival theories of industrial organization. It was a decade of the debt-laden, junk-powered corporate deal. Some 30,000 mergers, takeovers, and leveraged buyouts were consummated. Their annual value skyrocketed from an estimated $33 billion in 1980 to more than $250 billion by 1989, and the cumulative value during the decade exceeded one *trillion* dollars.

Debt fueled this razzle-dazzle deal making, and junk bonds provided an explosive additive. Total outstanding corporate debt more than doubled over the decade. Junk debt jumped even more dramatically—spurting by more than 2,500 percent. Interest payments required to service these staggering debt loads matched their frantic growth: More than half of corporate earnings were consumed by interest payments during the 1980s, compared to just 16 percent over the postwar period prior to 1970.

The administration in Washington responded to this massive merger movement with benign indifference, if not active encouragement. It declared a virtual moratorium on enforcement of the anti-merger provisions of the Clayton Act and the anti-monopoly provisions of the Sherman Act. "Merger activity in general," said William F. Baxter, President Reagan's first chief of the Justice Department's Antitrust Division, "is a very important feature of our capital markets by which assets are continuously moved into the hands of managers who can employ them efficiently," Interfering with mergers "would be an error of substantial magnitude." The President's Council of Economic Advisers concurred in this judgment. It testified "that mergers and acquisitions increase national wealth. They improve efficiency, transfer scarce resources to higher valued uses, and stimulate effective corporate management. They also help recapitalize firms so that their financial structures are more in line with prevailing market conditions. In addition, there is no evidence that mergers and acquisitions have, on any systematic basis, caused anticompetitive price increases."

Toward the end of the decade, however, it became increasingly apparent that these claims were overly optimistic. Entire industries were bankrupted by merger-induced debt. Notable among department stores are Campeau Corporation (including its Bloomingdale's, Burdines, Lazarus, and Jordan Marsh subsidiaries); Bonwit Teller and B. Altman; Garfinckel's; Carter Hawley Hale (including its Broadway, Emporium, and Weinstock's subsidiaries); Macy's; and Ames. Among airlines, the list includes Continental Airlines,

Eastern Airlines, Trans World Airlines, Braniff Airlines, and Northwest Airlines. Textile firms with decimated balance sheets include J. P. Stevens, Burlington, West Point-Pepperell, and Fieldcrest Cannon.

Observers—both within and outside the business community—pointed out that corporate size *per se* does not necessarily guarantee operational efficiency, technological progressiveness, or international competitiveness. Reflecting on the travails of General Motors, for example, Elliott M. (Pete) Estes, its former president, ruefully confessed: "Chevrolet is such a big monster that you twist its tail and nothing happens at the other end for months and months. It is so gigantic that there isn't any way to really run it. You just sort of try to keep track of it." GM's current chairman, John G. Smale, concurs in this diagnosis: "Bureaucracy seems to replace the free form of pure enterpreneurialism. And the size and complexity of the bureaucracy grow disproportionately with the growth of the business itself. Individual efforts become focused internally rather than externally, with more and more emphasis on internal processes and detailed analytical reports and presentations—and less emphasis on interaction with the customer and less urgency in anticipating and responding to changes in the competitive arena." No wonder that the conservative London *Economist* offers a Draconian remedy: break up this bureaucratic giant into separate, independent operating units that can function efficiently.

The *Economist* offers the same advice to IBM—i.e., to do what the Justice Department sought (unsuccessfully) to accomplish with its 1969 antitrust case: break up the company into separate units which could survive in the hurly-burly competition of the computer industry. Why? As the *Economist* puts it, "If bigger were better in research and development, IBM would be the best. It spends $6 billion a year on R&D. Its researchers won two Nobel prizes in the 1980s, and thickened IBM's portfolio of patents to over 33,000. Yet, despite all that, IBM's new technologies have time and again been beaten to market by those of smaller, nimbler firms." Says *Business Week*: "Competition and technological innovation from upstarts such as Sun Microsystems Inc. and Microsoft Corp. are forcing an inefficient supplier—and an enormous one at that—to trim down, be more competitive, and serve its customers better."

By the end of the decade, voluntary divestiture and downsizing had become standard operating procedure among *Fortune*'s 500. The bloom was off the rose of the synergies which ostensibly justified the creation of monstrous conglomerates. The newly rediscovered corporate strategy—promulgated by Big Business executives and Wall Street mavens—was deconglomeration! Why? To boost productivity and efficiency. In the words of Paramount's president, "Bigness is not a sign of strength. Indeed, just the opposite is true."

Transcending these developments, and dramatically underscoring their economic significance, is the defining event of the late-1980s: the stunning collapse of the Soviet Union and its satellite states. Above all else, the centrally planned Soviet state was founded on a faith in the virtues of economic giantism—a faith, as Engels put it, that economic society is best organized as "one giant factory." Lenin, like Marx, admired the turn-of-the-century American trusts and regarded them as superior economic organisms. He accepted without question the notion that "huge enterprises, trusts, and syndicates had brought the mass production technique to the highest level of development." Giant industrial organizations, he believed, unleashed forces for efficiency and modernization while, at the same time, "regulating in a planned way the production of commodities on which millions of people depend." The wise policy, Lenin believed, was not to resist industrial concentration but to actively promote it, which he and Stalin did by "liquidating" small firms and farms, and by consolidating them into centrally administered giants. Guided by this cult of bigness, the Soviet state thus became the ultimate conglomerate, operating every major sector of the economy, from food and crop production to the manufacture of shoes, refrigerators, and aircraft. The catastrophic failure of that system offers lessons relevant not only to the task of reconstructing the devastated economies of central and eastern Europe, but also to policy makers in the market economies of the United States and western Europe. The Soviet experience should remind us, once again, of the dangers inherent in concentrated economic and political power.

In the context of all these events, the ninth edition of this book seems felicitously timed. It offers a kaleidoscopic view of American industry—a collection of case studies illustrating different types of market structures, different histories and behavioral patterns, different performance records—with an emphasis throughout on international comparisons. Each industry case study offers the student of industrial organization a "live" laboratory for clinical examination, comparative analysis, and the evaluation of public policy issues and options. Each enables the student to engage as much in induction as in deduction. As such, we hope the book constitutes a useful supplement, if not a necessary antidote, to the economist's penchant for the abstractions of theoretical model building.

WALTER ADAMS & JAMES W. BROCK

CONTRIBUTORS

Walter Adams is Vernon F. Taylor Distinguished Professor of Economics at Trinity University (Texas) and former President of Michigan State University.

James W. Brock is Bill R. Moeckel Professor of Economics and Business at Miami University (Ohio).

William S. Comanor is Professor of Economics at the University of California, Santa Barbara.

Kenneth G. Elzinga is Cavaliers' Distinguished Teaching Professor and Professor of Economics at the University of Virginia.

Manley R. Irwin is Professor of Economics Emeritus at the University of New Hampshire.

Barry R. Litman is Professor of Telecommunications at Michigan State University.

Cecil Mackey is Professor of Economics and former President of Michigan State University.

Christian Marfels is Professor of Economics at Dalhousie University, Halifax, Canada.

Stephen Martin is Professor of Economics at the European University Institute, Florence, Italy.

Neil B. Niman is Associate Professor of Economics at the University of New Hampshire.

Stuart O. Schweitzer is Associate Professor of Community Medicine and Economics in the School of Public Health at the University of California, Los Angeles.

William G. Shepherd is Professor of Economics at the University of Massachusetts, Amherst.

Daniel B. Suits is Professor of Economics Emeritus at Michigan State University.

1

AGRICULTURE

Daniel B. Suits

I. INTRODUCTION

As the supplier of most of the food we eat and raw materials for industry, agriculture is clearly an important sector of the economy. But its importance extends even beyond this. In nations where farmers are unproductive, most of their workers are needed to grow food, and few can be spared for education, production of investment goods, or other activities required for economic growth. Indeed, among nations, one of the factors that correlates most closely with rising per capita income is the declining fraction of the labor force engaged in agriculture. In the poorest nations of the world, 50 to 80 percent of the population lives on farms, compared with less than 10 percent in western Europe and less than 1.8 percent in the United States.

In short, economic development in general depends in a fundamental way on the performance of farmers. This performance, in turn, depends on how agriculture is organized and on the economic context, or market structure, within which agriculture functions.

II. MARKET STRUCTURE AND COMPETITION

NUMBER AND SIZE OF FARMS

There are about 2 million farms in the United States today. This is roughly a third of the peak number reached seventy years ago, and as the number of farms has declined, the average size has risen. U.S. farms now average about 500 acres, but this average can be misleading. In fact, modern American agriculture is characterized by large-scale operations. Although only about 5 percent of all farms contain 1,000 or more acres, these include more than 40 percent of total farm acreage. Nearly a quarter of all wheat, for example, is grown on farms of 2,000 acres or more, and the largest 2.6 percent of wheat growers raise roughly half of all our wheat.

1

Size of farm varies widely by product, but even where the typical acreage is small, production is concentrated. Nearly 65 percent of tomatoes are grown on farms smaller than 500 acres, but the remaining 35 percent is marketed by the largest 9 percent of tomato growers. Broiler chickens are raised on still smaller farms, but more than 70 percent of all broilers are produced by the largest 2 percent of growers.

Farm size also varies with production technique, climate, and other factors. In the southern United States, 60 percent of cotton is grown on farms with fewer than 1,000 acres, whereas farms that small produce only a third of the cotton grown in the more capital-intensive western states.

In this age of large-scale commercial agriculture, the family farm, once the American ideal, is no longer common. Only about half of all present-day farmers earn their livelihood entirely from farm operations; the others must supplement their farming with other jobs. Moreover, large-scale agriculture is increasingly carried out by corporations. Although only 2 percent of all farms are incorporated, corporations own 12 percent of all land in U.S. farms and market 22 percent of the total value of all farm crops.

Corporate farms are especially important in states like California, where they operate a quarter of all acreage in farms and market 40 percent of the value of all field crops (including nearly 60 percent of all California sweet corn, vegetables, and melons). But even in a state like Kansas, over a third of all farm products are grown by corporations.

COMPETITION IN AGRICULTURE

Despite the scale and concentration of production, modern agriculture remains an industry whose behavior is best understood in terms of the theory of pure competition. Although production is concentrated in the hands of a small percentage of growers, the total numbers are so large that the largest 2 or 3 percent of the growers of any particular product still constitute a substantial number of independent farms. For example, although half of all the grain grown in the United States comes from the largest 2 percent of producers, this 2 percent consists of some 27,000 farms. Numbers like this are a far cry from those for manufacturing. There are fewer than 280 firms producing men's work clothing and only about 200 cotton-weaving mills, but both these industries are widely recognized as highly competitive. Thus, even if we ignore the competitive influence exerted by the thousands of smaller farms producing each crop, we are still talking about nearly one hundred times as many independent producers as can be found in the most competitive manufacturing industries.

In addition, the number and size of existing farms are only partial indicators of competitiveness. An important additional consideration is the ease with which new farms can enter to compete with those already producing. Not only are there no special barriers to setting up and operating a new farm, but many existing farms are adapted to a variety of crops and can easily shift production from one to another on the basis of the outlook for prices and costs.

As a result of this competitive structure, even large modern farms are too small a part of the total to influence price or total output through their own individual action. Each can decide only how much of which crops to grow and by what methods. When the combined result of these thousands of individual decisions reaches the market, the prices are determined by total volume in conjunction with demand.

DEMAND FOR FARM PRODUCTS

Another important element in the structure of agricultural markets is the nature of the demand for farm products. Potatoes are fairly typical and can be used as a convenient illustration.

1. DEMAND FOR POTATOES.   In Figure 1-1, the average farm price of potatoes in the United States is plotted vertically against the annual per capita consumption, measured horizontally. Each point represents the data for a recent year. The downward drift of the points from the upper left to the lower right confirms the everyday observation that people tend to buy more at low than at high prices. At the high price of $5.93, for example, the average consumption of potatoes shrank to 151 pounds per person in 1989, whereas at the low price of $3.71, the consumption expanded to 167 pounds in 1991. Of course, as a glance at Figure 1-1 reveals, price is not the sole influence on buying habits. Consumption during 1989 was somewhat greater and during 1991 somewhat smaller than would have been expected from the price of potatoes alone. Part of this variation can be traced to changes in buyers' incomes and part to changes in the prices of other foods that can be substituted for potatoes in the diet. Some of the variation is associated with changes in consumer tastes for potatoes, connected with the shifting popularity of such things as packaged mashed potatoes or "fries" at fast-food outlets.

By using the appropriate statistical procedures, it is possible to allow for the effects of many of these influences and to estimate the effect of price alone. The result is shown by the curve DD drawn through the midst of the observations. Such a curve, called a *demand curve*, represents the quantity of potatoes that buyers would be expected to purchase at each price, other influences held constant.

2. DEMAND ELASTICITY.   The response of buyers to changes in price is measured by the *elasticity of demand*, which expresses the percentage change in quantity purchased to be expected from a 1 percent rise in price.

The elasticity of demand for particular products is readily estimated from fitted demand curves, by selecting two prices close together and reading the corresponding quantities from the curve. The elasticity is then calculated by

FIGURE 1-1   Demand for potatoes.

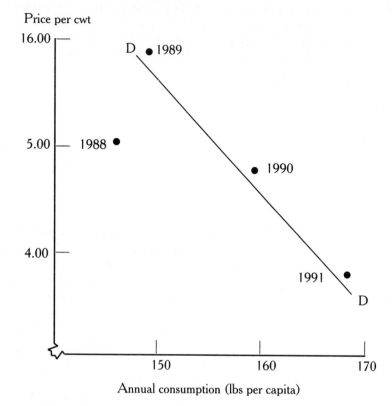

Annual consumption (lbs per capita)

the ratio of the percentage difference between the two quantities to the percentage difference between the two prices. Applying this procedure to the demand curve of Figure 1-1 yields an estimated elasticity of demand for potatoes of about -0.3.

We are rarely interested in such exact measurment of elasticity, but we often need a general idea of how elastic the demand for a given product is. For this purpose, it is convenient to place demand curves into broad catagories, using an elasticity of 1, called *unit elasticity*, as the dividing point. Demands with an elasticity smaller than 1 (in absolute value) are referred to as *inelastic demands*. Demand curves with an elasticity greater than 1 in absolute value are termed *relatively elastic*. Accordingly, the demand for potatoes, with an elasticity of -0.3, is *inelastic*. The demand for lettuce, estimated to have an elasticity of about -2.8, is *relatively elastic*.

3. DIFFERENCES IN ELASTICITY.   Buyers' response to price varies widely among products, depending on the characteristics of the product and the buyers' attitudes toward it. Things like potatoes, which most people view as food staples, have inelastic demands. Buyers feel they need a certain amount in their diet and so are reluctant to cut back when the price rises. By the same token, since they are already consuming about as much as they feel they need, they find use for only little more when the price falls.

By contrast, products viewed as luxuries exhibit relatively elastic demands, for their consumption can easily be reduced when prices rise, yet buyers are delighted at the chance to enjoy more when lower prices permit. Among farm products, the demands for fruits and fresh vegetables tend to be relatively elastic. The elasticity of demand for fresh peaches, for example, has been estimated at -1.49, five times that of potatoes. This high elasticity reflects the ease with which households can do without peaches when prices are high, and the welcome accorded the fruit when it becomes cheap.

Demand elasticity also depends on the product's relationship to others. Products that can be replaced by good substitutes tend to have relatively elastic demands. Even small increases in price lead large numbers of buyers to desert the product in favor of cheaper substitutes. This is one reason that demands for fresh vegetables tend to be relatively elastic. The high elasticity of demand for fresh peas, estimated to be -2.8, is largely due to the many other vegetables that can be eaten instead when peas are expensive.

These factors are reflected in the demand elasticities given in Table 1-1. We see that the demands for staple products like potatoes and corn tend to

TABLE 1-1 Elasticity of Demand for Selected Farm Products

Product	Elasticity of Demand	
	Price	Income
Cabbage	-0.25	n.a.
Potatoes	-0.27	0.15
Wool	-0.33	0.27
Peanuts	-0.38	0.44
Eggs	-0.43	0.57
Onions	-0.44	0.58
Milk	-0.49	0.50
Butter	-0.62	0.37
Oranges	-0.62	0.83
Corn	-0.63	n.a.
Cream	-0.69	1.72
Fresh cucumbers	-0.70	0.70
Apples	-1.27	1.32
Peaches	-1.49	1.43
Fresh tomatoes	-2.22	0.24
Lettuce	-2.58	0.88
Fresh peas	-2.83	1.05

n.a. = not available.

Source: Estimated by the U.S. Department of Agriculture.

be inelastic, whereas many individual fresh fruits and vegetables have highly elastic demands, partly because they are deemed to be less essential and partly because each has many close substitutes to which consumers can turn.

4. ELASTICITY OF DERIVED DEMANDS. A particularly important aspect of demand for farm products is that most are purchased from the farm by canners, millers, and other manufacturers who process the raw product before it reaches its final consumers. Even products sold for fresh consumption incur transportation, packaging, and retailing costs before they can be delivered to the table.

As shown in Table 1-2, only about 25 percent, on average, of the retail price of food items purchased in the United States consists of their original farm value. The other 75 percent is value added by processing and marketing, but these percentages vary widely among products. Farm value makes up about 65 percent of the retail price of products like meat and eggs, which reach the table fairly directly, but the farm value of the barley, rice, hops, and other ingredients constitutes only a small fraction of the retail price of a can of beer.

The small proportion of farm value in the retail price of food contributes to the low elasticity of demand for farm products. To see why this is so, consider frozen peas, a processed product with a demand elasticity of about -2. This means that a 5 percent reduction in the retail price of frozen peas tends to increase consumption by about 10 percent. But if frozen peas are typical, farm value constitutes only about 25 percent of the final retail price, and a reduction of 5 percent in the farm price of raw peas will produce no more than a 1.25 percent reduction in the retail price of frozen peas. Given the elasticity of -2, this will stimulate no more than about 2.5 percent greater sales of frozen peas, so the final result of the 5 percent reduction in farm price will be the purchase of only 2.5 percent more peas from farmers. This corresponds

TABLE 1-2 Shares in Final Retail Value of Food Products

	Billions of Dollars	Percentage
Final retail value	361.1	100
Processing and marketing costs		
Labor	123.7	34
Rail and truck transport	16.8	5
Power, containers, & other costs	119.8	33
Corporate profits	12.0	3
Farm value of products	89.0	25

Source: U.S. Department of Agriculture, *Agricultural Statistics, 1987* (Washington, DC: U.S. Government Printing Office, 1987).

to an elasticity of only -0.5 at the farm level, despite the highly elastic retail demand for frozen peas.

This relationship holds for all derived demands. In general, the smaller the farm share of the retail price is, the lower the elasticity of derived demand for the product tends to be, and demand at the farm level tends to be inelastic even when the retail demand for final product is relatively elastic.

5. COMMODITIES WITH SEVERAL USES. The elasticity of demand depends on what a product is used for. When products are used for more than one purpose, demand elasticity varies among the different uses, depending on whether buyers consider the use a "necessity" or a "luxury" and on whether the product has good substitutes for that use.

For example, wheat is used both for bread and for poultry and livestock feed. In its use for bread, wheat is a "necessity," and because of its gluten content, it has no good substitute. Indeed, recipes for most "rye," "corn," and other "nonwheat" bread call for the addition of wheat flour to impart cohesiveness to the dough. As a result, the demand for wheat by flour mills is quite inelastic. As a feed grain, however, wheat is readily replaced by corn, oats, sorghum grains, and other substitutes, and so the demand for wheat as feed grain is relatively elastic.

Statistical estimates bear out this difference. Researchers at the U.S. Department of Agriculture estimate the elasticity of demand for wheat to be ground into flour at only -0.2, but that for wheat to be used as feed grain at -3.

The overall elasticity of demand is the weighted average of elasticities for individual uses, with the weights proportional to the quantity purchased for each use. Because so much wheat is used for flour, its overall demand is quite inelastic, despite the high relative elasticity of its use in animal feed.

6. ELASTICITY AND ALLOCATION OF AVAILABLE CROPS. When the output of any crop declines, its price rises and buyers reduce their consumption, but not all uses are cut back equally. The greatest reduction occurs where the elasticity of demand is highest: in the less essential uses or in those in which the product can be readily replaced by close substitutes. Conversely, when the output increases, the price falls and buyers expand their consumption, but buy proportionally more for less essential uses or where the now cheaper product can replace expensive substitutes.

This principle can be seen in Table 1-3. Comparison of the allocation of a large wheat crop with that of a small crop shows that when the supply rose 7 percent, the consumption of wheat for human food increased only 3 percent, compared with a 50 percent increase in wheat used for animal feed. These proportions are about what would be expected from the difference in demand elasticity in the two uses.

TABLE 1-3   Allocation of Wheat in the United States (in Millions)

Year	Total Supply	Human Food	Annimal Feed
1985	952	678	274
1986	1111	700	411

Source: U.S. Department of Agriculture, *Agricultural Statistics, 1987* (Washington, DC: U.S. Government Printing Office, 1987).

OTHER FACTORS AFFECTING DEMAND

In addition to price, purchases are affected by population, incomes, prices of substitutes, consumer tastes, and—for things like soybeans that have important industrial uses—industrial technology. The influence of these factors is generally represented by shifts in the position of the demand curve. When, for any reason, consumers begin to buy more than they did before at given prices, this increased demand is represented by a shift of the demand curve to the right. A reduced demand is represented by a shift of the curve to the left. Curve DD in Figure 1-1, for example, shows the location of the demand for potatoes at the average for the period. The point representing 1989, however, lies on a curve shifted considerably to the right, and the point corresponding to 1991 lies on a curve shifted to the left.

INCOME ELASTICITY.   As with prices, the effect of income on buying can be expressed in terms of elasticity. The income elasticity of demand is the percentage change in quantity bought at given prices that occurs in response to a 1 percent rise in real income. For example, it is estimated that a 1 percent rise in real per capita income increases potato purchases by only 0.15 percent. This is expressed as an income elasticity of 0.15. (Unlike price elasticity, most income elasticities are positive, because the consumption of most commodities rises with income.)

An income elasticity of 1 is characteristic of a commodity whose purchase expands in proportion to income. The consumption of commodities with an income elasticity of demand of less than 1 expands more slowly than does income. An income elasticity below 1 is generally characteristic of staples like potatoes that even low-income families consume in quantity. The consumption of commodities with an income elasticity greater than 1 grows more rapidly than does income. An income elasticity greater than 1 is characteristic of luxury products whose purchase expands rapidly when rising income makes them affordable.

The income elasticity for many farm products is given in Table 1-1. Demands for basic staples like potatoes and onions have a low income elasticity, whereas cream, fruit, and fresh vegetables are characterized by a high income elasticity.

Rising income affects the composition of demand more than it does the total amount of food consumed. As incomes rise, families consume more expensive food, but less of other food. For example, families with high incomes eat, on the average, nearly three times as much sirloin steak as do families in lower brackets, but the high-income families' consumption of meat of all types is only 20 percent greater, and the difference in total food consumption is even less.

DEMAND ELASTICITY AND FARM INCOMES

A low elasticity of demand has important consequences for the behavior of farm incomes. Unlike most manufactured goods, which are priced first with production adjusted to whatever sales materialize, farm crops are grown first and then placed on the market for whatever prices they can bring. Because prices reflect the size of the crop, normal year-to-year variations in growing conditions generate year-to-year price fluctuations, whose magnitude depends on demand elasticity.

Inelastic demands make farm prices rise and fall more than in proportion to changes in output, so the total dollar value of a smaller crop is greater than that of a larger crop. This can be tested in Figure 1-1: According to the demand curve DD, the production of 140 pounds of potatoes per capita would bring a price of $6.18 per hundred pounds, making the crop worth about $8.65 per consumer, whereas the production of 170 pounds per capita would reduce the price to $3.82, and the crop would be worth only $6.48 per consumer. In short, an increase of 21 percent in the size of the crop would reduce its total value to farmers by about 25 percent.

For this reason, natural year-to-year fluctuations in growing conditions make farming very much a "boom or bust" proposition. Poor growing conditions in one area means severe losses for the farms affected but high incomes and prosperity for the others. Good growing conditions yield bumper crops but low prices and lower incomes for everybody.

III. SUPPLY

Just as demand represents the behavior of buyers in relation to prices, income, and other factors, *supply* represents the behavior of producers in relation to prices and costs. There is, however, an important difference between demand and supply, for although buyers tend to adapt quickly to new conditions, producers often require time to revise plans and production schedules or to acquire new facilities and equipment. For this reason, it is useful to distinguish three different situations.

HARVEST SUPPLY—THE VERY SHORT RUN

Once crops are mature and ready for market, the maximum total quantity available is fixed, and no action by farmers can generate an output beyond that total. Nevertheless, the maximum available is rarely harvested, for it seldom pays to strip fields so thoroughly that every last particle is collected. Some crops mature over a period of several weeks, and so growers must decide when the time is best for harvest and whether it is worthwhile to return to the fields for a second harvest a week or so later. Some crops can be harvested cheaply and quickly, but with greater loss of product than would be available from a slower, more careful harvest. Clearly, high prices at harvest time make it profitable to harvest a larger proportion of the potential crop, whereas low prices make careful harvest unprofitable and often lead to the outright abandonment of low-yield acreage when harvesting would cost more than the crop would be worth.

Although there is some flexibility in the quantity of a given crop that is marketed, the relatively low cost of harvesting (roughly 20 percent of variable cost) severely limits variation in marketing. Once a crop is ready for market, the harvested supply is extremely inelastic.

SHORT-RUN SUPPLY AND PRODUCTION COSTS

Short-run supply embodies the relationship between year-to-year variation in the quantities that farmers decide to plant and the prices they expect the crop to bring when it is harvested. These planting decisions depend on production costs.

1. COST STRUCTURE.   The production costs of farming—like those of any other business—are of two general types. Some costs are fixed regardless of output. Such *fixed costs* include taxes, interest on farm mortgage, depreciation of equipment, and similar expenses that are not affected by the amount of acreage planted and that would remain even if production were abandoned entirely.

Variable costs are zero as long as nothing is produced but rise sharply as soon as it is decided to undertake production. This sharp initial increase includes the costs of planning, acquiring materials, and other costs that would not be incurred if the farm were left idle. An important part of this initial variable cost consists of the labor of the farm owner and family members or the salaries of the managers of corporate farms.

Once these start-up costs have been incurred, output can be expanded by relatively small increases in outlay for seed, fertilizer, herbicides, labor, fuel, and similar variable costs. These costs rise slowly as more bushels are produced, but there is a limit to the output available from given facilities, and as this capacity limit is approached, further output is possible only with sharply rising variable costs. Additional output can be achieved only by more intensive care of the crop or by application of extra fertilizer or more of other inputs.

2. COSTS OF A CORN GROWER.   The variable costs of corn raised on an Iowa farm are given in Table 1-4. In keeping with modern farming methods,

TABLE 1-4 Variable Costs per Acre of Corn, Central Iowa Farms

Item	Quantity	Cost per Acre (1971 $)
Preharvest costs		
Labor (including owners)	3.78 hr	5.82
Seed	0.23 bu	3.22
Fertilizer and lime		
Nitrogen	100 lb	5.40
Phosphorus	22 lb	4.58
Potassium	19 lb	.99
Lime	0.23 ton	.91
Fuel, lubricants, repairs		3.95
Insecticides		2.18
Herbicides		2.24
Custom work		.85
Hail insurance		.30
Interest on operating expenses		1.01
Preharvest variable cost		31.45
Harvest costs		
Labor	2hr	3.08
Fuel, lubricants, repairs		1.81
Custom-hired harvesting and trucking		2.85
Other harvest expenses		1.98
Harvest cost		9.62
Variable cost per planted acre		41.07

Source: U.S. Department of Agriculture, Economic Research Service, *Selected U.S. Crop Budgets, Yields, Outputs and Variable Costs: North Central Region,* vol. 2 (Washington, DC: U.S. Government Printing Office, 1971).

the cost of labor is low, with chemical herbicides used in place of labor-intensive weed control. Commercial fertilizer is used to maintain high yields with little or no crop rotation. All together, the variable cost amounted to $41.07 per acre.

The source from which the data are taken gives no indication of fixed costs; however, on a national average, the fixed costs of farming amount to about a third of the variable cost. On this basis, we can estimate the total cost of corn grown in central Iowa at about $62.50 per acre.

3. AVERAGE AND MARGINAL COSTS. Total cost is translated into average and marginal costs in Figure 1-2. The *marginal cost* (MC) is the rate at which total cost rises as production is increased. Once start-up costs have been incurred and production is under way, additional corn can be raised for little more than the cost of the seed and materials needed to cultivate additional

FIGURE 1-2 Average and marginal costs of a corn grower (1971 dollars).

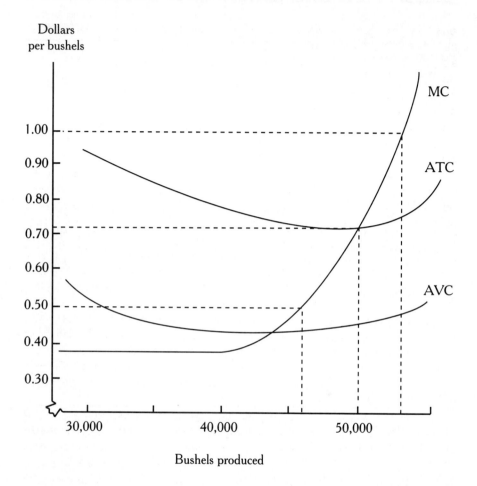

Bushels produced

acreage. This keeps the marginal cost low until production presses against the farm's capacity. As this capacity is approached, greater and greater outlays are needed to extract additional output, and so the marginal cost rises more and more sharply.

The *average variable cost* (AVC) is high at low levels of production because start-up costs are spread over limited output. As production expands, start-up costs are spread over more and more bushels of corn, pulling down the average variable cost per bushel. As production enters the range where the marginal cost rises steeply, the average variable cost stops falling and begins to rise.

Because fixed costs do not change as output expands, the average fixed cost is inversely proportional to the output, regardless of production. The *av-*

erage total cost (ATC) is merely the sum of the average variable and average fixed costs.

4. PROFIT MAXIMIZATION AND SUPPLY ELASTICITY.

According to a familiar proposition in competitive theory, the most profitable production plan for a competitive firm is the output that brings the marginal cost into equality with the expected price, provided only that the expected price is high enough to cover the minimum average variable cost at which the farm can operate. At an expected price of $0.50 per bushel, the most profitable output plan for the farm in Figure 1-2 would be about 46,500 bushels. If a price of $0.72 were expected, production would be profitably increased to 50,000 bushels, and a price of $1.00 would raise the most profitable output to 52,300 bushels.

The cost curves are consistent with the data in Table 1-4 and are typical of agricultural production. The marginal cost rises so sharply near capacity output that even wide year-to-year price swings exert little influence on the output of any individual grower, at least as long as he continues in operation. In other words, if the farm operates at all, it will produce very nearly the capacity output obtainable from land and facilities. As Figure 1-2 shows, even a 100 percent price increase—from $0.50 to $1.00—would induce the farmer to add only about 12.5 percent to his planned production. This output response amounts to a supply elasticity of only about 0.1.

But growers continue in production only as long as they expect prices high enough to cover their average variable costs. Fixed costs have already been sunk into the business and will continue whether or not anything is planted. Variable costs, on the other hand, are not incurred until the farmer decides to produce. Therefore, if there is no prospect of recovering all variable costs, it is better to keep the money.

According to Figure 1-2, the production of 52,300 bushels at a price of $1.00 would entail an average total cost of about $0.74 per bushel. This $0.26 per bushel difference would mean a profit of nearly $13,600 above cost. The production of 46,500 bushels at a price of $0.50 would entail an average total cost of about $0.73 per bushel. The difference of $0.23 per bushel amounts to a loss of about $10,700 for the year. Even so, this would be better than shutting down the farm entirely, for with no production at all, the entire fixed cost of $13,500 would be lost. Because the $0.50 price is above the minimum average variable cost (about $0.44 per bushel), the farmer will be $2,800 richer if he produces at a loss rather than shuts down his operation.

If, however, the expected price fell below $0.44 to, say, $0.40, the marginal cost would be equal to the price at an output of about 43,000 bushels and an average total cost of about $0.76. The operation would generate a loss of nearly $15,500, some $2,000 more than the farm would lose if it shut down entirely and settled for the loss of all fixed costs.

Because farms in production tend to operate close to capacity regardless of price, the principal supply response to falling prices occurs when some farmers find the crop no longer profitable. Similarly, the principal supply response to rising prices comes when farmers find prices moving back into the profitable range and so again take up production.

Variable costs differ widely among growers, depending on soil type, climate, length of growing season, and skill of the producer. During the same year that corn growers in central Iowa incurred an average variable cost of $0.47 per bushel, Nebraska farms produced corn at average variable costs ranging from $0.49 in the lowest cost districts to $0.71 in the highest-cost area, and a similar cost variation was characteristic of all other crops.

Figure 1-3 shows the striking variation in average costs among U.S. cotton growers. Each point on the curve shows the percentage of total cotton output contributed by growers whose average costs were lower than that indicated on the vertical axis.

The distribution of average variable costs among growers gives a good indication of the elasticity of the short-run supply of cotton. For example, practically all cotton grown in the United States that year came from growers with average variable costs below $0.40, but only 92 percent of growers had average variable cost below $0.30. We can therefore estimate that a 25 percent reduction in price from $0.40 to $0.30 would cause about an 8 percent reduction in cotton planting. This corresponds to a supply elasticity of about 0.3.

Another source of supply elasticity is the shifting composition of the output of multiproduct farms. For several reasons, many farms produce more than one crop. For example, many hog farmers also raise grain for feed. Some farmers raise crops that ripen at different times, in order to spread harvesting over a longer period and to use harvesting machinery more efficiently. Crop rotation is another reason for multicrop farming, and some farmers produce several crops to insure against natural calamities that might affect the yield of any particular crop and to protect against unfavorable market conditions for any one crop.

In any case, the proportions in which different crops are grown vary in response to the expected prices. The expectation of, for instance, high prices for soybeans relative to the price of corn leads farms that grow both to plant more soybeans and less corn. The expectation of cheap corn and high-priced hogs leads hog raisers to increase the number of hogs and to buy, rather than to grow, the extra feed. Such responses contribute to the elasticity of supply of the products involved.

MARKET EQUILIBRIUM

The interaction of short-run supply with demand governs the year-to-year behavior of production and prices.

Equilibrium price and output comprise a kind of "target" for the market, indicating the values toward which the actual price and quantity are continually pushed, but it should be understood that the year-to-year price and quantity are rarely observed at equilibrium values. For one thing, supply deals with

FIGURE 1-3 Percentage of total U.S. upland cotton crop raised by growers with average costs below those specified. Each point on the curve represents the percentage of U.S. cotton production (on the horizontal axis) raised by growers whose average costs were below the figure given on the vertical axis.

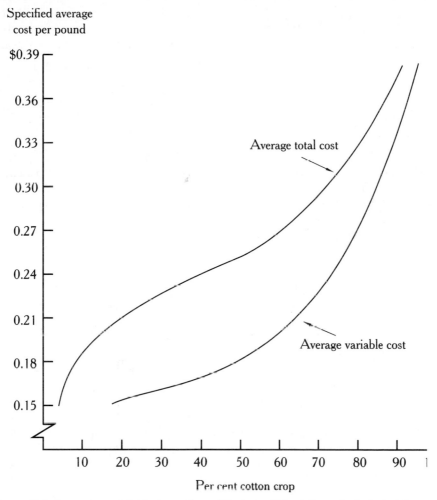

Source: U.S. Department of Agriculture, *Costs of Producing Upland Cotton in the United States* (Washington, DC: Economic Research Service, 1967).

production *plans* rather than actual outcome. The most a farmer can do is to plant and cultivate his crop in a manner calculated to earn the greatest profit under normal conditions. The actual outcome, however, depends on the vagaries of weather, insect damage, blight, and other growing conditions.

In addition, planting decisions are based on *expected* prices as they depend on expected demand, but consumers frequently do the unexpected. A shortage of substitutes, an unanticipated rise in popularity, or the introduction of new industrial uses may increase demand and prices above expectation. Likewise, an unexpected decline in demand may lead to prices below expectation. An example of the resulting year-to-year oscillation in price and output is shown in Figure 1-4.

LONG-RUN SUPPLY

THE ROLE OF AVERAGE TOTAL COST. Long-run supply depends on the ability of farmers to recover the total cost of production. Farms that cover variable costs but fail to recover all their fixed costs operate at a loss but can still remain in business by not repairing buildings and equipment and by digging into savings to meet family living expenses. Sooner or later, however, the time will arrive when buildings become unusable or equipment wears out. At

FIGURE 1-4 Price and output of onions (percentage of 1982–1991 mean).

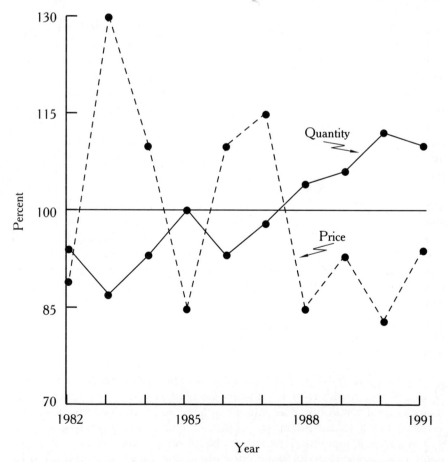

Source: Calculated from data in U.S. Department of Agriculture, *Agricultural Statistics, 1993.*

this point, farmers must decide whether to continue in operation. Unless the prospects are strong enough to promise not only the recovery of additional investment but a satisfactory profit as well, it is obviously better to shut down rather than to put more money into a losing proposition.

Long-run supply depends on the distribution of average total costs among growers and is considerably more elastic than short-run supply. This can be easily demonstrated in terms of Figure 1-3. We have already estimated the elasticity of short-run supply of cotton at about 0.3. When we make the same calculation for long-run supply, we find that although the price of $0.30 would be high enough to allow 92 percent of cotton farmers to cover their variable cost, only about 70 percent would be able to recover their entire total cost. This means that 22 percent of growers would operate at a loss. If this situation persisted, these farmers would ultimately have to stop producing cotton, leaving only the profitable 70 percent in operation. Thus the long-run result of a 25 percent decline in price would be a 30 percent decline in output. This is a long-run supply elasticity of 1.2.

RESPONSE TO SHIFTS IN DEMAND.   An important aspect of industrial performance is how effectively output is adjusted to changes in demand. Because of the three different supply situations that characterize agriculture, adjusting prices and outputs involves a sequence of events that may take many years to complete. Buyers whose demand has risen find themselves initially confronted by a very inelastic harvest supply. Although rising prices signal greater demand, growers can market only the result of production plans laid many months previously. The most they can do at harvest time is to strip fields with greater care than would have been profitable at lower prices and to harvest low-yield acreage that would otherwise have been abandoned.

Although rising prices have little immediate effect on output, they nevertheless perform two important functions. First, rising prices reallocate available supplies among alternative demands. Buyers who can do so are forced to resort to (now) cheaper substitutes or to go without the product entirely. Fewer purchases by these buyers, in keeping with their highly elastic demands, leaves more of the scarce crop available for essential uses. Second,   higher prices at harvest encourage growers to raise a larger crop next season by increasing the intensity of cultivation and by shifting some fields from other crops to the more profitable use. By the time of the next harvest, these efforts will have expanded output, and so prices will drop somewhat from the peak initially reached.

In the long run, as higher prices continue, production is further expanded by those farmers who invest in additional equipment and by new farms attracted to the profitable crop. This long-run expansion is accompanied by gradually declining prices but continues as long as prices remain above the average total cost. Expansion slows, however, and approaches a halt as prices fall to levels that just cover the average total cost of the least efficient, highest-cost farms engaged in production.

This adjustment process has a profound economic significance. The production of crops requires that land, labor, and other resources be diverted from manufacturing or other uses. Consumers who want the commodity signal their willingness to provide the necessary resources by paying high prices. But time is needed to transfer resources from one use to another. At harvest time, the most growers can do is to employ a few extra workers and a little extra gasoline to squeeze as much as possible from crops already in the field.

By the next season, however, growers can expand their production and bring prices down from the peak reached at the previous harvest. Over a period of years, resources can be shifted more efficiently, and the equilibrium price will settle at the lowest level permitting all growers of the new equilibrium quantity to cover the total costs of their operation.

The response to a reduced demand is just the reverse. A reduced consumer desire for a commodity is signaled by sharply falling prices, indicating that consumers would prefer fewer resources devoted to the crop. But most resources have already been irretrievably sunk into production. The most that growers can do at this point is to divert some small amount of resources away from harvest by abandonment of low-yield acreage and by less intensive harvest of the remainder. Few resources are immediately saved, but in planning for the next season, growers who find that they can no longer cover their average variable cost may shift their labor, fuel, and materials to other crops or release them for employment elsewhere. At this point, nothing can yet be done to shift resources already sunk into farm equipment and buildings. In the long run, however, as these wear out and are not replaced, labor and other resources that would otherwise be used in fabricating farm equipment and buildings are freed for employment in more valuable uses.

IV. IMPROVEMENT IN TECHNOLOGY

It is not enough for an industry merely to move resources in response to consumer demand. It must also keep the productivity of those resources as high as possible. The best way to raise productivity is by introducing new, more efficient methods.

Evaluation of the rate of technical improvement involves two distinct questions: (1) How rapidly do firms in the industry originate and develop new methods? (2) How rapidly do they adopt new methods as they become available?

On the first count, agriculture has a poor record. The intensely competitive structure and the relatively small scale of operation characteristic of farming simply do not lend themselves to research and development. Expensive laboratories and large research budgets, commonplace in many large industrial firms, are beyond the means of even the largest grower. If improvements had to wait until they could be developed on the farm, agricultural productivity today would not be far ahead of what it was a century ago.

Fortunately, however, we need not depend on the farm to develop its own technical improvements, as part of this job has been undertaken by the laboratories of state universities and agricultural experiment stations. To a far greater extent, however, improvements have arisen from the work of farm equipment firms, chemical manufacturers, and other suppliers to modern agriculture. For although farmers originate little themselves, they do provide a ready market for improvements once they have been developed and demonstrated. The result has been a rate of growth of productivity that has outstripped the rest of industry.

THE PROFIT INCENTIVE

The strong incentive to adopt better methods is the profit available to the first growers who introduce them. A grower who can cut the cost of his 30,000 bushels of corn from $28,000 to $23,000 immediately adds $5,000 to his annual net income. Since his individual contribution to the supply of corn is negligible, his actions cannot affect the price of corn. Until others take up the new method, the entire $5,000 is pure profit.

INNOVATION AND PRICES

Unfortunately for the grower, however, the new profitable situation carries within it the seeds of its own destruction. When other growers see the cheaper method, their own eagerness for greater profit leads them to imitate it. As the new method spreads, supply increases with a consequent fall in price to a new equilibrium at the new lower cost of production.

The long-run decline in price following technical innovation has a number of important consequences. First, it means that growers must be quick to change, for exceptional profits are available only temporarily to the first growers to introduce the new method. As other farmers follow suit, the rising supply lowers prices and wipes out the extra gains.

Falling prices also mean that, ultimately, growers have no effective choice about whether or not to adopt the new method. As prices approach the new lower-cost levels, farmers still using the old, higher-cost methods are no longer able to cover the total costs of their production. They thus must adopt the new method to survive, and if they hold back too long, they will be driven out of business by losses.

Above all, as prices fall to the new lower level, the entire gain from the new method is passed on to consumers. Those growers who first adopted the new method for the sake of extra profits and those who followed along in self-defense have combined in an action that has not only increased the productivity of resources but, with prices reduced to the new average total cost, has also eliminated extra profit and has delivered the entire cost saving to society at large.

BROILER CHICKENS: AN EXAMPLE OF INNOVATION

The continual improvement in production methods is one of the most striking features of American agriculture. A good example is the revolution in the production of broiler chickens, shown in Table 1-5.

It is probably difficult for modern readers to realize that only fifty years ago, chicken was too expensive for everyday use and was generally served only on holidays and special occasions. Broiler chickens were raised on farms where they ran freely in yards competing with one another for food, with heavy losses from accident, predators, and disease and with high labor costs for care. In those days it took 16 weeks and 12 pounds of feed to raise one 3.5 pound chicken, and labor costs ran as high as 8.5 hours per 100 pounds.

Then in about 1950, a revolution began in commercial broiler production: Chickens were raised indoors in individual cages. This eliminated wasteful competition among birds for feed, reduced disease and depredation, permitted automated delivery of feed, and substantially lowered labor costs. By 1980, labor costs had been cut to barely 1 percent of their level of forty years earlier, and it now took only 7 weeks to bring a bird to market weight. The lower costs expanded the supply, with an attendant drop in market price. Indeed, adjusted for inflation, the price of chicken fell from $0.46 per pound in 1934 to $0.08 in 1990. The fall in price, assisted by the increased demand growing out of a rising population and income, raised the per capita con-

TABLE 1-5   Production Costs, Output, and Price of U.S. Broiler Chicken, 1934–1990

| Year | Production Cost per 100 lb of Chicken | | Production (lb per cap.) | Price[a] ($ per lb) |
	Feed (lb)	Worker-hours		
1934	n.a.	n.a.	0.76	0.457
1940	420	8.5	3.13	0.394
1950	330	5.1	12.82	0.342
1960	250	1.3	32.76	0.164
1970	219	0.5	52.27	0.135
1980	192	0.1	68.48	0.109
1990	n.a.	n.a.	103.44	0.083

n.a. = not available.
[a] Price of chicken divided by Consumer Price Index to adjust for inflation.

Source: U.S. Department of Agriculture, *Agricultural Statistics, 1992* (Washington, DC: U.S. Government Printing Office, 1992).

sumption of chicken in 1990 to 140 times its level in 1934. Chicken, no longer a holiday dish, has become the cheapest meat in the store, and fried chicken has become a rival of hamburger in fast-food outlets.

Table 1-6 emphasizes that the story of broiler chicken is typical of what has happened to agricultural productivity. The continual search for more profitable methods has reduced the production cost of virtually every crop. Today, it takes only 10 percent as much labor to grow an acre of corn as it did seventy-five years ago, and the acre yields four times as much corn. Three times as much wheat can be raised per acre with one-sixth the labor, and a worker-hour of effort produces nine times as many pounds of hogs, thirteen times as much milk, and one hundred times as much turkey as could be produced seventy-five years ago.

TABLE 1-6 Productivity of Labor and Land in U.S. Agriculture: Selected Crops and Livestock, 1910–1986

Crop	1910– 1914	1945– 1949	1965– 1969	1982– 1986
Corn				
labor-hr per acre	35.2	19.2	6.1	3.1
bu per acre	26.0	36.1	48.7	109.3
Wheat				
labor-hr per acre	15.2	5.7	2.9	2.5
bu per acre	14.4	16.9	25.9	37.1
Potatoes				
labor-hr per acre	76.0	68.5	45.9	32.6
cwt per acre	59.8	117.8	205.2	283.9
Sugar beets				
labor-hr per acre	128.0	85.0	35.0	20.0
tons per acre	10.6	13.6	17.4	20.4
Cotton				
labor-hr per acre	116.0	83.0	35.0	5.0
lb per acre	201.0	273.0	505.0	581.0
Soybeans				
labor-hr per acre	n.a.	8.0	4.8	3.2
bu per acre	n.a.	19.6	24.2	30.7
Milk				
labor-hr per cow	146.0	129.0	84.0	24.0
cwt per cow	38.4	49.9	82.6	127.3
Hogs				
labor-hr per cwt	3.6	3.0	1.6	0.3
Turkeys				
labor-hr per cwt	31.4	13.1	1.6	0.2

n.a. = not available.

Source: U.S. Department of Agriculture, *Agricultural Statistics* (Washington, DC: U.S. Government Printing Office, appropriate issues).

TECHNOLOGY AND SCALE OF OPERATION

Most modern technical innovations reduce farming costs by replacing labor with machinery, which lowers the variable operating costs at the expense of higher interest, depreciation, taxes, and other fixed costs. But these higher fixed costs must be spread over a larger output if they are to reduce the average total cost.

When farms grow larger, what happens to the total number in operation depends on demand. Falling prices increase total consumption, in keeping with demand elasticity, and demand rises with population and income. If demand is sufficiently elastic or if population and income grow rapidly enough, the markets for output may expand enough to maintain or even to increase the number of farms in operation, despite the larger scale.

The U.S. history is shown in Table 1-7. Before 1920, demand grew faster than farm productivity, thereby increasing the number of farms despite the larger average size. As technical improvements accelerated and the population grew more slowly after 1920, the increase in size was accompanied by a smaller number of farms. In the last thirty years, the average size of U.S. farms has doubled, but the total number has fallen by 60 percent.

TECHNOLOGY AND DISPLACEMENT OF FARM LABOR

The rising productivity of farm labor means that any given amount of product can be grown with the help of fewer people. Unless consumption expands proportionately, rising productivity will reduce the number of people needed on farms. The history of declining farm labor is shown in the last column of Table 1-7.

It is the market that keeps the number of people engaged in agriculture in balance with productivity and demand. Increasing supply reduces farm

TABLE 1-7 Number and Size of Farms and U.S. Farm Employment, 1880–1990

Year	Number of Farms	Average Acreage per Farm	Farm Employment (thousands)
1880	4,008,000	133.7	10,100
1900	5,740,000	146.6	12,800
1920	6,453,000	148.5	13,400
1940	6,104,000	174.5	11,000
1960	5,388,000	215.5	7,100
1970	2,730,000	389.5	4,200
1990	2,140,000	461.0	2,891

Sources: U.S. Department of Commerce, Census of Agriculture (Washington, DC: U.S. Government Printing Office, appropriate years). 1990 data from U.S. Department of Agriculture, *Agricultural Statistics, 1992* (Washington, DC: U.S. Government Printing Office, 1992).

prices and the earnings of farmers below what workers receive in other industries, and consequently, farm people with marketable skills leave the farm for more promising jobs elsewhere. Those without skills for other employment find themselves trapped in low-income farming or are forced off the farm into the city where, without marketable skills, they are added to welfare rolls.

The fate of farmers is an excellent illustration of two important aspects of competitive markets. Competition generates an inexorable pressure to extract a greater output from available resources and passes on the productivity gains to the consumer in the form of lower prices and higher standards of living. But in doing so, the market operates without regard for the fate or feelings of the people involved. The supply and demand for farm labor are rigorously balanced by the competitive market, regardless of what happens to the farm families caught in the adjustment. Nowhere does the market take into account the human cost of this process.

OTHER SOCIAL COSTS OF MODERN AGRICULTURE

The increasing scale of operations needed by modern farming has brought profound changes to agriculture. More than before, farming is undertaken by large corporations, and the commercial farming of specialized crops is replacing the more diversified agriculture of the family farm.

Intense specialized farming increases the danger of soil exhaustion as single-crop operations replace crop rotation. In addition, contamination by runoff from chemical fertilizers and insecticides puts water supplies at risk and jeopardizes wildlife habitat. Many crop varieties have been adapted to the requirements of mechanical harvest, storage, and shipping rather than to improved nutritional value and flavor.

Some of these problems reflect more on the tastes and preferences of consumers than they do on farm technology. If people are willing to buy cheap tomatoes with the flavor and consistency of red baseballs rather than pay a little more for flavorful tomatoes that are too delicate for anything more than hand harvesting, competitive agriculture will provide for their preferences. On the other hand, growing consumer demand for food raised without chemicals has provided incentives for many farmers to forgo herbicides, pesticides, and artificial fertilizers.

Other such problems, however, are fundamental to the nature of competitive markets, for competitive producers pay attention only to those costs that they must pay themselves. Costs imposed on others count for nothing, and if farmers can improve their own position by shifting costs from themselves onto society at large, competition will force them to do so. The competitive market provides no mechanism to balance gains in output against higher social costs.

Irrigation systems, in some areas at least, offer an example of this kind of problem. Irrigation provides a stable and ample water supply and does for the fields what a lawn sprinkler does for a home garden. The adoption of the system increases the yield of the individual farm, yet widespread adoption of the technique depletes the water table. Nonetheless, the cost of the lower water table is borne by society at large. No bill is presented to the farmer for the wider consequences of his action. Individually, farmers can take account only of the private profit potential of the method, yet the collective result is a higher cost to society.

Similar external costs are incurred when water and wildlife habitat are polluted by agricultural chemicals and when consumer health is affected by antibiotics used in animal feed. The solution of these problems is beyond the power of the competitive market: The market cannot prevent farmers from using substances that pollute. Indeed, the very forces of competition compel them to use such substances. If their use is to be controlled, therefore, that control must come from outside the market.

Part of the answer lies in action by government to limit the type and amount of spray that farmers can apply and to monitor and regulate the use of the common water table. It would have been impossible for the agricultural industry itself to have abolished the use of DDT, despite the obvious damage it was inflicting on wildlife. Instead, it took action by the government to outlaw it.

A thorough discussion of environmental pollution far exceeds the scope of this chapter, but it is important to recognize that one aspect of the behavior of any industry is its impact on the environment. In this respect, unmodified competition as exemplified by agriculture has a poor record.

V. GOVERNMENT POLICY TOWARD AGRICULTURE

PRICE SUPPORTS

The severe dislocation wrought in agriculture by untrammeled technical innovation has generated overwhelming political pressure for government to intervene. This intervention has largely taken the form of efforts to hold up farm prices. After some ineffective initial attempts by earlier administrations, systematic U.S. farm price support was initiated as part of the New Deal of the early 1930s and, although repeatedly modified in detail, retains much the same general character today.

Any attempt to support prices must somehow define a target level at which to support them and then devise ways to enforce the supports. Targets for farm prices were initially defined in terms of *parity* prices. These are prices that bear the same relationship to the index of prices farmers pay (the *prices-paid* index) as they did during a selected base period. For example, if the farm price of cabbage had been $1.10 per hundred pounds during the base period when the prices-paid index was, say, 220, then the parity price of cabbage would be defined for any other year by dividing the prices-paid index by 200. When the prices-paid index reached 300, the parity price of cabbage would be 300/200, or $1.50 per hundred pounds. Base periods have been redefined from time to time, and more complicated formulas are now applied to the calculation of parity prices for individual commodities.

Once parity prices are defined, government then undertakes to maintain the prices of a number of key commodities at a specified fraction of parity. The fraction varies from commodity to commodity and from time to time, but regardless of the level of support, one difficulty is encountered. When prices are maintained at levels higher than the equilibrium prices defined by supply and demand, farmers will produce more than consumers are willing to buy. If the program is to succeed, government must somehow contend with this gap. The problem is illustrated in Figure 1-5. At the support price P, higher than the equilibrium price P^*, farmers produce and bring to market a total quantity Q_s, exceeding the amount Q_d that consumers are willing to buy at price P. If the farm price is to be maintained at P, something must be done about this gap.

RESTRICTION OF SUPPLY

There are three alternatives: (1) Government might induce farmers to restrict supply to $S'S'$ so that the desired support price becomes the new market equilibrium; (2) government might purchase the output $(Q_s - Q_d)$, which farmers cannot sell at the support price; or (3) the entire output Q_s could be sold on the market for whatever price it could bring (P'), and government could make up the difference $(P - P')$ by a cash subsidy paid directly to the growers.

Historically, government price-support programs have involved a mixture of the three techniques. The supply of peanuts, cotton, and a few other crops is directly restricted. Each commercial grower is given a certificate permitting the cultivation of a specified number of acres and is forbidden to plant more. These acreage allotments belong to the farmers to whom they were initially assigned, and if they no longer want to plant the crop themselves, they can rent the allotment to somebody else.

Supply has also been restricted by paying farmers to "set aside" part of their acreage, to divert part of their land to "soil-conserving" crops, or to place acreage in a "soil bank." In the early days of price supports, acreage diversion was the most important policy tool. At its peak in 1966, a total of 63.3 million acres was diverted from cultivation. This amounted to nearly 6 percent of all farmland in the United States and entitled farmers to more than $1 billion in government payments

More recently, outright payments to farmers for acreage diverted have been increasingly replaced by requiring farmers to limit cultivated acreage as a criterion for receiving other price-support payments.

FIGURE 1-5   Problems of price supports.

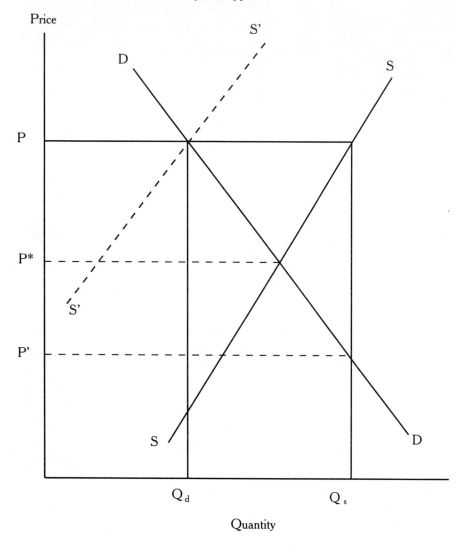

THE COMMODITY CREDIT CORPORATION

Any surplus that remains after the supply has been reduced is generally dealt with by loans issued to farmers by the Commodity Credit Corporation (CCC). The CCC is a government agency that, under the direction of the secretary of agriculture, designates each year the loan rate on each commodity to be supported. The loan rate is the amount per unit that the CCC is ready to lend to farmers who are in good standing with the price-control program. For example, when the loan rate for corn is $2.65, farmers are entitled to borrow $2.65 for each bushel of corn they grow, regardless of its actual market price. If too much corn is grown and the market price falls below $2.65, the borrowers are permitted to default on their loans and let the CCC take the corn. On the other hand, if the market price rises above $2.65, the farmers can repay their loans, reclaim their crop, and sell it at the higher price. Although the CCC was established as a lending institution, it provides a mechanism for the government to purchase farm output that cannot be sold at the price defined by the loan rate.

The CCC must, by law, extend loans on six "basic" commodities: corn, cotton, peanuts, rice, tobacco, and wheat and on a group of designated "non-basic" commodities that includes, among others, butter, honey, tung oil, rye, and wool. In addition, from time to time, as directed by the secretary of agriculture, the CCC extends price-support loans on an extensive and diverse list of crops ranging from staples such as dry peas, soybeans, and potatoes to almonds, cottonseed, and olive oil.

The volume of crops acquired by the CCC depends on loan rates in relation to production and demand. During 1991, the CCC acquired a total of $860 million of farm commodities, down from $6.7 billion in 1986.

Bumper crops can leave huge unsold surpluses in government hands. In some years the CCC has acquired more than 25 percent of the entire wheat crop. During one period thirty years ago, the CCC held what amounted to an entire year's wheat crop.

AGRICULTURAL SUBSIDIES

The third method of price support involves direct subsidy payments to make up the difference between the market price and the support level, but most price-support programs link direct subsidies to acreage limitation and CCC loans in a rather complicated package. In general, those farmers who agree to limit their acreage are offered subsidies designed to bring up to the full support level the price they will receive for a specified portion of their crop. Beyond the specified portion, their output is eligible for support by the CCC at loan rates set below the support level.

All told, the several price-support programs have absorbed huge government outlays. In 1991, the U.S. government paid $8.2 billion in direct subsidies, down from a peak of $16.7 billion in 1987.

OTHER PRICE-SUPPORT MEASURES

Not all prices are supported by governmental payments. The entire cost of the sugar support program is borne directly by consumers. Since the United States is a net sugar importer, sugar prices are supported by limiting imports to quotas assigned to forty-one sugar-producing countries.

The U.S. Department of Agriculture has estimated that the program has cost consumers about $2.3 billion a year, but the consequences of artificially high sugar prices extend well beyond the cost to consumers. Many wheat growers have found it profitable to switch their acreage to sugar beets. Perhaps the largest impact of the program has been the increasing switch of consumers— particularly beverage manufacturers—to sugar substitutes, especially high-fructose corn syrup (HFCS) and low-calorie sweeteners. As a consequence of this switch, sugar's share of the U.S. sweetener market has fallen from nearly 80 percent twenty years ago to barely more than 40 percent today.

The prices of a broad array of commodities, ranging from citrus fruit to nuts, are set by marketing orders issued by the secretary of agriculture. These orders enable producers to organize *marketing boards*, which are given wide powers to control the production and marketing of designated commodities. Aside from certain professional sports, marketing boards are the only unregulated legal monopolies permitted in the United States. The boards limit production, sometimes by restricting the quality or sizes of product that can be shipped. Some boards prop up prices by assigning quotas to individual producers and requiring that any additional output be placed "in reserve," usually to be exported at low prices. The political advantage of marketing boards is that they impose the entire cost of the program directly on consumers and avoid charges to the government budget.

The price of milk is supported by a combination of marketing order and CCC intervention. A marketing order by the secretary of agriculture sets the prices that dairies must pay farmers for fluid milk destined for human consumption. The higher this price is set, the less milk the dairies will buy, but the marketing order cannot control production. To deal with the result, all milk that farmers cannot sell at the established price is sold as "industrial-grade" milk at whatever price the market will bring. The demand for "industrial-grade" milk is supported indirectly by CCC loans extended to such manufactured dairy products as dry milk, butter, and cheese.

From time to time, the government passes on these accumulated products to poor families, to forestall extensive spoilage. The milk program has recently been modified to compensate dairy farmers who agree to reduce production below their normal output.

WHO BENEFITS FROM PRICE SUPPORTS?

Because the clear result of agricultural programs has been higher prices to consumers and higher taxes to taxpayers, one might well ask who benefits from this outlay. Presumably, the rationale for farm policy has been to relieve some of the suffering resulting from rising agricultural productivity. Yet oddly enough, practically nothing about governmental farm policy has benefited the people who need it the most.

The history of agriculture has been a continuous story of small farmers driven out and people pushed off their farms. But even most of those who remain on the farm receive few benefits from the program, because the nature of price supports directs the benefits toward the largest, most successful growers rather than toward the poor and weak who really need help.

A study made in 1982 indicated that the smallest 50 percent of all farms received less than 8 percent of total agricultural payments. This amounted to an average of only $250 per farm. In contrast, nearly 25 percent of payments went to the largest 5 percent of farms and represented an average of almost $7,000 received per farm. Payments to the few very largest farms ran to millions of dollars each!

This highly unequal distribution of benefits is a direct consequence of programs that pay for acreage diversion, that buy up unsold production, and that subsidize prices. The largest payments for acreage diversion naturally go to the farms with the most acres. When surpluses are purchased and prices are subsidized, the largest payments go to the farms with the most output. Moreover, this is only part of the picture, for consumers must pay higher prices for food. Nobody knows how much this amounts to—it has been estimated at many times the value of direct payments—but whatever it is, it is distributed among farmers in proportion to the size of their operations and so is concentrated in the hands of the largest producers.

AGRICULTURAL POLICY AND INTERNATIONAL TRADE

Since price supports maintain U.S. farm prices above levels in the world market, the government maintains the Export Enhancement Program to enable U.S. farm products to compete abroad. The program pays subsidies to make up the difference between U.S. prices and the lower prices in foreign markets. Some of these payments are made in cash, and some are made in kind from U.S. stores of surplus crops.

In addition, the government maintains a loan-guarantee program to enable nations with a poor credit standing to obtain loans to buy U.S. farm products. When such nations fail to repay, the U.S. taxpayer is left with the bill. The program provides for a total of $40 billion in guarantees, of which about $4.5 billion is in default.

Export subsidies, however, have done little to expand foreign purchases of U.S. farm products, partly because of competition by the European Economic Community which also heavily subsidizes exports. In addition, the U.S. program has been subject to great abuse. For example, some exporting firms have used subsidies to bribe foreign officials to get orders. In another case, eight tobacco exporters pleaded guilty to exporting cheap low-grade foreign tobacco to Iraq, which had obtained a guaranteed loan to purchase expensive U.S. tobacco. The difference in price enabled Iraq to use loans guaranteed by the United States (on which it has since defaulted) to help finance its military buildup for the Gulf War.

Even aside from abuses, most of the benefit of the export programs appears to accrue to a small number of very large agricultural exporting firms rather than to farmers.

The United States is by no means alone in its program to support farm prices. Most industrialized nations provide some sort of agricultural support. Canada relies on marketing boards to limit the production and maintain the prices of milk, poultry, and eggs. In some provinces, quotas assigned by the boards can be sold, and with the success of the program, the possession of a quota becomes a valuable farm asset.

A common rationale for agricultural protection is the desire to maintain national self-sufficiency in food. Mexico maintains self-sufficiency in corn by means of high tariffs that keep corn from the United States and Canada out of the Mexican market. The resulting high farm prices enable some 3 million Mexican farmers to produce corn, although most of the soil of Mexico is unsuited to the crop. The government buys the high-priced corn from the farm and then resells it to Mexican consumers at much lower prices. The subsidy costs Mexican taxpayers about $1 billion annually.

In order to guarantee that its farmers will grow enough rice to keep the nation independent of foreign supplies, the Japanese government forbids the import of rice. This holds the price of Japanese-grown rice at five times the world level and keeps rice production on land that would otherwise be devoted to more valuable uses. Indeed, rice paddies can still occasionally be seen nestled among high-rise office buildings in the middle of Japanese cities. The program costs Japanese consumers about $50 billion annually in higher food prices, more than $1,000 per nonfarm family.

Shortly after World War II, the industrial nations of Europe established the Economic Community (EC or "Common Market") to provide a Europe-wide free-trade area. This was designed to bring to the people of Europe the same economic advantages that the United States has always enjoyed with its continent-wide market. The EC was established at a time when Europe was a net importer of food, and its rules, set forth in the Common Agricultural Policy (CAP), were designed to stimulate European farm production.

The CAP established the European Agricultural Guidance and Guarantee Fund to provide common financing of price supports. The variety of products supported by the fund is wider than those under the U.S. program and includes beef, sugar, oil seeds, olive oil, wine, fruits, vegetables, protein crops, and some fibers. The fund establishes the target prices at which it will intervene and enter the market to purchase whenever the price of a farm commodity falls below its target. Direct support is combined with a system of flexible tariffs on imported farm commodities which are varied, sometimes daily, to maintain the price of imported farm products above the world market.

No production limitations are included in the program. As a result, output has outstripped what European consumers will buy at the high target prices. As much as possible of the resulting surplus is disposed of by exports, made possible by export subsidies that are set so as to make up the difference between the European farm price and the much lower price on the world market. For example, of the EC's milk production of 100 million metric tons, 25 million are disposed of at highly subsidized prices. A quarter of all butter production is subsidized, some for export and some for sale to the European baking industry. In 1970 the nations of the EC accounted for 10 percent of world sugar imports; today, the EC holds a 12.5 percent share of world sugar exports.

The continual rise in farm productivity has expanded European surpluses until today the fund's export subsidy absorbs nearly 40 percent of the total EC budget.

The current international patterns of food production and trade owe little to the competitive market and much to government interference. Among industrial nations worldwide, agricultural support programs cost more than $100 billion in governmental payments, and the total transfer to farmers from taxpayers and consumers has been estimated as nearly three times this amount. And because the programs stimulate an uneconomical use of resources, the cost to consumers is greater than the gains to farmers. The International Trade Research Consortium estimated that on a worldwide average, every $1.00 actually received by farmers costs consumers and taxpayers $1.34. That is, 25 percent of the cost of the program is sheer waste, a burden on consumers with no gain to farmers.

GATT AND AGRICULTURAL TRADE

Over the past fifty years, changing technology has transformed the world from a system of loosely linked, more or less independent economies into something approaching an integrated world market. This transition has been facilitated by the General Agreement on Tariffs and Trade (GATT), an international agreement established in 1948 to promote international trade. The member nations of GATT agree to reduce tariffs and to liberalize international trade by following rules developed in a series of "rounds." GATT has ninety-seven members today, accounting for about 90 percent of total world trade.

GATT rules are guided by three basic principles:

1. Any concession made to trade with any one nation must be extended to all members of GATT. This is known as the "most favored nation" provision.
2. Each nation must treat imported goods as favorably as it does its own domestic products.
3. Any protection accorded domestic producers must be by tariff rather than by import quotas or other nontariff restrictions.

GATT has made great strides in freeing industrial trade, but up to the present, agricultural products have been exempt from most GATT rules. Despite the general GATT provision for using tariffs rather than quotas, members are free to impose quantity restrictions on agricultural imports. Farm products are exempt from GATT's rules against export subsidies.

Occasional efforts to bring greater freedom to agricultural trade have met with little success. Indeed, the current round of GATT discussions, the "Uruguay round," is the first to concentrate on agriculture. Opened in 1986, the Uruguay round has yielded little progress. The United States, together with a group of agricultural exporters led by Australia and Canada, proposed new rules to subject farm production to market forces by removing artificial restrictions. Its most important proposals would have eliminated all agricultural subsidies that directly or indirectly affected international trade and would have removed import barriers. Income payments based on production would have been eliminated; any payments to farmers would have been made independent of output. Offending programs would have been phased out over a ten-year period.

Clearly, the proposed reforms would have required most industrialized countries, and particularly the EC, to alter their agricultural support policies. But this they have been reluctant to do. Indeed, the mere threat of reducing subsidies and of free trade in farm products brought protesting French farmers into the streets where they blockaded highways with their trucks and brought rail traffic to a halt by placing mounds of burning used tires along the tracks.

VI. CONCLUSIONS

Left to itself, agriculture is a highly competitive industry. As such, its performance is, in many respects, almost ideal. Indeed, it is hard to imagine a system better adapted to carry out the purely technical functions of producing and allocating products.

1. Although harvest is subject to the vagaries of random events, available supplies are rationed among users in accordance with consumer priorities as expressed by demand elasticities.
2. At each stage of production, the greatest amount is extracted from the available resources. In the very short run, when most costs have already been sunk into crops, relatively little can be done to adapt to consumer desires; however, given time, investment in production carefully follows up demand with matching shifts in output.
3. The industry has exhibited a remarkable history of rapidly increasing the productivity of the resources it employs and passing on the increase to the consumer in the form of lower prices. At the same time, land, labor, and other resources have been released for the production of other products.
4. All this has been accomplished with virtually no conscious collective planning, administrative direction, or political processes. Competitive pressure toward making improvements is inexorable. New methods are not debated; rather, they simply impose themselves on the industry.
5. The same competitive process, however, sweeps along its own way regardless of the fates of the people involved. The absence of social costs from the accounts of individual farms means that the search for cheaper methods can result in destroying the environment and imposing costs on society at large. Moreover, the same rapidly rising agricultural productivity that has made cheap food and higher living standards possible for con-

sumers has been accompanied by serious problems of human displacement. When productivity grows more rapidly than the demand for farm products, the people who have devoted their lives to agriculture are driven off the farm to fend for themselves as best they can.

6. Governments the world over have adopted special agricultural programs to alleviate the plight of farmers, but these programs have done practically nothing for the people who most need help. Instead, they have subsidized the largest farms, encouraged inefficient use of land and labor, and made economic chaos of the pattern of world trade in food and fiber.

It is the ultimate irony that the industry that, aside from controlling social costs, could best be regulated by the forces of a purely competitive market has instead been subjected by governments the world over to restrictions that prevent much of the benefit of that very competition from reaching consumers.

SUGGESTED READINGS

Avery, William P., ed. *World Agriculture and the GATT*. Boulder, CO: Lynne Rienner, 1993.

Bormann, F. Herbert, and Stephen R. Kellert. *Ecology, Economics, Ethics. The Broken Circle*. New Haven, CT: Yale University Press, 1991.

Burger, Kees, Martin Degroot, Jaap Post, and Vinus Zachariasse. *Agricultural Economics and Policy: International Challenge for the Nineties*. New York: Elsevier Science, 1991.

Carlson, Gerald A., David Zilberman, and John A. Miranowski. *Agricultural and Environmental Resource Economics*. New York: Oxford University Press, 1993.

Cramer, Gail L., and Clarence W. Jensen. *Agricultural Economics and Agribusiness*. New York: Wiley, 1982.

Demissie, Ejigou. *Small Scale Agriculture in America: Race, Economics, and the Future*. Boulder, CO: Westview Press, 1990.

Horwich, George, and Gerald J. Lynch. *Food, Policy and Politics*. Boulder, CO: Westview Press, 1989.

Lin, William, James Johnson, and Linda Calvin. *Farm Commodity Programs: Who Participates and Who Benefits?* U.S. Department of Agriculture, Agriculture Economics Report 474, Washington, DC, 1981.

Marks, Stephen V., and Keith E. Maskus, eds. *The Economics and Politics of World Sugar Prices*. Ann Arbor: University of Michigan Press, 1993.

Roberts, Ivan, Graham Love, Heather Field, and Nico Klijn. *U.S. Grain Policy and the World Market*. Canberra: Australian Government Publishing Service, 1989.

Tyres, Rod, and Kym Anderson. *Disarray in World Food Markets*. Cambridge: Cambridge University Press, 1992.

2

PETROLEUM

Stephen Martin

I. INTRODUCTION

Just as fossilized footprints mark the passage of a great dinosaur long after the dinosaur itself is gone, so the record of fluctuations in the price of crude oil marks passages in the world petroleum market. The traces of major political events can be seen in Figure 2-1: the Arab–Israeli War of October 1973, the fall of the shah of Iran in January 1979, Iraq's August 1990 invasion of Kuwait. Changes in market structure, which occur less abruptly, also underlie the movements depicted in Figure 2-1: the shift in ownership and control over Mideast crude-oil reserves from vertically integrated, Western-based international oil companies to local governments; the subsequent reduction in the growth of energy demand; and the development of new oil supplies outside OPEC's control.

We will examine the economic and political forces that have determined the performance of the world oil market and the U.S. submarket. Some of the questions that we shall address are, What industry characteristics have allowed the exercise of market power and by whom? What structural characteristics limit the exercise of market power? What have been the roles of the major oil companies and of the smaller, independent companies in the market? How have government policies in the consuming nations affected the market? What does industrial economics suggest concerning likely future market performance?

II. STRUCTURE

The petroleum industry is made up of four vertically related stages: production, refining, marketing, and transportation. Production involves the location and extraction of oil and natural gas from underground reservoirs; these may be so close to the surface that their oil seeps up through the ground, or they may require extensive drilling from platforms located miles offshore. The refinery segment manufactures finished products ranging from petroleum coke to motor gasoline and jet fuel. Wholesale and retail marketers distrib-

FIGURE 2-1 Selected crude-oil prices (dollars per barrel).

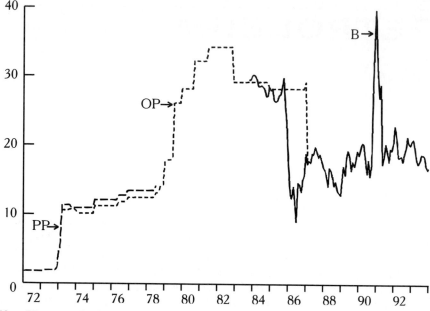

Note: PP = posted price, Saudi Arabian light (34x) crude.
OP = official price, Saudi Arabian light (34x) crude.
B = spot-market price, Brent blend.
  Source: Basic Petroleum Data Book, vol. 13, September 1993, sec. 6, Tables 10, 11, 14.

ute these products to consumers. Connecting these three vertical levels is a specialized transportation industry, which includes pipelines, tankers, barges, and trucks that move crude oil from fields to refineries and the finished product from refineries to marketers. In principle, these four vertically related segments might be supplied by independent firms, and for brief periods this has largely been the case. But throughout the history of the industry, the tendency has been for firms to integrate vertically and operate all along the line from production to distribution. These vertical links have been and promise to be critical to determining structure–performance relationships.

CRUDE OIL

1. DOMINATION BY THE INTERNATIONAL MAJORS. During the decade or so following World War II, the world oil market was dominated by the seven vertically integrated major oil companies. Five of these "Seven Sisters" were based in the United States;[1] the other two were British Petroleum and Royal Dutch/Shell. An eighth firm, Compagnie Française des Pétroles, was an additional player on the market.

Together, these eight firms controlled 100 percent of the 1950 world crude-oil production outside North America and the Communist bloc. Twenty years later, their combined share remained slightly above 80 percent.

The basis for this control was the system of joint ventures—partial horizontal integration—under which the vertically integrated majors divided ownership of the operating companies that exploited Middle East oil fields, the richest in the world (see Table 2-1).

This interconnecting network of joint ventures developed with the support of the international majors' home governments, each concerned, for reasons of national security, with ensuring the access of domestically based firms to crude-petroleum reserves. Thus, the U.S. Department of State induced American firms to take part in the 1928 "Red Line Agreement" that formalized control of the Iraq Petroleum Company, and the French government established the Compagnie Française des Pétroles to exploit its share of the Iraq concession.

The British government was similarly involved in the 1934 agreement that divided the Kuwait operating company between Gulf and British Petroleum. The U.S. government was instrumental in the 1948 reorganization of the Arabian–American Oil Company (Aramco) as a joint venture of Exxon, Texaco, Chevron, and Mobil to produce Saudi Arabian crude. The Iranian consortium—which delivered Persian oil fields into the hands of the seven majors, CFP, and a handful of American independents—was established after a CIA-backed coup returned the shah of Iran to power in August 1953 (reversing the nationalization of Iranian oil by the Mossadegh government). The fact that such a joint venture involving five American firms was contrary to U.S. an-

TABLE 2-1   Ownership Shares in Middle East Joint Ventures, 1970 (%)

	Aramco	Kuwait Oil Company	Iranian Consortium	Iraq Petroleum Company
Exxon	30		7	11.875
Texaco	30		7	
Gulf		50	7	
Chevron	30		7	
Mobil	10		7	11.875
Royal Dutch/Shell			14	23.75
British Petroleum		50	40	23.75
CFP			6	23.75
Others			5	5

Note: CFP = Compagnie Française des Pétroles.

Source: S. A. Schneider, *The Oil Price Revolution* (Baltimore: Johns Hopkins University Press, 1983), p. 40.

titrust law was set aside on the urging of the Department of State, for reasons of national security.

The continual contacts required for the management of these joint ventures resulted in a sharing of information and a communality of interest that is not characteristic of "arm's length" competition:

> [T]he international companies' vertical integration was complemented in practice by a degree of informal but effective *horizontal* integration. Their joint ownership of operating companies in the Middle East, and their voting rights under the complex operating agreements through which they controlled exploration, development and offtake there, gave them a unique degree of knowledge of each others' opportunities to increase crude offtake, and some leverage to influence each others' opportunities.[2]

The Mideast joint ventures were operated under restrictions that had the effect of ensuring output limitations. For example, partners in the Iraq Petroleum Company were obliged to file their requirements for crude oil five years in advance. Each partner thus gained definite information about the plans of every other partner. A firm that filed requirements for expanded output would thereby telegraph its plans to rivals, exposing itself to immediate retaliation.[3]

With its restrictive network of horizontal and vertical linkages, the world oil market for roughly the first decade after World War II was one in which "the seven companies controlled all the principal oil-producing areas outside the United States, all foreign refineries, patents and refinery technology: . . .they divided the world markets between them, and shared pipelines and tankers throughout the world; and. . .they maintained artificially high prices for oil."[4] Because the OPEC member states took control of the exploitation of oil deposits on their national territories, the international majors no longer dominate the world oil market. As we shall see, however, they remain major players on the world stage, and their operations continue to be characterized by a pattern of joint ventures.

During the postwar period, U.S. government regulations insulated the United States from the world market, although of course, the same suppliers dominated the world market and the U.S. submarket.

From the 1930s until the 1950s, controls on oil production by state governments (especially the Texas Railroad Commission) held crude-oil prices in the United States at artificially high levels. These prices proved attractive to foreign suppliers, and by 1948 the United States became a net importer of refined oil products. Three congressional investigations of the matter in 1950 conveyed to the oil companies a congressional preference for low imports. When domestic oil producers raised the price of U.S. crude oil in June 1950, the U.S. coal industry, with the support of the petroleum industry, sponsored a bill to place import quotas on petroleum. The Eisenhower administration set up "voluntary" import-restraint programs in 1954 and 1958, and when these proved ineffective, it imposed mandatory quotas in 1959.

These formal and informal restrictions on the flow of oil into the United States meant higher prices for U.S. consumers, perhaps by as much as $3 billion to $4 billion a year.[5] At a time when the price of crude oil on the eastern seaboard was about $3.75 a barrel, a cabinet task force estimated that the elimination of oil-import quotas would reduce the price of crude oil by $1.30 a barrel.[6]

Although the quotas had been justified on national security grounds, the effect of high U.S. prices in a shielded market was to encourage the extraction of relatively high-cost U.S. crude oil, thereby accelerating the depletion of U.S. reserves and conserving lower-cost reserves elsewhere in the world. In 1973, it became clear that U.S. national security would have been better served if the pattern of extraction had been reversed.

2. THE RISE OF THE INDEPENDENT OIL COMPANIES. The domination of the world oil market by the international majors set in motion a process of entry by new firms in search of profit. This process occupied the period from the mid-1950s through 1973 and marked a transition from a market dominated by the international majors to a market dominated by the governments of producing countries.

The first step was the 1954 Iranian consortium, when the United States government insisted that the majors make room for nine independents. Having gained a toehold in the Middle East, the independents sought to expand their access. Just as the majors had once been able to pit host nation against host nation by shifting production from country to country in order to resist pressure to expand output, so the host nations gradually gained the option of pitting the independent companies against the major companies.

In 1956, Libya granted concessions to seventeen firms. Independents subsequently accounted for half of Libyan output, and in due course, products refined from this oil found their way into European markets.[7]

The activities of Enrico Mattei, head of the Italian national firm Ente Nazionale Idrocarburi (ENI), had far-reaching consequences. He sought access to oil supplies in Iran and elsewhere and ultimately found it in the Soviet Union. After 1959, products refined from Russian oil joined the flow of independent oil onto world markets.

The increased flow of oil from these various independent sources created an excess supply at prevailing prices, despite the rapidly expanding demand. The result was downward pressure on prices, which the international majors could not resist. But the governments of the oil-producing nations collected taxes based on a "posted price" for oil, a price largely divorced from the reality of transactions in the marketplace (see Figure 2-1). This presented no problem for the majors as long as the market price of oil was rising. A falling market price combined with unchanging posted prices meant that an increasing share of profit went to the host countries in the form of taxes.

In August 1960, Exxon reduced its posted prices for oil, and in due course, the remaining major firms followed suit. This reduction in the posted prices for oil was no more than a reflection of reductions in transaction prices. And the reduction in transaction prices was the consequence of a more rapid expansion in supply than in demand. The more rapid expansion in supply than in demand was, in turn, a direct result of the actions of the host countries, which had granted the independents access to crude supplies as a way of breaking the grip of the Seven Sisters on the world oil market.

The reduction in posted prices was, therefore, an inevitable consequence of the actions of the host countries. But it appeared to them as a unilateral reduction in their own tax revenues imposed by international corporations. The reaction came at a September 1960 meeting of Saudi Arabia, Iran, Iraq, Kuwait, and Venezuela, when they agreed to establish the Organization of Petroleum Exporting Countries—OPEC.

The thirteen years that followed the formation of OPEC saw a long dance between the two loosely coordinated oligopolies, one of the international majors and one of the producing countries. At the start of this period, the balance of power lay with the companies; by the end, it lay with the countries.

Although the OPEC member states were beneficiaries of this shift in power, OPEC did not actively initiate it. The international companies had a long history of effective cooperation, and they were better at it than the producing countries were. By negotiating on a country-by-country basis, the major companies were able to prevent the countries from combining their bargaining power. Although OPEC was able to prevent further declines in the posted price, it was not able to reverse the reductions that had led to OPEC's formation.

The catalyst for change was the interaction of the independent companies and the revolutionary government of a relatively new oil province—Libya. Colonel Qaddafi's government took power in September 1969 and soon set about renegotiating the terms of Libyan oil concessions. As we have already remarked, these concessions involved the independent firms in an important way, and the independents were in a much weaker bargaining position, vis-à-vis the host countries, than the majors were. Any one of the integrated majors, faced with an unattractive proposal from a producing country, could credibly threaten to reduce output in that country and turn to supplies elsewhere around the world. But the independent companies often had no such alternative.

In August 1970, Occidental Petroleum Company agreed to Libya's demands for higher prices and higher taxes. This example inspired other oil-producing nations. In February 1971, the oil companies agreed in Teheran to the higher price demanded by the shah of Iran. The major oil companies revised the terms of their arrangement with Libya in April 1971.[8]

The oil-producing countries had demonstrated their ability to control the terms on which oil was lifted from their territories. But it was the major companies that, through their joint ventures, owned the operating companies (see Table 2-1). This, too, was to change.

Again, it was the radical states, rather than OPEC, that led the way. Algeria nationalized 51 percent of French ownership in Algerian reserves in February 1971; Libya nationalized British Petroleum's interests in November 1971; the Iraqi Petroleum Company was nationalized in June 1972. Long negotiations between Saudi Arabia, represented by Sheik Zaki Yamani, and Aramco followed, the result of which was that Aramco agreed to yield an initial 25 percent of its Saudi Arabian concession to Saudi Arabia.[9]

In the absence of intervening political developments, the transfer of world market control from the international majors to the producing countries would likely have continued at a gradual pace. The producing countries would have slowly replaced the international majors, with little change in market performance. In such a world, the Western "man in the street" might have remained blissfully unaware of the nature of the world oil market. But events unfolded rather differently.

3. THE OPEC EPISODE. The 1970s opened with the demand for oil increasing throughout the industrial world. Figure 2-2 shows the steady growth of the U.S. oil demand through the 1960s and early 1970s. Similar growth took place in Europe and Japan. In 1973, with simultaneous booms in North America, Europe, and Japan, the world's demand for energy—and oil—was at an all-time high.

At the same time, supply and the control of supply were increasingly concentrated in the low-cost Middle East. In 1970, proved reserves of crude oil in the Middle East were 333,506 million barrels, versus 67,431 million barrels of proved reserves in the Western Hemisphere and 54,680 million barrels in Africa.

The location and development of crude-oil reserves is a time-consuming process, particularly when oil fields are located offshore or in other hostile climates. Thus, the 1970s opened with a relatively small short-run supply of crude oil available from fringe, non-OPEC suppliers. These fringe suppliers faced substantially higher costs—development costs and operating costs—than did the Middle East producers.

This kind of market is illustrated in Figure 2-3(a). Fringe supply is small relative to market size; the residual demand left for the cartel comprises the bulk of the market. In such a market, a dominant firm or a perfectly colluding cartel would maximize profit by selecting an output that equates marginal revenue along the residual demand curve—the market demand curve after subtracting fringe output—to marginal cost. In Figure 2-3(a), the corresponding cartel output is q_1, which would sell at price p_1. Because fringe supply is small, this is very nearly the monopoly position.

Since the economic interests of the OPEC member nations diverge in fundamental ways, OPEC was far from being able to act as a monopolist or a perfectly colluding cartel. Instead, it was a political rather than an economic event that triggered coordinated action by OPEC and allowed it to take advantage of the demand–supply relationship depicted in Figure 2-3(a). It was in response to Western support for Israel during the Egyptian–Israeli War of October 1973 that Arab nations imposed production cutbacks and an embargo of crude-oil supplies to the West.

FIGURE 2-2 U.S. demand for refined oil products (thousands of barrels per day).

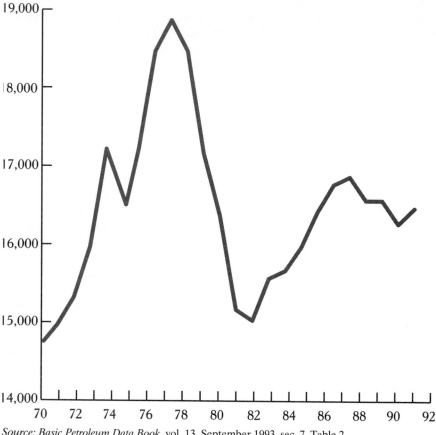

Source: *Basic Petroleum Data Book*, vol. 13, September 1993, sec. 7, Table 2.

The international oil companies were based in the West, and they had long benefited from the political support of their home governments, governments that sought to protect their perceived national security interest in a safe supply of oil. But the international majors administered the embargo of Western nations in accordance with OPEC directives, going so far as to provide Saudi Arabia with information on the shipment of refined oil products to U.S. military bases around the world.[10]

As the producing countries cut back the international majors' supplies of crude oil, the majors cut back supplies to the independent companies. With their survival threatened, the independents turned to the market for oil not tied up by long-term contracts—the relatively narrow spot market. They bid up the spot-market price of oil, and official OPEC prices soon followed. The immediate result was the 1973 rise in official prices shown in Figure 2-1.

FIGURE 2-3 Demand and supply in the world oil market.

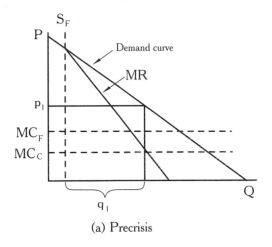

(a) Precrisis

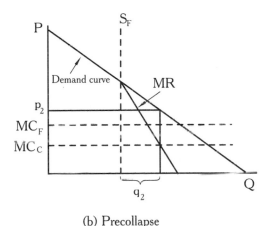

(b) Precollapse

From this price increase flowed longer-run changes. OPEC's revenue from the sale of oil rose from $13.7 billion in 1972 to $87.2 billion in 1974. The real U.S. gross national product, which grew 5.2 percent in 1973, fell 0.5 percent in 1973 and 1.3 percent in 1974.

Aside from the accelerated shift in the control of production to the producing nations, there were remarkably few structural changes during the period following the first price increase. The U.S. demand for oil fell slightly in 1974 and 1975 but then rose to new heights by 1978 (Figure 2-2). The share of imports in the U.S. market, and specifically imports from OPEC, peaked in 1977 but remained higher than in 1973 (Figure 2-4). The pattern of consumption and supply was much the same in other industrialized countries.

As shown in Figure 2-5, OPEC's share of world crude-oil production fell only about 5 percent over the period from 1973 to 1979. World production of

FIGURE 2-4 Import and OPEC shares of the U.S. petroleum market, 1960–1992.

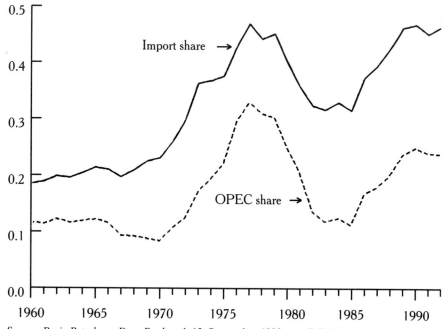

Source: *Basic Petroleum Data Book*, vol. 13, September 1993, sec. 7, Table 2, and sec. 14, Table 3.

FIGURE 2-5 Share of world crude petroleum production.

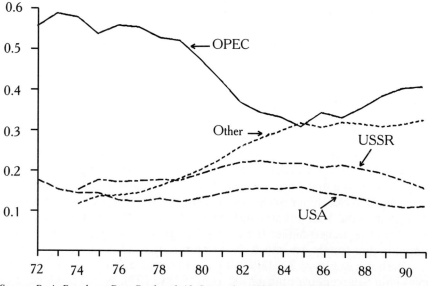

Source: *Basic Petroleum Data Book*, vol. 13, September 1993, sec. 4, Tables 1, 2.

crude oil grew throughout this period, but OPEC's production was essentially level: 30,989 thousand barrels per day in 1973, and 30,911 thousand barrels per day in 1979.

Because of the length of time needed to develop new petroleum reserves and to install energy-saving residential and industrial equipment, the underlying market conditions that greeted the fall of the shah of Iran in January 1979 were essentially the same as those that had greeted the Arab–Israeli War of 1973: peak demand, concentration of supply in the Middle East, and absence of spare capacity in the West.

The impact of the course of events on the market was similar. Supply was disrupted. Independent refiners had their crude supplies cut off, and desperate for crude oil, they turned to the spot market. The spot-market price shot up, and the official price, as shown in Figure 2-1, followed.

4. CULTIVATION OF CRUDE-OIL SUPPLIES OUTSIDE OPEC. The response to the second oil-price shock, however, was substantially different from the response to the first one. These differences occurred on both the demand side and the supply side of the market.

As shown in Figure 2-6, energy use in industrial economies declined slowly over the period from the first to the second oil-price shock but declined sharply thereafter. This illustrates the long response time required to realize changes in the demand for energy resources. It also indicates a long-term shift to greater efficiency in energy use, partly due to the greater real price of en-

FIGURE 2-6 Energy use per dollar of gross domestic product.

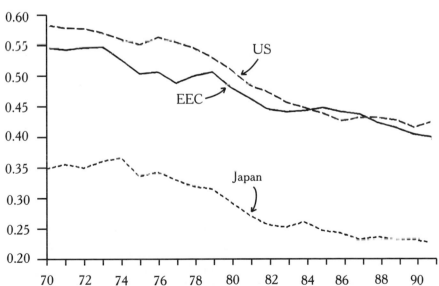

Note: Total primary energy supply (in millions of tons of oil equivalent) per billion dollars of gross domestic product (measured in 1985 U.S. dollars).
Source: International Energy Agency, *Energy Balance of OECD Countries*, various issues.

ergy and partly due to increasing concern about the impact of energy consumption on the environment. This shift will translate into slower growth in demand for energy from all sources, including petroleum.

There is a matching effect on the supply side of the market, described in Table 2-2. U.S. crude-oil production peaked in 1970 and has followed a downward trend from that date. This trend will continue, mitigated by new technology but not by the discovery or exploitation of new reserves. Long isolated from the world market by quotas, the United States has been thoroughly explored. But the output from Western Europe increased sharply over this period, as the North Sea oil fields of Britain and Norway came into production. North Sea oil output will peak in the mid-1990s, but it will remain an important source of natural gas.

Output in Latin America rose over this period, and this trend is likely to continue. Venezuela, despite being a charter member of OPEC, continues to expand its oil reserves. It markets a coal–water mixture that is a good substitute for fuel oil and is exempt from OPEC output quotas. In addition, Venezuela has aggressively acquired networks of refineries and service stations to ensure outlets for its oil. Other Latin American countries, including Peru and Colombia, continue to expand their oil industries; Mexico has increasingly cooperated with international oil companies to exploit its oil deposits.

Crude-oil output from the Third World, including China and various less developed countries (LDCs) in Africa, can be expected to increase, thereby reflecting a convergence of interest between the LDCs and the international majors. The international majors have been cut off from the Mideast oil fields

TABLE 2-2 World Crude Production by Area, 1974 and 1992 (Millions of Barrels)

	1974	1992
United States	3,203	2,625
Canada	617	592
Western Europe	142	1,745
Latin America	1,789	2,702
Asia	816	2,373
USSR	3,374	3,257
Africa	1,990	2,313
Middle East	7,987	6,376
Total	20,538	21,983

Source: American Petroleum Institute, *Basic Petroleum Data Book*, September 1993, Sec. 4, Table 1.

that were for generations the foundation of their dominant market positions. They will explore anywhere outside OPEC's sphere of influence for new reserves, which they can feed into their existing refining and marketing networks. They will do so as long as the new reserves can be developed at or below the spot-market price for oil.

At the same time, the LDCs know from bitter experience that it is their development efforts that are torpedoed by dependency on foreign sources of oil. For political reasons, the LDCs will encourage the development of local oil supplies even if the cost seems likely to exceed the spot-market price of oil. Their national security, and often the lives of their leaders, depends on it.

The same applies to the world's newest group of less developed countries, the members of the Commonwealth of Independent States. As the foundations of market economies are laid, crude-oil output from the former Communist bloc will decline during a transitional period. Once that difficult transition is past, petroleum deposits will be viewed as much as a source of foreign exchange as a source of energy and will be exploited accordingly. Indeed, independent oil companies are already rushing to establish ties in this part of the world despite the uncertain political and legal environments.

The international majors will, therefore, be welcome in the new oil provinces around the world. The supply of oil from the less developed countries will increase, and to some extent this increase will result from political rather than economic considerations.

Along with the substantial increases in crude-oil production outside OPEC has come the reduction in Middle East output indicated in Table 2-2. By mid-1985, OPEC found itself in the kind of market illustrated in Figure 2-3(b). After a decade of development efforts, the fringe supply was relatively large, but after a decade of conservation efforts, demand had grown much less rapidly than expected. The residual part of the market left for OPEC was substantially reduced, compared with the situation of Figure 2-3(a). In such circumstances, the best that OPEC could do would be to set a price p_2, much lower than p_1, and market a substantially smaller quantity q_2.

Just as the international majors' long domination of the world oil market created an incentive for the entry and expansion of independent oil firms, so OPEC's somewhat briefer period of control created an incentive for the development of new oil provinces. The entry of the independent firms undercut the resource base of the international majors, which responded by seeking new supplies of oil. The entry of new oil-producing countries has undercut the power base of OPEC member nations, and they have reacted by seeking secure outlets for their oil. The result has been a renewed trend to vertical integration, by OPEC member states and oil companies, with less horizontal concentration at all levels of the industry.

5. THE NORTH SEA MARKETS. Perhaps the most ironic consequence of OPEC's assertion of control over the world crude market is the development of active spot and futures markets for Brent crude blend, which has made the price of North Sea crude a bellwether for the industry and has ratified OPEC's inability to do more than slide down a contracting residual demand curve.

Sovereignty over North Sea oil deposits lies with Britain and Norway. Ownership of production entitlements is by a large number of international oil companies, through a complex web of joint ventures.[11] Concentration of production is high, although declining. Output is dominated by Exxon and Shell, whose combined market share was near 70 percent for much of the 1980s and remained over 50 percent in 1993.[12]

North Sea oil fields benefit from a location near the major consuming centers of Europe and from a product with desirable physical properties from the refining point of view. North Sea oil is a good economic substitute for U.S. and OPEC products. For these reasons and because the North Sea came "on line" at a time when the international oil companies were scampering to acquire access to oil outside OPEC's influence, the spot market for North Sea oil has assumed a central place in the interlocking world network of oil markets.

The economic importance of the North Sea market is far greater than its share of world oil output (never more than 6 percent). The prices of many other transactions are tied to those on the North Sea market. In times of real or perceived shortage, surges of demand on the North Sea futures market (the market for oil to be delivered one, two, or more months in the future) can drive up prices very rapidly and often down again just as rapidly (examine Figure 2-1 for the situation around the time of the Iraqi invasion of Kuwait).[13] The role of the price of Brent blend as a marker price for the world petroleum industry confirms OPEC's long-run inability to control the world oil market. But the picture is not entirely rosy: The spot and futures markets for North Sea oil are subject to speculative binges, and price fluctuations on these markets often have little to do with the fundamentals of supply and demand.

6. THE U.S. SUBMARKET. The U.S. crude market is relatively unconcentrated, by conventional measures. In 1991, the largest four U.S. crude producers had a combined share of 27.2 percent of U.S. crude production. The 1991 Herfindahl index for the U.S. crude industry was 0.02894, suggesting a level of concentration equivalent to an industry with roughly thirty-five equally sized firms.[14]

Students of industrial organization justify their interest in market concentration figures on the ground that such figures provide information about the likelihood that leading firms will come to recognize their mutual interdependence and act in a way that is likely to maximize their joint profit. Two factors suggest that the domestic concentration figures for the U.S. market understate the likelihood of recognition of interdependence.

The first is the common use of joint ventures among U.S. firms, which in a local sense replicates the use of joint ventures by international majors in the Mideast. For example, from 1970 to 1972, four of sixteen major U.S. oil companies made no independent bids for offshore oil leases but instead submitted a total of 863 joint bids with fourteen alternative partners. Chevron submitted 79 individual bids and 108 joint bids with nine alternative partners; Cities Service made 7 independent bids and 372 joint bids with four alternative partners. Among the sixteen majors, only Exxon made no joint bids during this period, although it made 80 independent bids.[15] A similar pattern of joint ventures occurs in the management of pipelines.

The partners in this overlapping network of joint ventures, to the extent that they profit from the same enterprises, have an incentive to avoid competition. The operation of the network of joint ventures provides each company with myriad bits and pieces of information about the market strategies of its fellows. No single firm could contemplate noncooperative behavior without expecting that this behavior would be promptly detected by other firms.

The second factor that suggests that U.S. domestic concentration figures understate the likelihood of noncompetitive behavior is the ongoing wave of mergers among major U.S. firms and between U.S. firms and international majors. During the 1980s, four of the eight majors took over other major firms, and two were themselves taken over by foreign firms. A similar consolidation is taking place among smaller U.S. oil companies. As oil companies the world over search for secure assets to petroleum reserves, some find it easier to look for reserves on Wall Street rather than in the ground.

The U.S. market, long protected by formal and informal trade barriers, is today part of the world market. The far-reaching history of joint ventures, cooperation, and collusion on the world stage suggests the likelihood of similar conduct in the United States.

REFINING

The refining segment of the petroleum industry transforms crude oil into a variety of final products, ranging from gasoline and liquefied petroleum gases to residual fuel oil and petroleum coke. Refining involves not only the distillation of crude petroleum into refined products but also a number of upgrading processes by which lower-value refined products can be transformed into higher-value products (lead-free gasoline, for example).

The most striking recent development on the world refining market has been the forward vertical integration of OPEC member nations into refining and, indeed, forward still further into marketing.

Through a subsidiary of the Kuwait Petroleum Corporation, Kuwait owns two European refineries with a capacity of 135,000 barrels per day, together with 4,800 retail gasoline stations in seven different European countries. Kuwait has acquired a 22 percent ownership of British Petroleum, much to the concern of the British government, and has sought refining assets in Japan.

Venezuela has employed joint ventures to acquire partial interests in refineries in Germany, Sweden, and Belgium and has acquired Citgo's refining and distribution operations in the United States.

Other OPEC members also have integrated forward into the U.S. market. In November 1988, Saudi Arabia acquired half-ownership of Texaco's U.S. refining and distribution network. The three refineries involved have a capacity of 615,000 barrels per day; the distribution network includes 11,450 retail gasoline stations. With this investment, the largest source of crude oil in the world moved to secure a market for its product. It also acquired an interest in maintaining profitability at the refining and distribution levels of the market, as well as the crude level.

The motives for this forward integration are partly political and partly economic. A move forward into refining is a way of broadening the local industrial base while taking advantage of existing assets and skills. At the same time, a refinery associated with a national oil company of an oil-producing state has an almost insuperable advantage when compared with an independent refiner. The real cost of crude oil to the integrated refiner is the cost of crude production (regardless of the transfer price from the crude division to the refinery division). But the cost of crude oil to an independent, nonintegrated refiner is the much higher market price for crude oil. Refining and distribution networks associated with producing nations will always be able to undersell independents. When there is a surplus of crude oil, it will be tempting—and profitable—to do so.

A consequence of this forward integration by oil-producing nations is a persistent excess capacity at the refining level, as illustrated in Table 2-3. Oil-producing nations integrate forward for economic and political reasons while (in some cases at least) restraining crude output. International majors retain the bulk of their refining operations, which are the heart of their distribution network, while seeking to develop access to crude oil outside OPEC's influ-

TABLE 2-3 Worldwide Refining Capacity (Thousand Barrels per Day)

	1973	1993	1992 Utilization Rate (%)
United States	13,671	15,209	86.7
Other Western Hemisphere	8,176	9,442	81.3
Western Europe	16,827	14,402	92.1
Middle East	2,757	4,924	84.4
Africa	825	2,939	81.2
Asia-Pacific	7,916	13,396	88.6
Communist Nations	9,110		
USSR/Eastern Europe		12,873	65.8
World	59,282	73,186	81.6

Source: American Petroleum Institute, *Basic Petroleum Data Book*, September 1993, Sec. 8, Tables 1 and 2.

ence. Refining capacity expands and crude reserves expand rapidly, but crude production grows slowly if at all. The result is excess refining capacity.

MARKETING

Marketing channels differ depending on the class of final consumer served. Nearly all jet fuel is sold by refining companies to their customers, who are principally the Department of Defense and commercial airlines. Most refiner sales of residual fuel are made directly to large utility and industrial customers, with the remainder going to large terminal operators and dealers.

In contrast, no more than one-third of distillate fuel oil is sold directly to customers from refiner-owned facilities, whereas two-thirds goes through independent marketers.

The retail market for gasoline, like all levels of the petroleum industry, has been severely affected by the structural changes that have occurred since 1973. In the United States, the number of gasoline service stations fell from 176,465 in 1977 to 111,657 in 1990.[16] In part, this reflects the shift from full-service to convenience-store distribution. And in another part, the decline in the number of outlets signals an increase in the scale of operation: "The stations that are disappearing are generally on the smaller side—30,000 to 40,000 gal./month. In their places are being built the big superstations —the 100,000-to 200,000-gal./month-plus pumpers that are more economical to operate and frequently are on a seven-day, 24-hour basis."[17] In part, this drop in the number of outlets reflects a high level of concentration of retail gasoline sales in local markets (in which the combined share of the largest four firms typically ranges from 40 to 60 percent). This concentration can be expected to increase as vertical integration from crude production through distribution (by OPEC members and major companies) clones the oligopolistic structure of the crude segment of the industry at levels closer to the final consumer.

These changes bode ill for the independent retail outlet. And because competition from independent retail outlets has always been an important factor in forcing competitive pressure vertically backward from the retail level, they bode ill for the consumer as well.

TRANSPORTATION

Pipelines are the most important mode of transportation in the United States, where large volumes of oil must be moved overland.[18] In any oil field, a network of small-diameter "gathering lines" collect crude oil from individual wells in an oil field and transmit the field's output to a larger-diameter "trunk line" for shipment to a refinery. Pipelines later move refined products to marketing centers.

Bulk shipment of oil in trunk lines is a classic example of a technology that exhibits decreasing cost per unit of output. Pipeline construction cost is roughly proportional to pipeline radius, whereas pipeline capacity is proportional to the square of the radius. If pipeline radius is doubled, pipeline capacity will increase by a factor of four. Construction and operating costs per unit of capacity fall, therefore, over the entire range of technically feasible pipelines. Petroleum shipment over pipelines is, for this reason, a natural monopoly.

It is not surprising, therefore, that the major oil companies dominate pipeline transportation. Nearly 90 percent of crude-oil pipeline shipments reported to the Interstate Commerce Commission in 1976 originated in lines that were owned or controlled by the sixteen major U.S. oil companies. At the same time, nearly 75 percent of refined-product shipments that originated in refineries (in contrast with those received from connecting carriers) went into pipelines owned by the major companies.

Pipeline transportation is inherently a local activity, moving from one point to another:

> Concentration in major crude oil transport corridors (i.e., specific pipeline markets) is extremely high. In the Texas–Cushing, Oklahoma corridor for example, the four largest pipeline companies together account for 76% of total crude carried. Three pipeline companies control all crude oil shipments in the Gulf Coast–Upper Mid-Continent corridor. . .in 1979, the four-firm concentration ratio for pipeline shipments in the nation's major crude carriers averaged 91%.[19]

This high degree of concentration is intensified by the use of joint ventures in pipeline management. As with joint ventures in offshore oil and Mideast exploration, such joint ventures are not conducive to "arm's length" competition.

The implications of this concentration of pipeline ownership and control are clear:

> A pipeline rate set well above the competitive cost of transporting crude oil. . .imposes no burden on the majors who own the pipeline. For them, the high price is simply a transfer of funds from the refinery operation to the pipeline operation. To the nonintegrated refiner, however, an excessive pipeline charge is a real cost increase that he cannot recoup elsewhere and that places him at a competitive disadvantage vis-à-vis his integrated competitors.[20]

Vertical integration from refining into pipeline transportation allows U.S. majors to apply a vertical price squeeze to refiners who are not integrated forward, much as vertical integration forward into refining allows crude oil-producing countries to apply a price squeeze to refiners who are not integrated backward into production.

III. CONDUCT AND PERFORMANCE

MODELS AND MARKETS

It is sometimes asserted that the price increases of 1973 and 1979–1982 reflect no more than the working of competitive forces. In this view—favored in particular by oil-producing nations—a price of oil at or near extraction cost fails to reflect the scarcity that current consumption imposes on future generations. A price substantially above the marginal extraction cost, in this view, is to be desired, because it encourages conservation and spreads consumption of a finite resource over a long time period.

This argument might explain the price increases observed in 1973 and 1979–1982. But it cannot explain the price declines since then. Oil is, after all, as much a finite resource as it ever was, and if future scarcity would produce a high price in 1982, it would, seemingly, produce a still higher price in 1993. Statistical tests do not support the argument that OPEC pricing is competitive.[21]

Some analysts have suggested that the world oil industry is driven by a single dominant firm—Saudi Arabia. According to the figures in Table 2-4, Saudi Arabia holds nearly 26 percent of the world's proven reserves (oil deposits that can be profitably extracted with current technology at current prices). It is widely believed that Saudi Arabia substantially understates its reserve hold-

TABLE 2-4   Estimated Proven Reserves of World Oil, 1993 (Billion Barrels)

Saudi Arabia	258.0
Iraq	100.0
United Arab Emirates	97.7
Kuwait	94.4
Iran	92.9
Venezuela	63.7
C.I.S.	57.0
Mexico	51.3
United States	24.7
China	24.0
Libya	22.8
Nigeria	17.9
Algeria	9.2
Norway	8.8
Egypt	6.2
World	997.0

Source: American Petroleum Institute, *Basic Petroleum Data Book,* September 1993, Sec. 2, Table 4.

ings. But this quantitative description does not capture the fact that Saudi crude is by far the least expensive in the world, with an estimated cost of 30 to 60 cents a barrel.

These reserve holdings mean that Saudi Arabia will be a factor on the world oil market as long as there is a world market for oil. If Saudi Arabia were to act as a wealth-maximizing dominant firm, it would in principle restrict output and raise the price above the cost of production. It would then gradually give up market share as other producers expand their output to take advantage of the opportunity for profit created by the price increase.[22]

A variation on this theme suggests that although no single OPEC member has sufficient control of reserves to exercise control over price, OPEC, as a group, is able to act as a collusive price leader. The predicted market performance is much the same as under the dominant-firm model. OPEC's share of the market should decline over time as independent producers respond to the incentive created by a price above the cost of production.

Figures 2-1 and 2-5 suggest that the dominant-firm and dominant-group models can help explain the world oil market. OPEC's share of world crude-oil production fell very slowly from 1973 to 1979. As already noted, this is a reflection of the long lead times for the discovery and development of oil reserves. OPEC's market share has dropped precipitously since 1979, bottoming out at 30 percent of the world market in 1985. Price, also, has fallen as OPEC's market share has fallen and the share of other producers has increased.

What the dominant-firm and dominant-group analyses fail to capture is the oligopolistic interactions that have influenced OPEC's behavior. When price is raised above the cost of production, individual OPEC member nations (and not just independent producers) have an incentive to increase their own output. The problem for OPEC, like that of any cartel, is to agree on a course of action (raising the price to some level) and then to secure adherence to the agreement. As is the case with any group, differences make for disagreements. The main difference that has plagued OPEC pertains to the rate of time preference for income.

OPEC member states differ substantially in the urgency with which they desire revenue from oil sales. Countries such as Saudi Arabia, Kuwait, and the United Arab Emirates have small populations, high GNPs per capita, and political circumstances that are well served by modernization at a slow pace. Their massive oil reserves ensure that they will earn oil revenue for the foreseeable future. Other OPEC members, however, such as Indonesia, Nigeria, and Algeria, have larger populations, smaller GNPs per capita, and substantially smaller oil reserves. Their best hope for economic development is through the maximization of short-run oil revenues. Political pressures reinforce this economic incentive. More than once, governments of OPEC nations have been overturned because of mismanagement of the oil sector, and the ousted leaders often do not survive to collect retirement benefits. Arguments between OPEC members who place great weight on short-run revenue and

OPEC members who place great weight on the long run have persistently complicated OPEC negotiations:

> A resolution of OPEC's impasse is badly needed by its members from the third world, like Indonesia, Nigeria, Venezuela and Algeria. These nations, with populations that are much larger than those of the Persian Gulf countries, are weary of seeing their oil revenues decline because of disputes in the Persian Gulf that have little to do with them.[23]

SAUDI ARABIA AND OPEC OUTPUT DYNAMICS

There is a fundamental difference between the ability of high-absorption and low-absorption OPEC members to encourage the market to behave as they would like it to do. High-absorption states can restrict their own output and urge their colleagues to follow suit. Total OPEC output will fall and prices will rise if, and only if, substantially all OPEC members reduce production. But a single low-absorption state, within the limits permitted by its developed reserves, can expand its output and deliver both oil and low prices to the market.

Although there is room for interpretation, this appears to be the role adopted by Saudi Arabia. By 1985, OPEC's market share had fallen to 30 percent, mostly on the strength of output cutbacks by Saudi Arabia, which enjoyed an OPEC quota of 4.353 million barrels per day but was estimated to be producing only 2.5 million barrels a day in September 1985. OPEC's official price remained at $28 a barrel, but the spot-market price for oil was no more than half the official price.

At this point, Saudi Arabia introduced a system of "netback pricing," under which the price paid for Saudi crude oil was determined by the market prices of the products refined from the crude. The immediate effect of the netback-pricing system was to eliminate risk for the purchaser of Saudi crude: If the price of refined products should fall, the price of crude would fall proportionately. The consequence was a sharp increase in the demand for Saudi oil, whose output had reached 6 million barrels a day by July 1986.

Other OPEC members soon adopted their own netback-pricing schemes, and oil prices fell as low as $6 a barrel. By August 1986, OPEC members, with the exception of Iraq, which held out for a quota equal to Iran's, reaffirmed their support for the quota schedule that they all were violating. A series of ineffective agreements followed. The pattern was as follows:

1. Agreement on a schedule of quotas that implies some reduction of current output by each OPEC member, but to a level above the previous quota.
2. Rumbling by Saudi Arabia that it will not act as a swing producer.
3. Production above quota levels by all members, with Saudi Arabia typically one of the least serious offenders.

4. Massive expansion of output by Saudi Arabia, resulting in a collapse of prices.
5. A meeting of OPEC ministers, who (go to Step 1 and begin the cycle again).

One analyst described the negotiations among the OPEC oil ministers as follows: "The good news is that there may be a deal. The bad news is that it will not be enforced."

THE IRAQ–KUWAIT NONCRISIS

In view of the critical impact of political crises on oil prices, one might have expected the Iraqi invasion of Kuwait to have induced a third oil crisis. But this was not the case, and in fact the lessons that OPEC members have drawn from oil market events related to the Gulf War seem likely to loosen OPEC's already shaky grip on the market.

Despite the brief spike in the spot market's Brent price to nearly $40 a barrel, the Iraqi invasion of Kuwait caused hardly a ripple in the supply of oil to world markets. Supplies from Iraq and Kuwait were abruptly cut to zero, but Saudi Arabia increased its output from 5.4 million barrels a day in August 1990 to 8.2 million barrels a day in November 1990, maintaining the overall supply at a comfortable level.[24]

Saudi Arabia perceived its own self-interest to lie in stable oil markets and prices that did not give too much encouragement to conservation efforts or the search for alternative fuels. Saudi Arabia was able to act in its own self-interest because it had excess capacity that allowed it to act unilaterally. But those OPEC members with significant petroleum reserves have not been slow to draw the conclusion that bargaining power within OPEC is related to production capacity. In the immediate aftermath of the Gulf War, Kuwait (naturally enough), Abu Dhabi, and Iran all set in motion substantial investment programs aimed at increasing their crude capacity. Not to be outdone, Saudi Arabia initiated an expansion of its capacity to more than 10 million barrels a day.

OPEC's medium-run influence on world oil markets will depend on the ability of OPEC member states *not* to use the capacity they are now installing. If history is any guide, however, they are not likely to be very successful.

IV. GOVERNMENT POLICY

U.S. ANTITRUST ACTIVITY

Antitrust policy has been the traditional approach to preserving competition in the United States. In 1911, the U.S. Supreme Court upheld a finding that the Standard Oil Company had violated the Sherman Antitrust Act while acquiring a dominant position in the refining, marketing, and transporting of petroleum. The Court imposed a structural remedy, ordering the parent holding company to divest itself of controlling stock interests in thirty-three subsidiaries.[25] This first case was also the last successful major case involving the U.S. oil industry.

To be sure, the oil giants have from time to time attracted the attention of antitrust enforcers. In 1940, a case so broad that it became known as the "Mother Hubbard" case was filed.[26] In a Justice Department civil suit, 22 major oil companies, 344 subsidiary and secondary companies, and the American Petroleum Institute were charged with violating both the Sherman and Clayton acts. The case was postponed, however, because of the onset of World War II and thereafter languished until it was dismissed in 1951 at the request of the Justice Department.

Throughout the postwar period, in fact, antitrust action against U.S. oil firms was suspended on grounds of national security. In the closing days of the Truman administration, the government accused the five U.S.-based international majors (along with British Petroleum and Royal Dutch Shell, which were beyond the jurisdiction of U.S. authorities) of seeking to restrain and monopolize crude oil and refined petroleum products, in violation of the Sherman Antitrust Act. But the State Department urged that the oil companies receive antitrust immunity for their cooperation in setting up the 1954 Iranian consortium, and this immunity weakened the cartel case, which dragged on for years. Exxon, Texaco, and Gulf eventually settled for consent decrees, and charges against Mobil and Socal were dismissed. The last parts of the case were dropped by the Justice Department in 1968.[27]

Much of the vitality of American antitrust law derives from the possibility of private enforcement. A private antitrust suit filed in 1978 by the International Association of Machinists and Aerospace Workers (IAM) sought to apply U.S. antitrust law to OPEC. OPEC declined to appear when the case was heard, and the Department of Justice refused to submit the amicus curiae brief requested by the district court. The district court declined to hear the case on technical grounds, among which was that the IAM did not have sufficient "standing" to sue OPEC, since IAM did not purchase directly from OPEC.

IAM appealed this dismissal to the circuit court of appeals, which upheld the lower court's refusal to hear the case on the ground that the case would interfere with U.S. foreign relations. Once again, conditions of national security short-circuited the application of the antitrust laws.[28]

GOVERNMENT POLICY RESPONSES TO OPEC

The fact that the oil crisis of 1973 was repeated just six years later is testimony to the fact that the Western governments generally were unable to develop adequate energy policies. The founding of the International Energy Agency (IEA) in November 1974 suggests government recognition of the importance of cooperation among consuming countries. However, France refused to join the IEA, apparently preferring bilateral government-to-government negotiations with oil-producing nations. Such government-to-government negotiations became common during the tight crude markets of 1978–1980, when security of supply was a matter of concern. Since that time, with crude oil in excess supply at official OPEC prices, sales with government involvement have declined. Supplies can be acquired on the spot market and generally at prices below the official OPEC levels.

Three interrelated aspects of the continuing debate over proper government policy toward the petroleum industry merit discussion. All reflect a failure to understand the way the markets work.

1. THE INTERNATIONAL ENERGY AGENCY.[29]   Twenty-one Western nations are members of the International Energy Agency. They are pledged to share their oil supplies if it is determined that a shortage of oil has occurred or is likely to occur. *Shortage* is defined in terms of a physical interruption in supply, and it seems clear that the IEA's focus is on embargoes imposed for political reasons. Price increases, however, have been an important aspect of past oil shocks. It would be desirable to alter IEA procedures so as to facilitate a response to a sharp price increase as well as to a sharp supply decrease.

2. THE STRATEGIC PETROLEUM RESERVE.   In 1975, Congress established the Strategic Petroleum Reserve (SPR) as insurance against future interruptions in foreign supplies. Targets, set in later detailed plans, were for 500 million barrels in storage by 1980 and 1 billion barrels by 1985. By March 1988, 545 million barrels were actually stored in salt caverns around the Gulf of Mexico. Modest sales from the SPR were made during the Gulf War. In view of the increases in supply from Saudi Arabia, such sales probably had little direct effect on oil markets. But it seems likely that the knowledge that the SPR existed had a calming effect on world oil markets. It would be desirable to adopt a policy of quietly building up the Strategic Petroleum Reserve during periods of low prices, as insurance against shortages and price spikes.

3. LOW PRICE OR ENERGY INDEPENDENCE?   The U.S. government has been unable to come to grips with the constraints imposed by the market on its policy options. What is striking is that OPEC member nations suffer from a symmetric analytical failure.

Figure 2-7 shows a stylized demand curve for the world oil market. In the long run, a monopolist or complete cartel can be anywhere on the demand curve. The constraint imposed by the market is that price and quantity are inversely related: A high price means a low quantity demanded, and a low price means a high quantity demanded. It is clear from endless public statements that the OPEC oil ministers want desperately to be at a place like X in Figure 2-7. They want to charge a high price for oil and sell a great deal of oil, but over the long run the market will not let them do so.

FIGURE 2-7   Wishful thinking at home and abroad.

Price

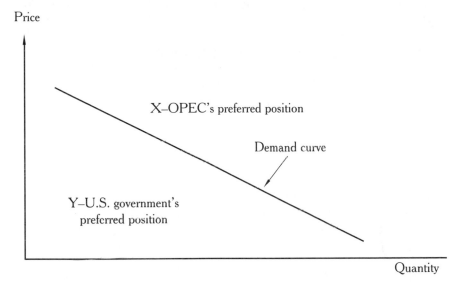

Unfortunately, it is not proximity to rich oil reserves that leads to fuzzy thinking. The debate on U.S. policy toward the petroleum industry has long been bedeviled by a similar failure to come to grips with the limitations imposed by the market.

U.S. policymakers have long recognized the ills associated with a high oil price. Supply shocks mean short-run bursts of inflation, which eventually raise interest rates. A high oil price raises the cost of all energy-intensive activity, by consumers and by industry. Economic growth is slowed, and the U.S. trade balance—because the high price applies to a good for which the U.S. is a net importer—is made worse.

One might think, therefore, that low oil prices would be welcomed by U.S. policymakers. But no:

> The U.S. energy secretary. . .said Monday that the price of oil, currently at $13 to $14 a barrel, was too low and would not help U.S. energy interests in the long term. . . .
> She agreed that in the short term the low price of oil represented a boost for the American economy. But in the long run she insisted that it posed problems from both an economic as well as an environmental perspective by stimulating demand in the nation that already ranks as both the world's largest oil consumer as well as importer.[30]

A recurring position in the U.S. policy debate would very much like to have the United States at a point like Y in Figure 2-7: a low price for oil, but we don't buy much of it.

Some of this is simply putting parochial interests ahead of national interests, as when the governor of Texas sits on the lap of Santa Claus and asks for

"$25-a-barrel oil."[31] More important, however, is the failure to think through the implications of dependence on domestic supplies for long-run national security.

NATIONAL SECURITY: THE SHORT RUN

The current consumption of U.S. oil makes the nation's future dependence on foreign oil worse, barrel for barrel. The only way to avoid this is permanent tariffs or quotas that artificially raise the price of oil in the United States and distort input/consumption choices in less efficient and productive ways. It is not in the national security interest of the United States to be protected from foreign oil that is cheaper than domestic oil.

The specter is raised of unending Middle East reserves and vulnerability to a cutoff of oil supplies from a politically unstable region of the world. The response to this contingency points to the increase in supplies elsewhere, an increase that will intensify as the price of oil rises. The expansion of oil reserves outside OPEC implies that U.S. security interests lie in an open world market.

There is no *long-run* U.S. security interest in avoiding the current consumption of cheap foreign oil. There is, however, a *short-run* U.S. security interest in neutralizing rises in oil prices due to sporadic supply interruptions, and that is a problem that can be addressed through the proper use of the Strategic Petroleum Reserve.

NATIONAL SECURITY: THE LONG RUN

Secure energy supplies need not lie in oil reserves around the world. They may also lie in alternatives to conventional oil and natural gas—shale oil, coal gasification, solar power, nuclear fusion, and others. Experience shows that this sort of research cannot be left entirely to market forces. The costs of commercial-scale plants are far higher, and the development times far longer, than commercial enterprises can support. If government investment in a Strategic Petroleum Reserve to maintain short-run energy security is appropriate, then government support for the long-term—twenty- or fifty-year—research needed to ensure long-run energy security is also appropriate.

V. PUBLIC POLICY CHALLENGES

The oil price increases of 1973, 1979, and 1981 induced a permanent change in world demand for energy. The demand for energy will grow as the world economy grows, but much less rapidly than would have been the case if the OPEC-administered price increases had never taken place. As the price of oil falls, the use of oil relative to other energy sources will increase, but industry will maintain its flexibility in order to reduce its use of oil in the event of oil price increases. The demand side of the world oil market, therefore, is permanently changed in a way unfavorable to any would-be cartel.

There is a corresponding change on the supply side of the market: the expansion of fringe or non-OPEC supply as oil companies and less developed countries seek secure oil supplies. It will take a prolonged period of low oil prices—five or six years—to shrink the non-OPEC supply and move the market toward, but not to, the precrisis situation of Figure 2-3(a). Some of the increased fringe supply is politically rather than economically motivated and will not leave the market even in the face of extended low prices.

A critical change in structure is the expanded number of players. For generations, the world oil market was dominated by the Seven Sisters, the vertically integrated majors. For somewhat more than a decade, the world oil market was dominated by the thirteen OPEC member nations. For the foreseeable future, events on the world oil market will reflect the actions of the OPEC member nations (integrating forward toward the final consumer), the international majors (integrating backward through the development of non-OPEC reserves), the new supplying nations, and the independent oil companies. These various firms and nation-firms will collude when they can and compete when they must. The greater number of suppliers reduces the likelihood of successful long-run collusion.

Nonetheless, when the growth of demand and the reduction in fringe supply present an opportunity to OPEC or an expanded "world OPEC," it will be taken. Periodically, supply will be cut back and oil prices will rise. That is, in the future, there will be more oil price shocks.

Governments will—if they can —mitigate the effects of these shocks. The policy to do so will use the market as a trigger to release reserves over the short run and to share reserves among industrialized countries and to supplement the market to develop long-run alternative sources of supply. Policies that fail to use the information provided by the market in the short run and that rely on the market to develop new technology over the long run will exacerbate future oil shocks.

NOTES

1. Three of these were survivors of the 1911 breakup of the Standard Oil Trust: Standard Oil of New Jersey, now Exxon; Standard Oil of New York, now Mobil; and Standard Oil of California, now Chevron.

2. J. E. Hartshorn, *Oil Trade: Politics and Prospects* (Cambridge: Cambridge University Press, 1993), p. 117. The economic and political tensions among the international cartel partners are described in Walter Adams and James W. Brock, "Retarding the Development of Iraq's Oil Resources: An Episode in Oleaginous Diplomacy, 1927–1939," *Journal of Economic Issues*, March 1993, pp. 69–93.

3. M. A. Adelman, *The World Petroleum Market* (Baltimore: Johns Hopkins University Press, 1972), pp. 84–87. Relevant here is Stigler's theory

of oligopoly: The more rapidly that cheating is likely to be detected, and therefore subject to retaliation, the less likely cheating is to occur. See George J. Stigler, "A Theory of Oligopoly," *Journal of Political Economy*, February 1964, pp. 44–61, reprinted in George J. Stigler, *The Organization of Industry* (Homewood, IL: Irwin, 1968), pp. 39–63.

4. Anthony Sampson, *The Seven Sisters* (New York: Bantam Books, 1975), p. 147, describing the findings of the Federal Trade Commission report *The International Petroleum Cartel*.

5. S. A. Schneider, *The Oil Price Revolution* (Baltimore: Johns Hopkins University Press, 1983), p. 46.

6. Senate Subcommittee on Antitrust and Monopoly, Cabinet on the Judiciary, *The Petroleum Industry: Part 4, The Cabinet Task Force on Oil Import Control* (Washington, DC: U.S. Government Printing Office, March 1970).

7. Sampson, *Seven Sisters*, pp. 174–175.

8. Ibid., pp. 253–272.

9. Ibid., pp. 278–282. On the eventual end of this process, see Youssef M. Ibrahim, "A U.S. Era Closes at Aramco," *International Herald Tribune*, April 6, 1989, p. 13.

10. Louis Turner, *Oil Companies in the International System* (London: Allen & Unwin, 1983), p. 136.

11. Exxon and Shell, for example, are equal partners in all their North Sea assets.

12. Paul Horsnell and Robert Mabro, *Oil Markets and Prices* (Oxford: Oxford University Press, 1993), pp. 25–28.

13. See Steven Butler, "Oil Traders Devise Strategies for the 21st Century," *Financial Times*, May 25, 1990; and Horsnell and Mabro, *Oil Markets and Prices*.

14. The inverse of the Herfindahl index gives the number of equally sized firms that would produce the level of concentration measured by the Herfindahl index. See M. A. Adelman, "Comment on the 'H' Concentration Measure as a Numbers Equivalent," *Review of Economics and Statistics*, February 1969, pp. 99–101. The concentration and Herfindahl figures are computed from output data reported in the *Basic Petroleum Data Book*, vol. 13, September 1993.

15. Walter Adams, "Vertical Divestiture of the Petroleum Majors: An Affirmative Case," *Vanderbilt Law Review*, November 1977, pp. 122–123.

16. *Basic Petroleum Data Book*, vol. 13, September 1993, sec. 15, Table 2.

17. *1984 National Petroleum News Factbook Issue*, p. 103.

18. Pipelines assume an increasingly important role in the Middle East, where they provide a way of moving oil that avoids the Persian Gulf and (some) of the hazards of war.

19. Walter Adams and James W. Brock, "Deregulation or Divestiture: The Case of Petroleum Pipelines," *Wake Forest Law Review*, October 1983, pp. 711–712.

20. Ibid., pp. 1134–1135.
21. James M. Griffen, "OPEC Behavior: A Test of Alternative Hypotheses," *American Economic Review*, December 1985, pp. 954–963.
22. Darius Gaskins, "Dynamic Limit Pricing: Optimal Limit Pricing Under Threat of Entry," *Journal of Economic Theory*, September 1971, pp. 306–322; Norman J. Ireland, "Concentration and the Growth of Market Demand," *Journal of Economic Theory*, October 1972, pp. 303–305.
23. Youssef M. Ibrahim, "Iran Threat to Increase Oil Output," *New York Times*, November 22, 1988, p. D1.
24. Youssef M. Ibrahim, "Saudi Oil Output Is Highest in Decade," *International Herald Tribune*, November 5, 1992, p. 2.
25. See Bruce Bringhurst, *Antitrust and the Oil Monopoly* (Westport, CT: Greenwood Press, 1979).
26. *U.S.* v. *American Petroleum Institute et al.*, Civil No. 8524 (D.D.C., October 1940).
27. B. I. Kaufman, "Oil and Antitrust: The Oil Cartel Case and the Cold War," *Business History Review*, Spring 1977; Sampson, *Seven Sisters*, pp. 150–159.
28. Irvin M. Grossack, "OPEC and the Antitrust Laws," *Journal of Economic Issues*, September 1986, pp. 725–741.
29. See Douglas R. Bohi and Michael A. Toman, "Oil Supply Disruptions and the Role of the International Energy Agency," *Energy Journal*, April 1986, pp. 37–50; David R. Henderson, "The IEA Oil-Sharing Plan: Who Shares with Whom?" *Energy Journal*, October 1987, pp. 23–31; and George Horwich and David Leo Weimer, eds., *Responding to International Oil Crises* (Washington, DC: American Enterprise Institute for Public Policy Research, 1988).
30. Erik Ipsen, "U.S. Bemoans Cheap Oil," *International Herald Tribune*, October 26, 1993, p. 11. Such views appear to prevail regardless of political affiliation. In a widely publicized April 1986 visit to Riyadh, then Vice-President George Bush was quoted as saying, "[T]he interest in the United States is bound to be cheap energy if we possibly can. But from our interest, there is some point where the national security interests of the United States say, 'Hey, we must have a strong, viable domestic industry'." (UPI, reported in the *New York Times*, April 7, 1986, p. 1).
31. John Badan, "A Welfare Plan for U.S. Oil," *Wall Street Journal*, December 18, 1987, p. 22.

SUGGESTED READINGS

Adelman, M. A. *The World Petroleum Market*. Baltimore: Johns Hopkins University Press, 1972.
Crémer, Jacques, and Djavad Salehi-Isfahani. *Models of the Oil Market*. Chur: Harwood Academic Publishers, 1991.

Hartshorn, J. E. *Oil Trade: Politics and Prospects*. Cambridge: Cambridge University Press, 1993.

Heal, Geoffrey, and Graciela Chichilnisky. *Oil and the International Economy*. Oxford: Clarendon Press, 1991.

Horsnell, Paul, and Robert Mabro. *Oil Markets and Prices*. Oxford: Oxford University Press, 1993.

Mabro, Robert, Robert Bacon, Margaret Chadwick, Mark Halliwell, and David Long. *The Market for North Sea Crude Oil*. Oxford: Oxford University Press, 1986.

Sampson, Anthony. *The Seven Sisters*. New York: Bantam Books, 1976.

Schneider, Steven A. *The Oil Price Revolution*. Baltimore: Johns Hopkins University Press, 1983.

Turner, Louis. *Oil Companies in the International System*. 3rd ed. London: Allen & Unwin, 1983.

Yergin, Daniel. *The Prize*. New York: Simon & Schuster, 1991.

3

AUTOMOBILES

Walter Adams and James W. Brock

Automobile production is one of the most influential segments of the American economy. The 8.2 million new cars sold in the United States in 1992 represented buyer outlays of some $140 billion. The industry employs more than 800,000 people directly and is at the epicenter of a vast constellation of allied fields, accounting for 77 percent of all U.S. rubber consumption, 60 percent of malleable iron purchases, 40 percent of machine tools, 25 percent of glass consumption, and 20 percent of semiconductor purchases. The automotive supplier–manufacturing–assembly base comprises more than 4,000 plants in forty-eight states and the District of Columbia, with its finished products distributed through 30,000 dealer franchises employing another 900,000 people.

Dominated by three of the world's largest firms, automobile production also is one of the most concentrated of all American industries. This, in conjunction with their widespread economic impact, means that the performance of the Big Three is inevitably a matter of vital national concern. It is all the more significant, then, that automobiles have come to epitomize the "American disease" of an industry wracked by foreign competition, layoffs of hundreds of thousands of workers, shutdowns of scores of plants, and financial losses of tens of billions of dollars.

I. HISTORY

The automobile as we know it today first took shape in the 1890s. The early pioneers experimented with gasoline engines, steam engines, and electric motors as sources of propulsion. By 1900, they had sold approximately 4,000 cars. Production expanded rapidly thereafter, reaching 187,000 automobiles by 1910. Entry into the industry was relatively easy. The manufacturer of automobiles was primarily an assembler of parts produced by others. The new entrepreneur needed only to design a vehicle; announce to the public its imminent availability; and contract with machine shops and independent producers for the engines, wheels, bodies, and other components.

The next decade marked the emergence of the Ford Motor Company as the dominant producer. Believing the demand for new cars to be price elastic, Henry Ford's goal was to provide an inexpensive car capable of reaching a large potential market. Standardization, specialization, and mass production, he felt, were the keys to lowering manufacturing costs, and constant price reduction the key to tapping successively larger layers of demand. "Every time I reduce the charge for our car by one dollar, I get a thousand new buyers," Ford said. His strategy seemed simple enough: to take lower profits on each vehicle and thereby achieve larger volume. As Ford saw it, successive

> price reductions meant new enlargements of the market, and acceleration of mass production's larger economies, and greater aggregate profits. The company's firm grasp of this principle. . .was its unique element of strength, just as failure to grasp it had been one of the weaknesses of rival car makers. As profits per car had gone down and down, net earnings had gone up and up.[1]

By 1921, Ford's Model T (which had remained largely unchanged for nineteen years) accounted for more than half the market.

The 1920s witnessed a shift of preeminence from Ford to General Motors (GM), the latter a consolidation of formerly independent firms (Chevrolet, Oldsmobile, Oakland, Cadillac, Buick, Fisher Body, Delco). GM adopted a two-pronged strategy: First, contrary to Ford's emphasis on a single model, GM offered a broad array of products to blanket all market segments. Its motto was "a car for every purse and purpose." Second, again contrary to Ford's strategy, GM elected to modify its cars each year with a combination of engineering advances, convenience improvements, and styling changes. GM felt that annual model changes, despite the expense, would stimulate replacement demand and increase sales, and indeed, this strategy catapulted the company into unchallenged industry leadership for half a century.

In this third era, the groundwork was also laid for the high concentration that became the hallmark of the auto industry. Figure 3-1 depicts the evolutionary process that produced the triopoly that dominates U.S. production.

Starting in the mid-1950s, successive waves of imports challenged the domestic oligopoly. By the 1970s, imports had captured more than 25 percent of the U.S. market and triggered repeated efforts by the Big Three—in collaboration with the United Auto Workers—to obtain government protection from foreign competition. In the 1980s, in partial response to the domestic industry's lobbying campaigns, foreign firms (primarily Japanese) began to construct plants in the United States; by 1992, the combined output of these "transplants" equaled that of Ford Motor Company.

FIGURE 3-1 Historical evolution of the auto industry's structure.

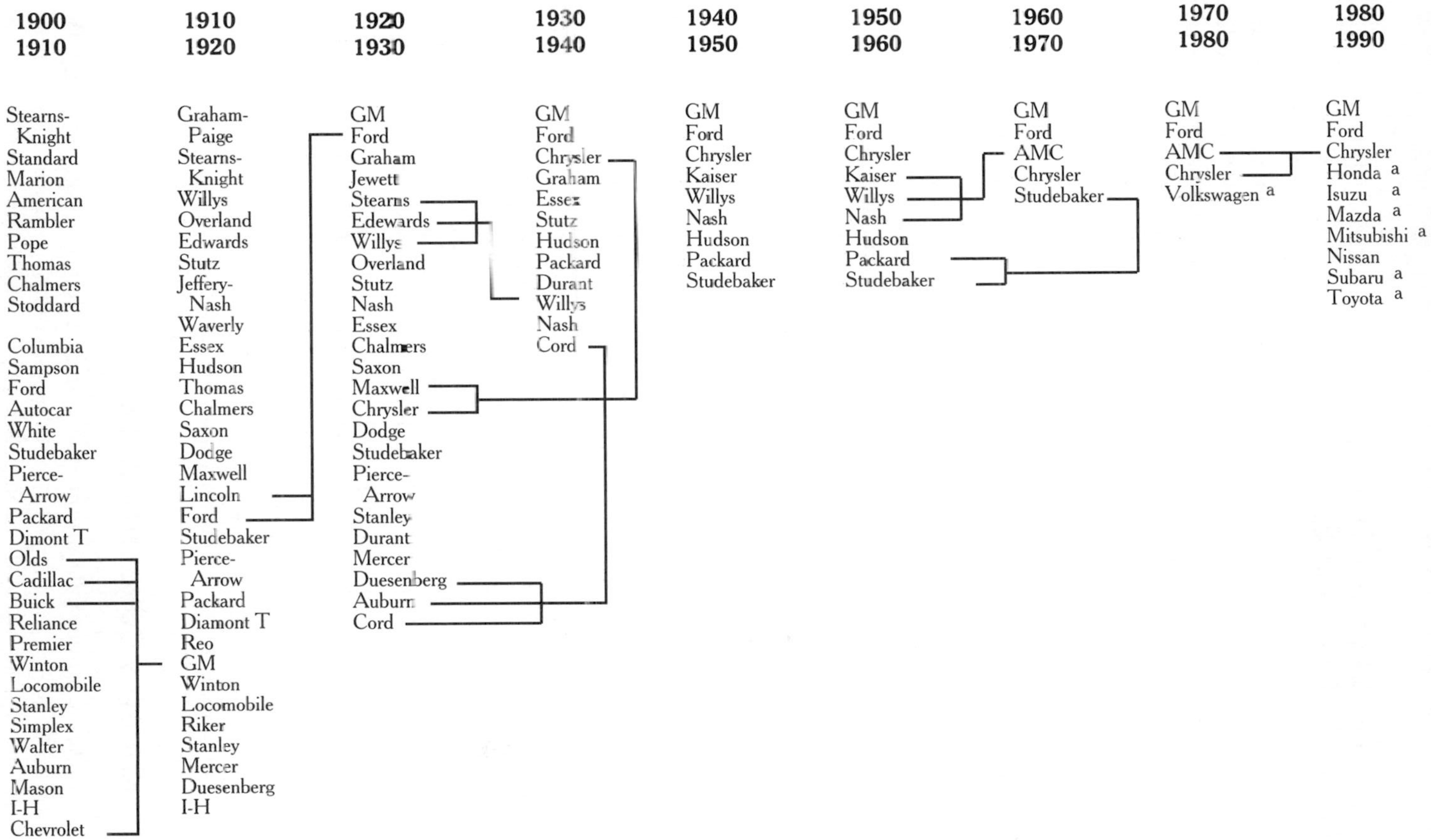

1900 1910

Stearns-
Knight
Standard
Marion
American
Rambler
Pope
Thomas
Chalmers
Stoddard

Columbia
Sampson
Ford
Autocar
White
Studebaker
Pierce-
  Arrow
Packard
Dimont T
Olds
Cadillac
Buick
Reliance
Premier
Winton
Locomobile
Stanley
Simplex
Walter
Auburn
Mason
I-H
Chevrolet

1910 1920

Graham-
Paige
Stearns-
Knight
Willys
Overland
Edwards
Stutz
Jeffery-
Nash
Waverly

Essex
Hudson
Thomas
Chalmers
Saxon
Dodge
Maxwell
Lincoln
Ford
Studebaker
Pierce-
Arrow
Packard
Diamont T
Reo
GM
Winton
Locomobile
Riker
Stanley
Mercer
Duesenberg
I-H

1920 1930

GM
Ford
Graham
Jewett
Stearns
Edewards
Willys
Overland
Stutz
Nash
Essex
Chalmers
Saxon
Maxwell
Chrysler
Dodge
Studebaker
Pierce-
  Arrow
Stanley
Durant
Mercer
Duesenberg
Auburn
Cord

1930 1940

GM
Ford
Chrysler
Graham
Essex
Stutz
Hudson
Packard
Durant
Willys
Nash
Cord

1940 1950

GM
Ford
Chrysler
Kaiser
Willys
Nash
Hudson
Packard
Studebaker

1950 1960

GM
Ford
Chrysler
Kaiser
Willys
Nash
Hudson
Packard
Studebaker

1960 1970

GM
Ford
AMC
Chrysler
Studebaker

1970 1980

GM
Ford
AMC
Chrysler
Volkswagen [a]

1980 1990

GM
Ford
Chrysler
Honda [a]
Isuzu [a]
Mazda [a]
Mitsubishi [a]
Nissan [a]
Subaru [a]
Toyota [a]

*U.S. plants opened by foreign-based producers.

67

II. INDUSTRY STRUCTURE

The most important structural features of the U.S. automobile industry are buyer demand and the nature of the product; the number of rival sellers and their relative size (concentration); the extent of economies of scale; and barriers to new competition.

DEMAND AND THE NATURE OF THE PRODUCT

The demand for new cars is influenced by a variety of factors. First, the demand for new cars is predominantly a replacement demand. Because the purchase of a new car can be deferred, market demand is quite volatile. Second, because the purchase of an automobile constitutes a major investment for the average household (at a median price of $18,000 in 1992), the demand for new cars is highly sensitive to macroeconomic conditions, including employment, income trends, and interest rates. Thus, during recent business cycles, annual U.S. new-car sales varied dramatically, rising from 9.2 million units in 1983 to 11.5 million in the boom year 1986 and then plunging to 8.2 million in 1992, with the 1993 economic recovery compelling the industry once again to scramble to expand production capacity in an effort to meet sharply rising consumer demand. Third, an important determinant of demand is, of course, price. Although the aggregate demand for new cars is slightly price elastic, the demand for a particular model is much more price sensitive because of the availability of close substitutes.

Noteworthy is the long-term shift in the composition of new-car demand in favor of smaller cars and more utilitarian vehicles. The share of small cars (compacts and subcompacts) rose from 25 percent of all new-car sales in 1967 to 52.4 percent by 1984; over the same period, the share of large cars declined from 51 to 21.4 percent. This shift is partly explained by fuel shortages and the skyrocketing price of gasoline in the 1970s. Yet the trend in demand toward smaller cars began well before the first oil embargo of 1973. Gasoline supplies and prices may have amplified this trend, but other factors have been exerting a long-run influence, such as smaller family size and the growing prevalence of multiple-car households, whose second car is generally smaller and more utilitarian in purpose. This family/utilitarian trend has accelerated in recent years, with substantial increases in sales of minivans and light trucks.

INDUSTRY CONCENTRATION

Domestic automobile production is dominated by a tight triopoly. Because of the growing competition of imports, domestic sales are less concentrated, but the impact of this competition is attenuated by the trend toward cross-national ownership positions and joint ventures between the American oligopoly and its major foreign rivals, as well as the Big Three's persistent political pressure for government import restraints.

1. CONCENTRATION OF DOMESTIC PRODUCTION. The manufacture of automobiles in the United States is, as Table 3-1 shows, highly concentrated. From a high of eighty-eight firms in the early 1920s, the number of domestic producers had dwindled to four by the mid-1970s.

General Motors is, and during most of the post–World War II era has been, the largest firm in the industry, frequently producing more cars than the rest of the industry combined. Indeed, GM is the largest industrial corporation both in the United States and in the world: As Table 3-2 shows, GM's total sales revenues in 1992 equaled those of the two largest Japanese auto producers (Toyota, Nissan) combined. Only nineteen nations have gross national products exceeding GM's annual sales revenues! Ford, the nation's and the world's third largest industrial firm, accounts for approximately one-quarter of domestic production, and Chrysler, for approximately 10 percent. (American Motors, long a distant fourth, was acquired by Chrysler in 1987; Chrysler subsequently also acquired Jeep.) Table 3-3 provides details concerning the Big Three's North American assembly plants.

In recent years, the number of domestic producers has increased, with foreign firms constructing U.S. assembly plants. As Table 3-4 shows, Japanese automakers have launched seven new auto assembly operations. These production facilities represent a substantial development, comprising a combined production volume of 1.4 million cars in 1992 and accounting for 25 percent of U.S. new-car production in 1992 (up from less than 1 percent in 1983). Yet because many of these operations are joint ventures between American producers and their foreign rivals (or involve foreign firms in which U.S. pro-

TABLE 3-1 U.S. Auto Production: Market Shares and Concentration, 1913–1992 (%)

Year	General Motors	Ford	Chrysler	Other U.S. Producers	Share of Top Three
1913	12	40	[a]	48	
1923	20	46	2	32	68
1933	41	21	25	13	87
1946–1955	45	24	19	12	88
1956–1965	51	29	14	6	94
1966–1975	54	27	17	2	98
1976–1985	59	24	13	4	96
1992	42	24	9	25	75

[a] Chrysler not yet in existence.

Source: Lawrence J. White, *The Automobile Industry Since 1945* (Cambridge, MA: Harvard University Press, 1971); *Ward's Automotive Yearbook*, various years.

TABLE 3-2 World's Top Ten Auto Producers, 1992

Rank	Company	Vehicles Produced (million units)	Sales Revenues (billions)
1	General Motors	7.1	$ 132.8
2	Ford	5.8	100.8
3	Toyota	4.2	79.1
4	Volkswagen	3.5	56.7
5	Nissan	3.0	50.2
6	Chrysler	2.2	36.9
7	Peugeot	2.1	29.4
8	Renault	2.0	33.9
9	Honda	1.9	33.4
10	Mitsubishi	1.8	25.5

Source: Ward's Automotive Yearbook, 1993.

ducers hold sizable ownership stakes), they do not represent the entry of genuinely independent new competitors.

Nevertheless, the Big Three collectively continue to dominate the industry. They are extensively integrated not only horizontally but also vertically and internationally. GM, for example, produces an estimated 70 percent or more of the parts and components from which its automobiles are assembled; in-house parts production at Ford is about 50 percent; and Chrysler's internal supply of parts and components is on the order of 30 percent. During the latter 1980s, the Big Three vertically integrated in a new direction—forward into distribution—by buying stakes in a number of rental car firms: GM bought minority stakes in Avis and National; Ford acquired ownership stakes in Hertz and Budget Rent-a-Car; and Chrysler purchased the Thrifty, Dollar, General, and Snappy rental car concerns. Because rental agencies are the single largest buyers of new cars, the Big Three thus acquired substantial captive outlets for their automobiles.

GM, Ford, and Chrysler are also extensively integrated internationally. Ford, for example, is the largest auto producer in Australia, Britain, Mexico, and Argentina; the second largest in Canada and Australia; and the third largest in Brazil and Spain. General Motors is the second largest auto firm in western Europe, Canada, Australia, and Germany and the third largest in Britain. In fact, if the U.S. sales of cars, minivans, and light trucks manufactured in Canada are counted as imports (as befits the definition of the term), then Chrysler, General Motors, and Ford currently rank as three of America's top four importers of vehicles—ahead of Nissan, Honda, and Mazda. The recent adoption of the NAFTA (North American Free Trade Agreement) is likely to accelerate this trend toward the internationalization of the industry.

2. FOREIGN COMPETITION. The import share of U.S. auto sales grew from 0.4 percent in the immediate post–World War II decade to 21 percent in the

TABLE 3-3 Big Three Assembly Plants

Company	Plant Location	Annual Capacity (000 units)	1993 Models Assembled
General Motors	Arlington, TX	200	Buick Roadmaster Cadillac Brougham Chevrolet Caprice
	Bowling Green, KY	80	Corvette
	Detroit, MI	240	Buick Riviera Cadillac Allante Cadillac Eldorado Cadillac Seville Oldsmobile Toronado
	Doraville, GA	240	Cutlass Supreme
	Fairfax, KS	240	Pontiac Grand Prix
	Flint, MI	240	Buick LeSabre Oldsmobile 88
	Lake Orion, MI	240	Cadillac DeVille Oldsmobile 98
	Lansing, MI	250[a]	Buick Skylark Pontiac Grand Am Oldsmobile Achieva
	Lordstown, OH	500	Chevrolet Cavalier Pontiac Sunbird
	Oklahoma City, OK	280	Chevrolet Century Cutlass Ciera
	Oshawa, Ont.	240[b]	Chevrolet Lumina Buick Regal
	Ramos Arizpe, Mex.	150	Buick Century Chevrolet Cavalier Cutlass Ciera
	Spring Hill, TN	325	Saturn
	Ste. Therese, Queb.	200	Camaro, Firebird
	Van Nuys, CA	150	Camaro, Firebird
	Wentzville, MO	250	Buick Park Avenue Oldsmobile 88 Pontiac Bonneville
	Willow Run, MI	150	Buick Roadmaster Chevrolet Caprice

TABLE 3-3 Big Three Assembly Plants (cont'd)

Company	Plant Location	Annual Capacity (000 units)	1993 Models Assembled
			Oldsmobile Custom Cruiser
	Wilmington, DE	225	Chevrolet Beretta Chevrolet Corsica
Ford	Atlanta, GA	250	Taurus, Sable
	Chicago, IL	250	Taurus, Sable
	Cuautitlan, Mex.	150	Tempo, Thunderbird, Cougar, Grand Marquis, Topaz
	Dearborn, MI	200	Mustang
	Hermosillo, Mex.	180	Escort, Tracer
	Kansas City, MO	240	Tempo, Topaz
	Lorain, OH	240	Thunderbird, Cougar
	Oakville, Ont.	250	Tempo, Topaz
	St. Thomas, Ont.	220	Crown Victoria, Grand Marquis
	Wayne, MI	250	Escort
	Wixom, MI	250[c]	Continental, Mark VIII, Lincoln Town Car
Chrysler	Belvidere, IL	240	Fifth Avenue, Imperial New Yorker, Dynasty
	Bramalea, Ont.	250	Concorde, Intrepid, Vision
	Detroit, MI	5	Viper
	Newark, DE	240	Spirit, Acclaim, LeBaron
	St. Louis, MO	275	Caravan, Voyager, Town & Country
	Sterling Heights, MI	240	Daytona, Shadow, Sundance
	Toledo, OH	225	Jeep Cherokee, Wrangler

TABLE 3-3 Big Three Assembly Plants (cont'd)

Company	Plant Location	Annual Capacity (000 units)	1993 Models Assembled
	Toluca, Mex.	175	LeBaron, Spirit, Shadow, Acclaim, Sundance, Phantom New Yorker
	Windsor, Ont.	250	Caravan, Voyager

ª Two plants, each with capacity of 250,000.
ᵇ Two plants, each with capacity of 240,000.
ᶜ Combined capacity, two plants.

Source: *Ward's Automotive Yearbook*, 1993; *Automotive News*, 1993, *Market Data Book*.

years 1976 to 1983. In 1992, imports made up 24 percent of U.S. new-car sales, with Japanese makers accounting for three-quarters of these. However, this share is somewhat misleading because it includes "captive" imports produced abroad by foreign firms but marketed in the United States under Big Three nameplates. Examples include the Dodge Colt (produced by Mitsubishi in Japan), Chevrolet's Geo Storm (produced in Japan by Isuzu) and Geo Metro (produced in Japan by Suzuki), and Pontiac's LeMans (produced in South Korea by Daewoo).

Foreign producers, led by the Japanese, have been a critical, if not the only, source of effective competition for the U.S. oligopoly in the post–World War II period. Their competitive success has been especially significant as a force for deconcentrating the American automobile market. Thus, although GM, Ford, and Chrysler together accounted for three-quarters of U.S. production in 1992, import competition reduced their collective share of sales to 64 percent on the U.S. market (representing individual firm shares of 35, 21, and 8 percent for GM, Ford, and Chrysler, respectively).

Initially, foreign producers focused their efforts on the low-price, small-car segment of the market. Then, beginning in 1981, after the Big Three obtained government restraints on the number of Japanese cars imported, Japanese producers began to move upscale into the midsize region of the market—ironically, the traditional mainstay of the Big Three. Japanese and European producers also began to win larger shares of the high-price luxury end of the market (Lexus, Infiniti, Mercedes Benz, BMW). As a result, the Big Three now confront greater foreign competition across the entire spectrum of the American market.

TABLE 3-4 Transplants: Foreign-Owned U.S. Auto Production Operations

Company	Owner (s)	Location	Year Established	Employment	1992 (000 units)	Models Produced
AutoAlliance	Ford (50%) Mazda (50%)	Flat Rock, MI	1987	3,500	169	Mazda: MX-6, 626 Ford: Probe
Diamond–Star	Mitsubishi[a]	Normal, IL	1988	3,100	140	Chrysler: Laser, Talon Mitsubishi: Eclipse, Galant
Nissan Motor	Nissan	Smyrna, TN	1983	5,900	171	Altima, Sentra, compact trucks
Honda	Honda	Marysville, OH	1982	10,200	458	Accord, Civic
Subaru–Isuzu	Subaru (51%) Isuzu (49%)[b]	Lafayette, IN	1989	1,900	58	Isuzu: Rodeo, compact trucks Subaru: Legacy
Toyota Motor	Toyota	Georgetown, KY	1988	3,975	240	Camry
NUMMI	GM (50%) Toyota (50%)	Fremont, CA	1984	4,250	181	Chevrolet Geo Prizm Toyota Corolla, compact trucks
Total				32,825	1,417	

[a] Begun with Chrysler as co-owner; Chrysler sold its stake to Mitsubishi in 1991.
[b] GM holds 38% ownership stake in Isuzu.

Needless to say, however, import quotas and the perennial threat of protectionism jeopardize the otherwise salutary effect that foreign competition has had in eroding concentration in the U.S. auto market.

3. CROSS-OWNERSHIP AND JOINT VENTURES. Another significant feature of concentration in the industry is the burgeoning web of cross-ownership and joint ventures between U.S. and foreign automakers (see Figure 3-2). More

FIGURE 3-2 Recent Big Three partnership arrangements.

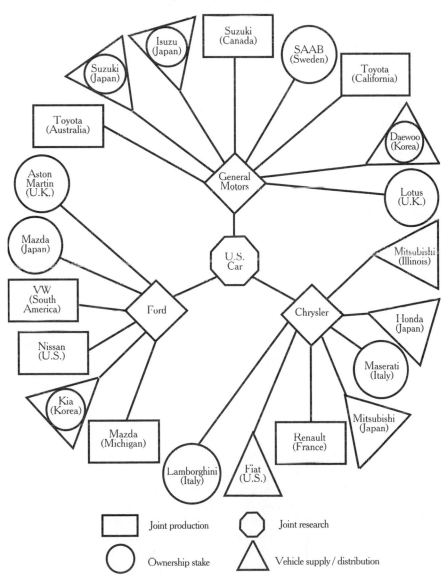

recently, these partnerships have proliferated and have come to include (1) emerging South Korean auto producers (Hyundai, Kia, Daewoo, Samsung); (2) joint research ventures among the Big Three themselves, such as the United States Council for Automotive Research (USCAR), which comprises a growing number of projects (including materials and composites, battery systems, and vehicle recycling); and (3) a joint effort between the U.S. government and the Big Three for the purpose of developing less-polluting, lightweight autos capable of traveling 60 to 80 miles per gallon.

Given the industry's inordinately concentrated structure, these joint ventures are especially problematic because they expand interfirm communication, coordination, and cooperation and because they thus threaten to further undermine independent decision making. Given the crucial competitive role of imports, joint ventures linking the U.S. oligopoly with its major foreign rivals are of particular concern.

THE QUESTION OF ECONOMIES OF SCALE

To what extent is this observed level of concentration necessitated by economies of large-scale operation? Are the economies of scale so great as to require the industry's tight triopoly structure?

In fact, economies of scale are not as extensive as might be commonly assumed. Engineering estimates put the minimum efficient scale of production in the 200,000- to 400,000-unit range. Although large in an absolute sense, these estimates imply that a firm with a 3 to 6 percent share of U.S. auto production would be big enough to reap all significant economies of scale or, conversely, that the industry could conceivably support fifteen to thirty efficient producers (at 1992 production levels). Some analysts suggest that the risks and vagaries of the market might require a firm to produce two distinct lines of automobiles, rather than one, thereby approximately doubling these minimum efficient scale estimates.[2]

An examination of actual plant-capacity levels corroborates the conclusion that there are definite limits to scale economies in auto production. As Table 3-3 indicates, the Big Three assemble cars at multiple locations rather than concentrating all their assembly operations in only one or two giant plants. Moreover, Table 3-3 also indicates that their plants are typically designed for capacity volumes in the 250,000-unit range, a figure quite comparable with the foregoing minimum efficient scale estimates. In addition, recent evidence suggesting that less vertical integration (and greater reliance on outside suppliers) lowers costs would further reduce the minimal optimal scale required for efficient production.[3]

GM's persistent woes seem to indicate not only that economies of scale are limited but also that there are serious diseconomies of scale. Measured by annual dollar sales, GM is one-third bigger than its next largest rival, Ford, three and a half times larger than Chrysler, and bigger than the two largest Japanese auto firms (Toyota, Nissan) combined. But GM's size seems to be a liability rather than an asset: It has the lowest productivity and the highest per-car production costs in the industry. Despite modernization investments of billions of dollars, GM's production costs are variously estimated to be $200 to $2,000 higher per car than those of Ford, Chrysler, and the Japanese.[4] By its own admission, GM has been forced to turn to joint ventures with smaller foreign rivals in a struggle to learn how to make cars economically. In addition, GM's "Saturn" project—a separate, stand-alone auto-manufacturing division with its own production, procurement, and marketing operations—is obviously intended to escape the bureaucratic morass of GM's disproportionately large size. Indeed, if the Saturn project makes sense, then GM's size does not—an increasingly prevalent conclusion. "The basic question nagging this biggest, most diverse, and most integrated of car companies," *Business Week* concludes, "is whether [GM] is just too big to compete in today's fast-changing car market."[5] Perhaps GM's main operating divisions (Buick, Cadillac, Chevrolet, Oldsmobile, Pontiac) would be closer to the optimal production scale if they were divested and operated independently.

BARRIERS TO ENTRY

Barriers to the entry of new competition, another important element of market structure, are substantial in the automobile industry.

According to the Department of Transportation, the cost of constructing a new auto assembly plant would amount to more than $200 million.[6] The production of major components and parts would drive costs even higher for a new entrant: The combined construction costs for four basic types of facilities (engine plant, transmission plant, parts and components plant, final assembly plant) would exceed $1 billion.

A new entrant must not only produce automobiles but also market them to consumers—another substantial obstacle to new competition. In 1991, the Big Three spent $2 billion on advertising their vehicles ($1.1 billion by GM, $518 million by Ford, $415 million by Chrysler). With these high levels of advertising by incumbent firms, a newcomer would have to incur heavy expenses merely to announce its arrival. Exclusive advertising contracts with television networks (according to which one of the Big Three is the only auto firm permitted to advertise for certain sports) may compound entry barriers. Moreover, the Big Three's recent acquisition of rental car agencies may raise entry barriers further yet, by reducing the opportunities for a new entrant to introduce its cars to the driving public via this avenue.

Finally, in addition to producing and advertising its cars, a new entrant would have to assemble a dealer system for distribution and service. Given the extensive dealer systems in place for the Big Three firms, this would be a daunting task: GM, Ford, and Chrysler cars are sold through some 34,000 dealer franchises.

It is not surprising, therefore, that since the early 1950s, few new firms have commenced domestic production of automobiles and those that have are established companies that have produced cars abroad and imported them into the United States for years.

III. INDUSTRY CONDUCT

Market behavior in the auto industry conforms to what would be expected in a tightly knit oligopoly. Strategic decisions with respect to both price and nonprice rivalry are marked by a notable degree of parallelism.

PRICING

As economic theory predicts, the general pricing pattern in autos is characterized by the recognition of mutual interdependence among a few large firms. The Big Three understand that their interest as a group is best served by avoiding serious price competition among themselves. The *Wall Street Journal* explains:

> Auto makers can maximize profits because in the oligopolistic domestic auto industry the three major producers tend to copy each other's price moves. One auto executive notes that if one company lowered prices, the others would follow immediately. . . .As a result, price cuts wouldn't increase anybody's market share, and "everybody would be worse off," he says.[7]

General Motors has traditionally served as the price leader. It typically initiates general rounds of price increases; it can, by refusing to follow, deter either of its main American rivals from launching price alterations; and its prices typically establish points around which its rivals gather. The prices adopted by the Big Three appear at times to represent the outcome of a tacit bargain arrived at through a delicate process of communication and signaling: One of them will begin by announcing its "preliminary" planned price hikes, and the others will follow by disclosing their own "tentative" or "anticipated" price changes. Once they have revealed their hands to one another, they then announce their final prices which, not surprisingly, tend to be quite similar.

In the less concentrated market segments populated by larger numbers of rivals, however, pricing patterns are distinctively different. In small cars, for example, there is no clearly identifiable price leader, and so pricing behavior exhibits the variability and unpredictability characteristic of more competitive markets. More generally, the disciplining role of imports in constraining the Big Three's pricing power is tellingly revealed during times when currency exchange rate changes compel foreign producers to boost substantially their U.S. prices, price hikes that the Big Three are quick to follow (as during the late 1980s and again in 1992–1993).

NONPRICE RIVALRY

With price competition eschewed where possible, rivalry in the industry is channeled primarily into nonprice areas. Most prominent among these are styling, model changes, and advertising. Yet nonprice conduct, too, is marked by mutual oligopolistic interdependence, both generally (in product uniformity among the Big Three) and specifically (in mutual restraint by the Big Three in exploring new market segments).

1. STYLING AND MODEL CHANGES. Product styling and annual model changes are the predominant mode of nonprice rivalry in the domestic industry. For the most part, these changes are cosmetic style variations rather than fundamental technological breakthroughs. From the industry's viewpoint, they have the advantage of stimulating the replacement demand for new cars while avoiding the competitive uncertainty and "market disruption" that aggressive technological rivalry would entail.

Yet as in the case of pricing, the traditional result of leadership/followership and "protective styling imitation" has been a surprising degree of uniformity in the offerings of the Big Three. A notable exception has been the minivan, a niche that Volkswagen pioneered in the 1950s with its microbus, which the specter of imminent bankruptcy compelled Chrysler to risk exploiting in the early 1980s and which has since grown to be a major market segment. The advent of foreign competition and the willingness of foreign producers to gamble on new ideas clearly have triggered a proliferation of distinctively different models and market niches in recent years.

2. ADVERTISING. Advertising constitutes a second dimension of nonprice rivalry in the industry, with the Big Three ranked among the ten largest advertisers in the nation. GM is the industry's largest advertiser, spending $1.1 billion on advertising in 1992; Ford and Chrysler spent $518 million and $415 million, respectively.

3. MUTUAL INTERDEPENDENCE AND OLIGOPOLISTIC RESTRAINT: THE CASE OF SMALL CARS. The historic resistance by the Big Three to the introduction of small cars poignantly illustrates the recognition of oligopolistic interdependence and mutual restraint in nonprice rivalry.[8]

At the conclusion of World War II, the small, lightweight, inexpensive automobile was seen as a prime means for expanding the postwar car market in a manner analogous to Henry Ford's Model T decades earlier. The United Auto Workers, for example, urged Detroit to build a small car, citing an opinion survey conducted by the Society of Automotive Engineers that revealed that 60 percent of the public wanted the industry to produce a small car. In May 1945, General Motors and Ford revealed that they were considering the production of small cars. The following year, Chrysler announced that "if the market exists and if other companies have a low-priced car, Chrysler will be ready with something competitive."

However, the Big Three did not seriously undertake to produce and market such a car until the 1970s, at least for the American market. (A small, lightweight car developed by GM was marketed in Australia in 1948 by a GM subsidiary; Ford's light car, the Vedette, appeared the same year in France.) Attempts were made to meet successive import surges with the introduction of the "compact" car in the late 1950s and the "subcompacts" of the 1960s. But these efforts seemed halfhearted and dilatory. In 1962, for example, Ford canceled the planned introduction of its "Cardinal," featuring a front-mounted, front-wheel drive layout similar to the X-cars, Escort, and Omni of twenty years later.

Lawrence White explains this antipathy to small cars in terms of oligopolistic firm behavior. General Motors, Ford, and Chrysler, he contends, each seemed to recognize that vigorous entry into small cars by any one of them would trigger entry into the field by the others. Further, each seemed to believe that the demand for small cars was not great enough to permit profits acceptable to the group if all simultaneously elected to enter this segment.

> Twice, one or two of the Big Three pulled back from plunging ahead with a small car when the market did not look large enough for all three. . . .A sizable niche might have been carved out at the bottom of the market by a Big Three producer in 1950, or again, with a "subcompact" in 1962 or 1963. Room-for-all considerations, however, appeared to rule this out.[9]

Reinforcing this "room-for-all" Weltanschauung was the apparent desire by each of the Big Three to protect the group's profits in large cars by withholding the small car as an inexpensive option. As White explains,

> In this behavior, the Big Three definitely recognized their mutual interdependence, since in the absence of retaliation by rivals a single firm contemplating the production of a small car should have expected to gain more profits from stealing the dissatisfied customers from other firms than he would lose from dissatisfied customers of his own large cars. . . .But the Big Three mutually contemplating a small car could only see lost profits from reduced sales of large cars.[10]

Foreign competitors, of course, were not immobilized by such considerations because they had no established positions to protect. They thus broke through the logjam of tacit restraint and forced the domestic oligopoly to confront the challenge of building small, lightweight, fuel-efficient automobiles. "In the absence of the press of competition from imports," White concludes, "it is likely that the big Three might never have provided small cars to the market."[11]

SUMMARY

Perhaps H. Ross Perot, founder of the EDS Corporation, erstwhile director of General Motors, and independent presidential candidate, best summarized the domestic industry's oligopolistic behavior since World War II:

General Motors and the entire American automobile industry had a big respite from competition. . . .[I]t got so bad that [the Big Three] tried to get divisions to compete with one another—Chevrolet compete with Pontiac, Oldsmobile with Buick, and so on. . . .Now we've got a whole generation of people who think that's what competition is. And I don't like that, and I say "Fellows, that's intramural sports." I said, "You don't even tackle there, you just touch the guy. . . .You don't even play with pads.". . .Now the Japanese have showed up, and they're competing professionally. . . .First board meeting. . .I gave 'em my immigrant's view of General Motors. And I said, "You don't understand competition."[12]

IV. INDUSTRY PERFORMANCE

For decades, industry defenders have claimed that the automobile market's highly concentrated market structure was necessitated by the efficiencies of large-scale production, the expense of modern technological innovation, and the need for effective planning. Testifying before a congressional committee in 1974, GM spokesmen insisted that General Motors' "size has been determined by the product itself, the requirements of efficient manufacture, distribution and service, as well as market demand. . . .Its present size is a result of its ability to satisfy consumer demand more effectively than its competitors."[13]

In the light of the domestic oligopoly's performance over the past two decades, such claims today seem tenuous: Chrysler's dependence on government financial support to rescue it from bankruptcy in 1979–1980; record-breaking financial losses by the Big Three in 1980 ($4.2 billion), 1991 ($7.5 billion), and 1992 ($30.1 billion); the closing of at least ten Big Three assembly plants between 1987 and 1990 (representing a combined reduction of 1.7 million vehicles in annual capacity); and layoffs of more than 300,000 workers.

PRODUCTION EFFICIENCY

One measure of the degree to which production inefficiency came to afflict the Big Three is evident in Table 3-5, comparing labor productivity trends for U.S. and Japanese auto producers between 1960 and 1983. These statistics reveal, first, that giant firm size does not guarantee efficiency in production and, second, how far the domestic oligopoly fell behind the state-of-the-art performance by others, especially the Japanese.

Confronted by withering foreign competition, the Big Three struggled through the 1980s to raise their productivity and to reduce their bloated cost structures. Some of their efforts have succeeded: Chrysler's labor productivity approximately doubled between 1980 and 1992; Ford's Atlanta plant may now be one of the world's most efficient; and by some estimates, Ford and

TABLE 3-5 Vehicles Produced per Worker: U.S. Versus Japenese Firms

	1960	1970	1983
General Motors[a]	8	8	11
Ford[b]	14	12	15
Chrysler[a]	11	11	16
Nissan	12	30	42
Toyota	15	38	58

[a] Worldwide.
[b] United States.

Source: Michael A. Cusumano, *The Japanese Automobile Industry* (Cambridge, MA: Harvard University Press, 1985), pp. 187–188.

Chrysler may now produce some of their cars at costs approximately equal to those of their most efficient foreign rivals. General Motors, however, continues to stumble under the burden of high production costs and low productivity. Indeed, after investing billions of dollars to refurbish its production plant and equipment during the 1980s, GM's newer plants are even less efficient than its older ones, and GM remains 30 to 40 percent less productive than its smaller American rivals, Ford and Chrysler, and only one-seventh as productive as Toyota.

PRICING

The pricing dimension of the industry's performance is marked by two troublesome problems. First, the long-run record of automobile prices is one of sustained, sizable price escalation. The average selling price of new cars in the United States rose 445 percent over the 1967–1992 period (from $3,300 to $18,000), a rate four times that of inflation in the economy generally. Measured differently, the fraction of family earnings required to buy a new car, on average, climbed from one-third in 1973 to nearly one-half in 1992, giving rise to a serious affordability problem.[14] Whether the industry's sharply increased reliance on leases to move cars (leases accounted for a record one-quarter of new-vehicle transactions in 1993) represents a price reduction to overcome this affordability problem or, alternatively, whether it offers a means for masking price hikes remains to be seen.

Second, and characteristic of a tightly knit oligopoly, the industry's prices behave in a perverse, administered fashion over the course of business cycles. In particular, the Big Three's prices tend to be rigid in a downward direction during recessions and, instead, are often raised in the face of sales declines.[15] Such perverse pricing exacerbates instability, not only in autos (by further depressing sales), but also in the economy at large (given the impact of automobile production on American industry generally).

DYNAMIC EFFICIENCY

Dynamic efficiency encompasses product innovation, an area in which the domestic industry's performance is notable in at least four respects.

First, the rate, breadth, and depth of product innovation were great in the era before World War II, when the field was populated by numerous independents. Innovation competition was intense, and new people with new ideas could put their concepts (the bad with the good) into commercial practice.

Second, with the demise of a vibrant independent sector and with the consolidation of the industry into a tight oligopoly, the pace of genuine technological innovation slackened. Innovations like front-wheel drive, disc brakes, fuel injection, utilitarian minivans and fuel-efficient subcompacts, and four-wheel steering languished in the hands of the Big Three. According to David Halberstam,

> Since competition within the industry was mild, there was no impulse to innovate; to the finance people, innovation not only was expensive but seemed unnecessary. . . . Why bother, after all? In America's rush to become a middle-class society, there was an almost insatiable demand for cars. It was impossible not to make money, and there was a conviction that no matter what the sales were this year, they would be even greater the next. So there was little stress on improving the cars. From 1949, when the automatic transmission was introduced, to the late seventies, the cars remained remarkably the same. What innovation there was came almost reluctantly.[16]

Third, while the domestic oligopoly slumbered, foreign producers took the lead in exploiting the frontiers of automotive technology. According to veteran industry observer Brock Yates, foreign firms "continued to move ahead with fuel injection, disc brakes, rack and pinion steering, radial tires, quartz headlights, ergonomically adjustable bucket seats, five-speed manual transmissions, high-efficiency overhead camshaft engines, independently sprung suspensions, advanced shock absorbers, and strict crash-worthiness standards."[17]

Fourth, such competition from abroad has compelled the Big Three to become more innovative. And they have recorded some notable gains, for example, the success of Ford's aerodynamic Taurus program (which in 1992 edged out the Honda Accord as the top selling model in America), GM's Saturn project, and the quality improvements shown in Table 3-6. Yet the Big Three continue to lag on the innovation front, with foreign producers—especially the Japanese—continuing to hold the lead in developing and commercializing such advanced innovations as all-wheel steering, electronically controlled suspensions and transmissions, and lightweight ceramic engine componentry. The magnitude of the domestic oligopoly's challenge is evident in Table 3-7, showing the results of a recent in-depth study of the innovation process at American and Japanese firms. These researchers found that Japanese automotive R&D projects "were completed in two-thirds the time and

TABLE 3-6 Automotive Quality: Average Number of Problems per 100 New Cars Sold

	1980	1990
General Motors	108	40
Ford	100	35
Chrysler	89	31
Honda	34	14
Nissan	47	15
Toyota	24	16

Source: Consumer Reports, 1980 and 1990 annual edition.

with one-third the engineering hours of the non-Japanese projects. In absolute terms, the Japanese used an average of 2 million fewer engineering hours and typically completed a project more than a year and a half earlier." They pointedly note that these "differences have significant implications in an industry in which engineers may be a constrained resource, a model's life may be only four to five years, and market demands are continually changing."[18]

SOCIAL EFFICIENCY

Social efficiency examines how well an industry serves the broader public interest.

1. SMOG AND AUTOMOTIVE AIR POLLUTION. By the early 1960s, the typical American automobile spewed approximately one ton of pollutants per year into the nation's atmosphere, and motor vehicles accounted for an estimated 60 percent of all air pollution. At first, the industry denied the existence of the

TABLE 3-7 Automotive R&D Projects: Japan Versus U.S.

	Japan	United States
Number of R&D projects studied	12	6
Engineering hours consumed per project (average, in thousand hours)	1,155	3,478
Lead time (average, in months)	43	62
Average number of body types	2.3	1.7
Ratio of unique parts (%)	82	62
Cost of parts developed by outside suppliers (%)	70	19

Source: Adapted from Kim B. Clark, W. Bruce Chew, and Takahiro Fujimoto, "Product Development in the World Auto Industry," *Brookings Papers on Economic Activity 3* (1987): Table 1, p. 741.

problem. "[W]aste vapors are dissipated in the atmosphere quickly and do not present an air-pollution problem," Ford Motor Company told Los Angeles County supervisors in 1953. "The fine automotive powerplants which modern-day engineers design do not 'smoke'."[19]

Later, as automotive air pollution worsened and as national concern about the problem heightened, the domestic auto companies—under the guise of a "joint venture"—maneuvered to eliminate competition among themselves in developing and commercializing pollution control technology. In an antitrust suit filed in 1969 and not contested by the industry, the Justice Department charged that domestic auto firms "conspired not to compete in research, development, manufacture, and installation of [smog] control devices, and did all in their power to delay such research, development, manufacturing, and installation."[20]

When the government promulgated auto-emission regulations in the 1970s, the Big Three insisted that the regulations were impossible to meet, even though Honda and other foreign producers introduced redesigned engines combining high performance with low exhaust emissions and without the need for costly, complicated catalytic converters. When the Big Three finally responded, however, they chose the catalytic converter as the foundation for their approach, an approach that the National Academy of Science characterized as "the most disadvantageous with respect to first cost, fuel economy, maintainability, and durability."[21]

Only under extreme government pressure does the domestic oligopoly act, as in 1992, when in response to tough new air-quality standards enacted in California, Ford unveiled modified cars able to meet the standards four years ahead of time, at an additional production cost of only $100 per car and with no major technological changes required.

2. **AUTOMOTIVE SAFETY.** In 1965, a Senate committee reported,

> 49,000 persons lost their lives in highway accidents, 1,500,000 suffered disabling injuries, and an equal number suffered non-disabling injuries. . . . Since the introduction of the automobile in the United States, more Americans have lost their lives from highway accidents than all the combat deaths suffered by America in all our wars.[22]

Although a variety of factors influence automobile safety (road design, weather conditions, reckless and drunk driving), it is incontrovertible that the design of the automobile itself plays a key role. Nevertheless, for a long time the industry seemed casually indifferent to the problem. Patents awarded to the auto companies in the 1920s and 1930s for such safety features as padded dashboards and collapsible steering wheels were shelved for decades until their incorporation was mandated by government decree. As automobiles became progressively more dangerous in design over the postwar period, the industry insisted that safety should be optional, supplied only in response to consumer demand. Yet it steadfastly refused to make available the safety information and options essential to informed and free consumer decision making. The industry spent millions extolling raw horsepower and rocket acceleration while hiding behind its slogan "Safety don't sell."

Eventually, the protracted decades-long battle over safety produced results. The Big Three conceded the importance of safety belts. And in 1990—after stubbornly resisting air bags for two decades and actively dissuading consumers from purchasing them[23]—the Big Three discovered that "Safety *does* sell." They raced to equip their new cars with these life-saving devices, thereby providing themselves a significant advantage over many of their foreign rivals. "What is tragic and ironic" about the industry's traditional resistance to safety, one expert notes, "is that in the early 1980s, publication of crash test information was the only area in which the domestic car companies were beating the Japanese."[24]

The explanation may once again be rooted in oligopolistic interdependence and mutual restraint. As the former GM president Alfred Sloan once put it,

> I feel that General Motors should not adopt safety glass for its cars. I can only see competition being forced into the same position. Our gain would be purely a temporary one and the net results would be that both competition and ourselves would have reduced the return on our capital and the public would have obtained still more value per dollar expended.[25]

3. AUTOMOTIVE FUEL CONSUMPTION. The fuel economy of automobiles decisively affects the nation's petroleum consumption and significantly influences national dependence on geopolitically volatile foreign oil supplies.

Here again, however, the domestic industry initially considered neither the fuel efficiency of its products nor the looming problem of finite petroleum supplies to be particularly pressing concerns. It ignored warnings, even when sounded by some of its own officials. Instead, the industry seemed to absolve itself of responsibility. Asked in 1958 what steps his division was taking in the area of fuel economy, the general manager of GM's Buick division replied, "We're helping the gas companies, the same as our competitors."[26] Only months before the first OPEC oil embargo and the energy crisis of 1973, GM's chairman suggested more rapid licensing of nuclear electric power plants as a good means for dealing with America's energy problem. One month before the overthrow of the shah of Iran in 1979 and the onset of the nation's second oil crisis in six years, GM publicly took the position that auto "fuel-economy standards are not necessary and they are not good for America."[27]

As a consequence of the industry's nonchalant attitude, the fuel efficiency of U.S. automobiles steadily worsened from 1958 to 1973. Ironically, this made the industry especially vulnerable to the flood of fuel-efficient imports that swamped the American market in the wake of gasoline shortages and skyrocketing fuel prices. Under the press of government regulation, the Big

Three have since nearly doubled the fuel economy of their cars. Yet in 1991, in the wake of the Gulf War and America's emergency efforts once again to secure its foreign oil supplies, the domestic oligopoly resisted pleas that it continue to improve the fuel economy of its cars. Once again, the Big Three insisted that additional gains in fuel efficiency were technologically impossible—even though in 1991, at the same time that the Big Three were gearing up their political lobbying apparatus, Honda unveiled its Civic VX—a car able to get 55 miles per gallon, with more horsepower and a roomier passenger compartment than Chevrolet's Cavalier.

V. PUBLIC POLICY

Public policy toward the automobile industry can be examined in three areas: antitrust, protection from foreign competition, and government regulation of automotive smog, safety, and fuel economy.

ANTITRUST

Most of the antitrust suits in the industry have been tangential and peripheral in nature and have never squarely challenged the industry's concentrated structure. For example, the government charged GM and Ford with collusive pricing, but only in the fleet market for new cars sold to businesses and rental car agencies. In another case, it charged the auto companies with unlawfully conspiring to eliminate competition, but only in the pollution control field. Of late, the government has abandoned even these pusillanimous antitrust forays; instead, it has permitted a rash of joint ventures and cross-ownership arrangements among the Big Three and their major foreign rivals, which pose serious anticompetitive problems. Moreover, in what might be considered "failing-company" antitrust, the government engineered the Chrysler bailout in 1979/1980, in part to prevent the automobile triopoly from degenerating into a duopoly.

PROTECTION FROM FOREIGN COMPETITION

Recognition of and respect for mutual oligopolistic interdependence among the Big Three became solidified in the post–World War II era. This, together with the protection afforded by formidable entry barriers, insulated the domestic industry from effective competition. Noncompetitive conduct, including tacit vertical collusion between management and organized labor and steady price–wage–price escalation, flourished in this noncompetitive milieu.

Foreign competition, initially dismissed as an anomaly, eventually began to disturb this oligopolistic bonhomie, inducing the domestic industry to seek government protection from imports. Beginning in 1974, management and labor began to lobby for protection. Their efforts were crowned with success when in 1981, the Reagan administration persuaded the Japanese to accept "voluntary" import quotas. These "temporary" quotas, ostensibly designed to give domestic producers "breathing space," were renewed in 1983 and formally expired in 1985, only to be replaced by a quota system promulgated by the Japanese government.

Predictably, the quotas substantially drove up the price of Japanese imports which, in turn, enabled the domestic oligopoly to push through sizable price boosts of its own. New-car prices (foreign and domestic) are variously estimated to have risen by $800 to $1,000 per car as a result of the quotas, with additional dealer markups on some Japanese models reaching as high as $2,600. In the aggregate, "protection" from foreign competition is estimated to have cost American buyers nearly $16 billion in artificially inflated prices. In addition, the quotas effectively cartelized Japanese production by allocating, via government fiat, quota shares among Japanese competitors for the U.S. market. And in an ironic twist, numerical quota restrictions impelled Japanese producers to take steps to circumvent them by launching production operations in the United States (prompting the Big Three to demand that government tighten the quotas to "offset" growing Japanese production in the United States), as well as by upgrading their offerings and invading the midsize and luxury segments of the market. As a result of both developments, Japanese producers have come to pose a much greater competitive threat to the Big Three.

The Big Three's political war against foreign competition continued in 1991 when they demanded that the government restrict imported minivans— this even though the Big Three account for 85 percent of U.S. minivan sales, one of their targets (Mitsubishi) had not yet even sold minivans in the United States, Chrysler is the nation's largest importer of minivans (from Canada), and Ford and Chrysler had previously contracted to market Japanese minivans under their own domestic nameplates!

REGULATION: SAFETY, POLLUTION, AND FUEL ECONOMY

Have government attempts to regulate automotive safety, pollution, and fuel consumption been too costly compared with the benefits obtained? Is it true, as the industry maintains, that excessive regulation "adds unnecessary costs for consumers, lowers profits, diverts manpower from research and development programs, and reduces productivity—all at a time when our resources are desperately needed to meet the stiff competition from abroad"?[28] Or is it a fact, as Henry Ford II conceded, that

> we wouldn't have had the kinds of safety built into automobiles that we have had unless there had been a Federal law. We wouldn't have had the fuel econ-

omy unless there had been a Federal law, and there wouldn't have been the emission control unless there had been a Federal law.[29]

Is it true, as the industry contends, that such government regulations deny freedom of choice to consumers? Or do the Big Three opportunistically exploit "free consumer choice" only when it suits their purposes, but with no concern about denying freedom of choice to consumers when it does not serve their self-interest (as in the case of preventing consumers from obtaining imported cars, safer cars, or less polluting cars)? Is it true, as the industry claims, that government regulations prevent it from obeying consumer sovereignty by supplying the products that car buyers prefer? Or is it the case that if the industry acted more responsibly, it might minimize the need for (and intrusiveness of) direct government regulation?

Some economists contend that less direct forms of regulation—for example, "incentive-based" measures, featuring taxes or "fees" imposed on the sale of unsafe, polluting, gas-guzzling cars—would be preferable. But is it realistic to assume that government would impose fees if to do so would jeopardize the financial viability of a General Motors, a Ford, or a Chrysler? Would the government seriously consider shutting down GM, for example, if the firm simply refused to pay the "fees"? Conversely, given their size and political clout, would not a threat by any of the Big Three to shut down almost inevitably compel the government to grant delays, extensions, exemptions, and so on, as the industry has repeatedly proved? Whether direct or "incentives based," public policy here as elsewhere seems bedeviled by the labor–industrial bigness complex that dominates the U.S. auto industry.

VI. CONCLUSION

Protected by their oligopoly control of a continental-sized domestic market during the booming postwar period, America's Big Three auto firms convinced themselves of their innate superiority. They thus became blinder and ever more vulnerable to the competitive storm building across the Pacific. When the full force of that storm hit, the economic and social damage was devastating. Denial and disbelief were followed by a decade-long struggle for recovery.

The American experience thus may well be a precursor of the challenges looming ahead for established European automotive firms, hitherto operating in markets tightly protected by their governments. Efforts by the world's industrialized nations to break down trade barriers, including NAFTA and Europe '92, clearly pose the threats—and opportunities—of greater global competition in automobiles. Whether other nations have the political will to subject their automotive manufacturers to such competition, whether their firms will embrace these competitive challenges or resist them, and whether transnational joint ventures among automobile firms will facilitate such competition or neutralize it are some of the unanswered questions confronting students of this industry.

NOTES

1. Ford Allen Nevins, *The Times, the Man, the Company* (New York: Scribner, 1954), p. 493.
2. Lawrence J. White, "The American Automobile Industry," in *The Structure of American Industry*, ed. Walter Adams, 6th ed. (New York: Macmillan, 1982).
3. Michael J. Smitka, "The Invisible Handshake: The Development of the Japanese Automotive Parts Industry," *Business and Economic History* 19 (1990): 163–171.
4. Alex Taylor, "The New Drive to Revive GM," *Fortune*, January 13, 1992, p. 53; Harbour & Associates, *The Harbour Report: Competitive Assessment of the North American Automotive Industry, 1989–1992* (1992); Economic Strategy Institute, *The Future of the Auto Industry* (1992).
5. *Business Week*, March 16, 1987, p. 110.
6. U.S. Department of Transportation, *The U.S. Automobile Industry: 1980* (Washington, DC: U.S. Government Printing Office, 1981), p. 66.
7. *Wall Street Journal*, August 3, 1983, p. 1.
8. This account is drawn from Lawrence J. White, "The American Automobile Industry and the Small Car, 1945–1970," *Journal of Industrial Economics* 20 (1972): 179; Brock Yates, *The Decline and Fall of the American Automobile Industry* (New York: Vintage Books, 1984), chap. 3; and Paul Blumberg, "Snarling Cars," *New Republic*, January 24, 1983.
9. White, "American Automobile Industry," p. 191.
10. Ibid., p. 180.
11. U.S. Congress, Senate, Subcommittee on Antitrust and Monopoly, *Hearings on S. 1176*, 93rd Cong., 2nd sess., 1974, p. 1957.
12. Interview, *Washington Post*, July 7, 1985.
13. Senate Subcommittee on Antitrust and Monopoly, *The Industrial Reorganization Act: Hearings Before the Subcommittee on Antitrust and Monopoly, Part 4: Ground Transportation Industries*, 93d Cong., 2nd sess., 1974, p. 2468.
14. Mary Connelly, "Dilemma of the '90s: 'I Can't Afford It'," *Automotive News*, July 5, 1993, p. 3.
15. For example, see Joseph B. White, "GM Follows Ford by Boosting Car Prices," *Wall Street Journal*, August 15, 1990, p. B1; John K. Teahen, "Big 3 Hike '92 Prices $540 Despite Sales Drought," *Automotive News*, August 12, 1991, p. 1; William J. Hampton, "GM's Price Hikes: Foresight or Folly?" *Business Week*, April 14, 1986, p. 36; Amal Nag, "Car-Price Increases May Backfire," *Wall Street Journal*, October 7, 1985, p. 25.
16. David Halberstam, *The Reckoning* (New York: Morrow, 1986), pp. 244–245.

17. Yates, *Decline and Fall of the American Automobile Industry*, p. 149.
18. Kim B. Clark, W. Bruce Chew, and Takahiro Fujimoto, "Product Development in the World Auto Industry," *Brookings Papers on Economic Policy* 3 (1987): 740–742.
19. Senate Subcommittee on Air and Water Pollution, *Hearings: Air Pollution—1967, Part 1*, 90th Cong., 1st sess., 1967, p. 158. Some of the Big Three were sufficiently concerned about the air pollution problem to begin researching it as early as 1938. See "Smog Control Antitrust Case," *Congressional Record*, May 18, 1971, pp. 15626–27 (House ed.).
20. "Smog Control Antitrust Case," p. 15627.
21. National Academy of Sciences, "Report by the Committee on Motor Vehicle Emissions, February 12, 1973," reprinted in *Congressional Record*, February 28, 1973, pp. 5832, 5849 (Senate ed.).
22. Senate Committee on Commerce, *Report: Traffic Safety Act of 1966*, 89th Cong., 2nd sess., 1966, pp. 2–3.
23. Albert R. Karr and Laurie McGinley, "Auto Shoppers Encounter Stiff Resistance When Seeking Air Bags at Ford Dealers," *Wall Street Journal*, July 31, 1986, p. 23.
24. Quoted in Max Gates, "NHTSA Under Gun for Simple Format on Crash Tests," *Automotive News*, July 26, 1993, p. 8.
25. Quoted in Senate Select Committee on Small Business, *Hearings: Planning, Regulation, and Competition—Automobile Industry*, 90th Cong., 2nd sess., 1968, p. 967.
26. Quoted in John Keats, *The Insolent Chariots* (New York: Lippincott, 1958), p. 14.
27. Senate Committee on Commerce, *Hearing: Automotive Research and Development and Fuel Economy*, 93d Cong., 1st sess., 1973, p. 564; Ed Cray, *Chrome Colossus* (New York: McGraw Hill, 1980), p. 524.
28. House Committee on Government Operations, *Hearings: The Administration's Proposals to Help the U.S. Auto Industry*, 97th Cong., 1st sess., 1981, p. 129.
29. Quoted in Senate Subcommittee for Consumers, *Hearings: Costs of Government Regulations to the Consumer*, 95th Cong., 2nd sess., 1978, p. 87.

SUGGESTED READINGS

Books and Reports
Automotive News. *Market Data Book* (annual).
Bianchi, Patrizio. *Industrial Reorganization and Structural Change in the Automobile Industry*. Bologna: CLUEB, 1989.
Bollier, David, and Joan Claybrook. *Freedom from Harm*. Washington, DC: Public Citizen, 1986.
Crandall, Robert W., et al. *Regulating the Automobile*. Washington, DC: Brookings Institution, 1986.
Cray, Ed. *Chrome Colossus*. New York: McGraw-Hill, 1980.

Ford, Henry. *My Life and Work*. New York: Doubleday, 1926.

Halberstam, David. *The Reckoning*. New York: Morrow, 1986.

Harbour & Associates. *The Harbour Report: Competitive Assessment of the North American Automotive Industry, 1979–1989*. Troy, MI: Harbour & Associates, 1990.

———. *The Harbour Report: Competitive Assessment of the North American Automotive Industry, 1989–1992*. Troy, MI: Harbour & Associates, 1992.

Ward's Automotive Yearbook (annual).

Womack, James P., Daniel T. Jones, and Daniel Roos. *The Machine That Changed the World*. New York: Rawson Associates, 1990.

White, Lawrence J. *The Automobile Industry Since 1945*. Cambridge, MA: Harvard University Press, 1971.

Wright, J. P. *On a Clear Day You Can See General Motors*. Grosse Pointe, MI: Wright, 1979.

Yates, Brock. *The Decline and Fall of the American Automobile Industry*. New York: Vintage Books, 1984.

Articles

Fisher, Franklin M., Zvi Grilches, and Carl Kaysen. "The Costs of Automobile Model Changes Since 1949." *Journal of Political Economy*, October 1962.

Hampton, William J., and James R. Norman. "GM: What Went Wrong?" *Business Week*, March 16, 1987.

Milner, Edward W., and George E. Hoffer. "Has Pricing Behavior in the U.S. Automobile Industry Become More Competitive?" *Applied Economics* 21 (1989).

Perot, H. Ross. "How I Would Turn Around GM." *Fortune*, February 15, 1988.

Taylor, Alex. "U.S. Cars Make a Comeback." *Fortune*, November 16, 1992.

Teahen, John K. "20 Years of Price Hikes," *Automotive News*, September 30, 1991.

STEEL

*Walter Adams**

I. INTRODUCTION

For almost three-quarters of a century, the steel industry was a quintessential prototype of a tight oligopoly whose concentrated structure led to oligopolistic behavior and resulted in lackluster economic performance.

Until the 1960s, a handful of vertically integrated giants dominated the industry. Their well-honed system of price leadership and followership was marked by a consummate insensitivity to changing market conditions. Their virtually unchallenged control over a continent-sized market made them lethargic bureaucracies oblivious to technological change and innovation. Their insulation from competition induced the development of a cost-plus mentality, which tolerated a constant escalation of prices and wages and a neglect of production efficiency. Eventually all this made the industry's markets vulnerable to invasion by newcomers, both domestic and foreign. When these newcomers finally appeared, the industry found itself in disarray and desuetude. It was saddled with a jumble of largely obsolete and poorly located plants and high production costs, and worst of all, it was without the expertise to deal with competitive challenges.

In 1951, Benjamin Fairless, the president of United States Steel, boasted that

> Americans don't like to take second place in any league, so they expect their steel industry to be bigger and more productive than the steel industry of any other nation on earth. It is; but what many Americans do not know is that their own steel industry is bigger than those of all other nations on earth put together.

This hubris was unwarranted. Forty years after Fairless's smug statement, the American steel industry was substantially smaller than Japan's and accounted for less than 14 percent of the world's steel output.

---

* I am deeply indebted to Professors Joel B. Dirlam and Hans Mueller for their substantial contributions to this chapter.

A brief look at the industry's history throws some light on these dramatic changes.

HISTORY

Before the formation of the U.S. Steel Corporation in 1901, the industry was the scene of active and, at times, destructive competition. Competition for market shares was vigorous, often taking the form of aggressive price cutting. Companies that failed to adopt the best technology or anticipate market shifts fell by the wayside. Unlike the British steel industry, which was becoming conservative and defensive, innovative American managers and an industrious work force created a dynamic industry that was highly cost-competitive in international markets.

In this early period, various gentlemen's agreements and pools were organized in an effort to control the production of steel rails, billets, wire, nails, and other products, but the outstanding characteristic of these agreements was the "frequency with which they collapsed." Their weakness was that inherent in any pool or gentlemen's agreement: "60 per cent of the agreers are gentlemen, 30 per cent just act like gentlemen, and 10 per cent neither are nor act like gentlemen."[1]

With the birth of U.S. Steel, a "trust to end all trusts," these loosely knit agreements were superseded by a more stable form of organization. Under the leadership of Judge Gary, its first president, U.S. Steel inaugurated the famous "Gary Dinners," which were a transparent form of collusion among competitors. Cooperation replaced competition, and U.S. Steel held a price umbrella over the industry, which was high enough to accommodate its much smaller, even marginal rivals.

This policy of "friendly competition," of course, was not without cost. It permitted the corporation's rivals to expand and to gain an increasing share of the market. Over time, it led to the transformation of an asymmetrical oligopoly, dominated by a giant firm, into an industry with a more balanced oligopolistic structure. Friendly competition had one other fundamental consequence. Eventually, when the oligopoly felt entrenched enough to pursue a policy of constant price escalation—in good times and bad—in periods of declining as well as rising costs—it attracted newcomers.

The challenge to the steel oligopoly, starting in the 1960s, came from two principal sources: foreign competition and the appearance of the minimills. Against the former, the industry's primary defense was to plead for government protection in the form of a variety of trade restraints. Against the latter, the primary response was the gradual abandonment of those industry segments in which the minimills had established themselves as the low-cost producers.

By the 1980s, the steel oligopoly seemed moribund, a collection of helpless giants begging for government relief from self-inflicted injury.

II. MARKET STRUCTURE

Before analyzing the structure of the American steel market, it is important to note some of the characteristics of that market. First, there are about ten thousand distinct iron and steel products, including pig iron; semifinished steel (billets, blooms, and slabs); rolled products (bars, rods, structurals, hot- and cold-rolled sheets, and plates); and steel products with a very high unit value—forgings and castings.

Second, even very narrowly defined products are often further differentiated according to metallurgy, physical properties, and surface conditions. In addition, differences in quality of the more sophisticated products (such as cold-rolled and coated sheets) have become important enough to make steel buyers quite selective concerning the reputation of their suppliers, both domestic and foreign.

Third, the term *market* connotes the interaction of buyers and sellers in a geographical trading area. Steel—a relatively heavy good—is differentiated by the location where it is offered for sale. A steel user in San Francisco is economically (that is, in terms of freight costs) closer to Japan than to Pittsburgh. The geographical area of the United States is, therefore, not necessarily synonymous with the "home market" of the American steel industry.

Mindful of these caveats, the structure of the steel market in the United States consists of a group of large integrated companies, the so-called reconstituted mills, minimills, imports, and a specialty steel segment (producing primarily alloy, stainless, and tool steels).[2] The growing relative importance of minimills and imports over the last three decades is portrayed in Tables 4-2 and 4-3.

THE INTEGRATED SECTOR

The integrated companies produce most of their steel from scratch, in sprawling mills that are from two to seven miles long. They use a step-by-step manufacturing process that was developed in the nineteenth century and has not changed very much since then:

> [C]oke and iron ore go into a blast furnace and out comes liquid pig iron. The liquid pig iron goes into a steelmaking furnace and out comes liquid steel. The liquid steel is poured into pots to make solid ingots of steel or, in a more recent innovation, it is poured into a continuous-casting machine, which extrudes a solid slab of steel. The ingots or slabs are allowed to cool to room temperature. Then they are reheated, red-hot, and driven through a rolling mill a half a mile long, and out comes a finished steel product [plates, hot- and cold-rolled sheets, heavy structurals, and so forth]. The method requires machines that cost billions of dollars, it burns huge amounts of energy, and it depends on the mass labor of thousands of steelworkers. That is to say, the

manufacture of steel is capital intensive, energy intensive, and labor intensive.[3]

The minimum efficient scale of a modern integrated plant ranges from 6 million to 8 millions tons of raw steel capacity.

Historically, this segment of the industry has been shaped by mergers and consolidations. The formation of U.S. Steel in 1901 was a "combination of combinations" that consolidated 180 formerly independent plants. With its 65 percent control of the industry, U.S. Steel was a Brobdingnagian giant dwarfing all of its remaining competitors. Its size and dominance, however, did not derive from natural economies of scale, but from mergers and combinations.

In the 1920s, "independents" such as Bethlehem, Republic, and National consummated a series of major mergers that enabled them to grow rapidly, not only in terms of absolute company size, but also in terms of market share.[4] The gains in market share were made largely at the expense of U.S. Steel and transformed the industry from an asymmetrical into a symmetrical oligopoly.

In the period after World War II, the merger trend was temporarily stalled by the government's antitrust objections to Bethlehem's acquisition of Youngstown, a merger that would have combined the second and sixth largest steel producers. In the late 1960s, however, when the merger movement resumed momentum, several integrated firms began to buttress their positions through horizontal consolidations. The following combinations took place:

1. 1968: Wheeling Steel (tenth largest) and Pittsburgh Steel (sixteenth largest).
2. 1971: National Steel (fourth largest) and Granite City Steel (nineteenth largest).
3. 1978: LTV (seventh largest) and Youngstown (eighth largest).
4. 1983: LTV (third largest) and Republic (fourth largest).

Today, the Big Six comprising the integrated steel oligopoly have the following market shares: USX, 10.8 percent; Bethlehem, 10.3 percent; LTV, 9.3 percent; National, 6.1 percent; Inland, 5.1 percent; and Armco, 3.7 percent.

Other structural changes in the integrated steel sector are noteworthy. First, the diversification movement that began in the 1960s led to a number of acquisitions outside the steel industry. National, for example, acquired an aluminum firm, a fast-growing drug distributor, the nation's largest oil distribution company, and a California savings and loan association; Armco bought an insurance holding company; and—in the largest diversification move of all—U.S. Steel acquired the Marathon Oil Company. (Incidentally, its ownership of Marathon Oil means that USX now derives more revenues from petroleum than from steel.) Several of these acquisitions have been unprofitable, if not disastrous (such as Armco's venture into the reinsurance business), and some have subsequently been spun off.

Second, under the increasing pressure from minimills and imports, the integrated producers embarked on a major restructuring program. During the 1980s, they closed more than 450 antiquated steelmaking facilities, some dating back to the last century. They cut production capacity by more than 25 percent (from 154 million to 116 million tons). They sold or divested themselves of major segments of their business as they sought to focus resources on the product area in which they were able to compete most effectively, that is, sheet products. In the process, they reduced their workforce by 60 percent (from 340,000 in 1980 to 125,000 by the end of the decade). So drastic was this restructuring that eleven of the seventeen integrated producers operating in 1980 had either passed into Chapter 11 bankruptcy or disappeared altogether. Needless to say, the impact on venerable steel communities—such as the Monongahela Valley; Buffalo, New York; and South Chicago—was devastating both in human and economic terms.

Nevertheless, the massive plant closings, accompanied by the investment of $35 billion in the modernization of existing plant and equipment, had positive results. According to the International Trade Commission, the industry's real production costs declined by 28 percent, and labor productivity increased by 60 percent. Technological advances and changes in cost structures meant that by 1991, a ton of steel that would have cost $669 to produce using the technology and cost structure of a decade earlier could now be produced for $480.[5]

Moreover, an entirely new group of producers, the so-called reconstituted mills, emerged from this era of restructuring. These were mills that had been sold off by traditional producers or mills that had gone through bankruptcy proceedings and emerged with lower costs, especially such "legacy" costs as pension and group insurance liabilities, payments for leased equipment, environmental liabilities, and interest burdens.[6] One example is California Steel Industries, a joint venture between Kawasaki of Japan and a Brazilian natural resources company, which took over the former rolling and finishing facilities of Kaiser Steel in Fontana, California. Another is UPI, a fifty–fifty partnership between USX and Pohang Iron and Steel of Korea, which took over U.S. Steel's former rolling mill in Pittsburgh, California.[7]

Finally, since the middle 1980s, the largest Japanese steel producers began making massive investments in the American steel industry. These investments included the acquisition of existing plants, the formation of joint ventures, and participation in technology transfer agreements (see Table 4-1), and they accounted for approximately 25 percent of the modernization expenditures by U.S. mills during the period. As a result, today one in six American steel workers is employed on a project directly or indirectly financed by a Japanese mill. There is a striking irony here: In a few short years, those major Japanese steel companies that had been the fiercest competitors of the U.S. producers have become partners with most of the major integrated steelmakers in the United States. The consequences of this partnership for competitive behavior remain to be seen.

TABLE 4-1   Major U.S.–Japanese Joint Ventures

Joint Venture (Founded)	Shareholder (Share %)	Facility	Type of Products (TPY/Thousands)
National Steel Corp. [1984]	NKK (70) National Intergroup (30)	Integrated steel production	Hot-rolled sheets Cold-rolled sheets Coated sheets Tin plate (Total 4,960)
Armco Steel Co. LP [1989]	Kawasaki Steel Corp. (50) Armco (50)	Integrated steel production	Hot-rolled sheets Cold-rolled sheets Coated sheets (Total 4,800)
USS/KOBE Steel Co. [1989]	Kobe Steel Ltd. (50) USX (50)	Integrated steel production	High-quality bars Tubular products (Total 2,400)
Inland Steel [1989]	Nippon Steel Corp. (13)	Integrated steel production	Sheet products (Total 4,170)
I/N Tek [1987]	Nippon Steel Corp. (40) Inland Steel (60)	Cold-strip mill Sendzimir mill	Cold-rolled sheets (Total 1,000)
I/N Kote [1989]	Nippon Steel Corp. (50) Inland Steel (50)	Galvanized coating line	Hot-dipped galvanized sheets (Total 500) Electrogalvanized sheets (Total 400)
L-S Electro Galvanizing Co. [1985]	Sumitomo Metal Ind. (40) LTV Steel (60)	Coating line	Electrogalvanized sheets (Total 400)
L-S II Electro Galvanizing Co. [1989]	Sumitomo Metal Ind. (50) LTV Steel (50)	Coating line	Electrogalvanized sheets (Total 400)
Wheeling–Nisshin [1984]	Nisshin Steel Ltd. (80) Wheeling–Pittsburgh (20)	Coating line	Aluminized/galvanized sheets (Total 270)
Two ventures			Galvanized sheets (Total 240)
PROTEC Coating Co. [1990]	Kobe Steel Ltd. (50) USX (50)	Coating line	Hot-dipped galvanized sheets (Total 600)

Source: Japan Iron & Steel Federation.

THE MINIMILL SECTOR

Since the early 1960s, minimills have made dramatic inroads into the market position of the integrated steel giants; at the same time, they have captured a share of the market once held by imports. By 1990, for example, minimills accounted for 100 percent of domestic shipments of reinforcing bars, 85 percent of merchant bars and light structurals, 83 percent of wire rod, 50 percent of large structural shapes, and 36 percent of special-quality bar. Overall, they now account for a bigger share (almost 25 percent) of the domestic steel market than imports do.

Minimills are relatively small, nonintegrated companies, typically with a raw steel capacity of 500,000 tons to 1.5 million tons.[8] Unlike their integrated rivals, they do not produce steel from scratch. Therefore, they have been untrammeled by the immense capital investment that the integrated companies had to put into the huge coke ovens and blast furnaces that make up the clanging cathedrals of the Rust Belt. Instead, the minimills convert scrap directly into finished steel products. They operate modern electric furnaces, continuous casters, and rolling mills to achieve maximum efficiency. According to the International Trade Commission,

> The current cost of a state-of-the-art, 4-million metric ton-per-year integrated facility in an OECD country is estimated to be between $4 billion and $8 billion ($1,000 to $2,000 per ton of capacity), while a state-of-the-art minimill (500,000 metric tons) costs about $250 million (or $500 per ton of capacity). Even for existing plants, capital requirements are lower for smaller facilities. The annual reinvestment requirements for an integrated plant have been estimated at $40 per ton of capacity compared to minimill requirements of $15 per ton.[9]

Also unlike their integrated competitors, the minimills are led by a new breed of entrepreneurs who eschew chic executive dining rooms and disdain fleets of company jets and limousines. Occupying spartan headquarters operated by skeleton corporate staffs, they share the rigors and rewards of the marketplace with their employees, by means of productivity incentives and profit-sharing programs. In minimills, says a trade magazine, "personalities and personal management styles dominate and stand in sharp contrast to the bureaucracies that long throttled competition, decision-making, and accountability in the integrated sector. And labor is regarded as a resource, not merely a cost."[10]

In spite of their advantageous labor and capital costs, the minimills were initially hampered by technological limitations that confined them to low value-added product lines, such as bars, small shapes, and wire rods. By the mid-1980s, however, when they found themselves with little market share to conquer (except from one another), they decided to break out of their traditional niche by way of technological innovation. Nucor, the largest and most enterprising of the group, built a new plant to compete head-on with Bethlehem, USX, and Inland in producing large I-beams. Drawing on the innovative skills of a German equipment supplier, Nucor also built revolutionary new plants to produce hot- and cold-rolled sheets—the bread-and-butter products that had been the traditional preserve of the steel majors.

Today, Nucor's costs are an estimated 14 percent below those of the lowest-cost foreign producer and 18 percent below the costs of U.S. integrated producers.[11] (Whereas Big Steel boasts that its plants can now produce a ton of steel in three to four man-hours, Nucor's ultramodern Crawfordsville plant regularly rolls out a ton in an average of fifty-four minutes.)

Noted a Nucor plant manager: "We're Big Steel's worst nightmare, and we're not going away." Added UBS Securities analyst Charles Bradford: "There is nothing to stop the minis from taking major pieces out of Big Steel's sheet market except the time it takes to build plants."[12]

IMPORTS

Steel imports increased from a trickle of mostly low-quality products in the 1950s to a flood of high-grade steel in the mid-1960s. From 1970 to 1992, imports fluctuated widely between a low of 13.8 percent to a high of 26.4 percent of the U.S. market. Currently, imports enjoy a market share of less then 20 percent (see Table 4-2).

TABLE 4-2   Steel Shipments, Imports, Exports, Apparent Consumption, and Import Penetration in the United States, 1960, 1970, and 1980–1992

Period	Shipments	Imports	Exports	Apparent Consumption	Import Penetration (%)
		Thousand short tons			
1960	71,149	3,359	2,977	71,531	4.7
1970	90,798	13,364	7,062	97,109	13.8
1980	83,853	15,495	4,101	95,247	16.3
1981	88,450	19,898	2,904	105,444	18.9
1982	61,567	16,663	1,842	76,388	21.8
1983	67,584	17,070	1,199	83,455	20.5
1984	73,739	26,163	980	98,922	26.4
1985	73,043	24,256	932	96,367	25.2
1986	70,263	20,692	929	90,026	23.0
1987	76,654	20,414	1,129	95,939	21.3
1988	83,840	20,891	2,069	102,622	20.4
1989	84,100	17,321	4,578	96,843	17.9
1990	84,981	17,169	4,303	97,847	17.5
1991	78,846	15,845	6,346	88,345	17.9
1992	82,241	17,075	4,288	95,027	18.0

Source: USITC publication 2436, September 1991; American Iron & Steel Institute.

The level of imports and their share of the U.S. market tend to fluctuate in accordance with world demand and supply conditions and, most important, with changes in the value of the U.S. dollar. A strong dollar makes imports "cheap" in the domestic market; a weak dollar has the opposite effect. In 1984, for example, when the dollar was strong (U.S. $1 = ¥ 237), imports from Japan reached a peak of 6.4 million metric tons. By 1991, after the dollar had plunged in value (U.S. $1 = ¥ 134), imports from Japan fell to 2.8 million metric tons, a drop of 60 percent. Indeed, after 1988, the weakness of the dollar made the U.S. market so unattractive to imports that many countries failed to fill the quota to which they were entitled under existing voluntary restraint agreements.[13] The weak dollar, it seems, was a more powerful deterrent to imports than were the severe quota restrictions.

Aside from price considerations, the market share of imports varies from one region to another. According to a study by the Department of Commerce, "Imports serve to meet shortfalls in the U.S. capacity for particular regions, products, and time periods. They are largest relative to demand in regions where there are capacity shortfalls. For those regions with capacity overhangs, import penetration is typically well below national averages."[14] A comparison of import statistics by destination bears out these conclusions.

Imports also meet shortfalls in particular products. A 1980 survey by the International Trade Commission showed that customers had switched from domestic to foreign suppliers in nearly half of all the instances because a domestic supply was not available.[15] More than 13 percent did so because they could not find domestic products of the required quality, and 7 percent preferred multiple sourcing. Only 28 percent switched to imports because of price considerations. In the same year, a General Accounting Office survey of domestic-steel users found that for products other than those supplied by minimills, the salient motives for buying imported steel were better quality, assured availability, and marketing help.[16] Although the big steel producers claim to have caught up with their foreign rivals (or even surpassed them) on this score, a 1993 investigation by the International Trade Commission reveals that many deficiencies in product quality and customer service still persist.[17]

SUMMARY

Since World War II, the U.S. steel industry has experienced a profound structural transformation. The power of U.S. Steel, which once dominated the industry, has greatly diminished, and the market share of the large integrated producers has been significantly eroded. In the relatively short period between 1979 and 1989, the market share of the major integrated producers declined dramatically, from 73 percent to 42 percent. At the same time, primarily because of their lower costs and lower prices, the minimills increased their share from 8.2 to 20 percent, and the reconstituted mills attained a 29 percent share (see Table 4-3). Since the import share of the U.S. market did not change significantly, this means that it was the rise of the minimills and the reconstituted

TABLE 4-3 U.S. Market Share: Major Steel Mills, Reconstituted Mills, and Minimills

	1979			1989		
	Number	Shipments[a]	%	Number	Shipments[a]	%
Major Mills	8	73.4	(73)	5	35.7	(42)
"Reconstituted"	0	0	(-)	12	24.1	(29)
Other Integrated	20	17.7	(18)	8	5.8	(7)
Minimills	48	8.2	(8)	33	17.0	(20)
Specialty	10	1.0	(1)	9	1.5	(2)
Total	86	100.3	(100)	67	84.1	(100)

[a]Shipments in millions of net tons.

Source: Paine Webber, cited in Japan Steel Information Center, "Steel Trade Cases Offer No Solution to Global Steel Problems," 1993, p. 7.

mills that explains the loss of market share by the integrated producers during the 1980s. This, as we shall see, has had profound implications for public policy.

III. MARKET CONDUCT

Business conduct or behavior is closely related to industry structure. A tight, oligopolistic structure is generally expected to lead to nonaggressive pricing behavior. An absence of competitive pressures may also lead to a lack of progressiveness which, over the longer term, will result in substandard efficiency and lackluster performance.

THE "ADMINISTERED PRICE" ERA

From the turn of the century until the early 1960s, the U.S. Steel Corporation was the acknowledged price leader of the steel industry, and for the most part, the other oligopolists followed in lockstep.[18] The pricing discipline observed by the large American steel producers, even during severe recessions, became the envy of foreign steel producers, which often experienced great price instability during periods of weak demand. Although this "administered pricing" helped create an environment of stability and predictability, it had harmful effects over the longer term: It exposed the industry to external challenges that eventually resulted in a significant erosion of its oligopoly power.

During the administered price era, steel prices were characterized by remarkable rigidity and uniformity. Unless impelled by sharp increases in direct costs or dangerous sniping by rivals, U.S. Steel generally preferred to resist both price increases and decreases and to sacrifice stability "only when the decision [was] unavoidable."[19]

After World War II, this price policy was transformed into one of "upward rigidity," that is, a pattern of stair-step price increases at regular intervals. At times this upward flexibility was achieved in the face of declining demand. This rigidity in steel prices was reinforced by price uniformity. With or without resort to the conspiratorial basing-point system, the integrated giants matched one another's prices with monotonous consistency. They maintained a lockstep price uniformity by punishing any major mill that showed deviationist tendencies.

THE "COMPETITIVE" ERA

Beginning in the mid-1960s, this policy of price uniformity and upward rigidity came under pressure, primarily because of competition from minimills and imports. In wire rods, for example, Paula Stern (then a member of the International Trade Commission) found it "obvious why the integrated producers lost so much of the market to the minimills. . . .The efficiency of their technology, management, and cost control techniques enable minimills to keep their prices low." According to Stern, price data showed

> that the average delivered price paid for rod from nonintegrated firms was well below that price paid for rod from integrated firms in most regions and in most of the period of investigation. But even more significant is the fact that these efficient U.S. mills were able to sell wire rod at a price that, on average, was below the average price of imported wire rod.[20]

Like the minimills, imports have had a moderating influence on the administered pricing policy of the domestic oligopoly. Between 1959 and 1969, the Council on Wage and Price Stability found that "there were limiting forces which operated to prevent U.S. steel companies from increasing prices and maintaining the previous higher profit levels. *Chief* among these was import competition."[21] The Federal Trade Commission came to the same conclusion.[22]

THE "PROTECTIONIST" ERA

Predictably, the domestic oligopoly did not view these competitive challenges with complacency. Although it could do little to curb the competition of minimills, it launched an orchestrated political campaign to neutralize the threat of foreign competition. The result was a succession of "voluntary" import quotas imposed on foreign producers.

1. THE VOLUNTARY RESTRAINT AGREEMENT (VRA), 1969–1974. Under this agreement, steel imports from Japan and from the European community each were limited to 5.8 million tons annually, compared with their then current levels of 7.5 million and 7.3 million tons, respectively. The agreement also provided for an annual growth factor of 5 percent in the allowable quotas.

The price effects of the VRA were dramatic. According to one study, between January 1960 and December 1968, a period of nine years, the composite steel price index rose 4.1 points—or 0.45 points per year—indicating the moderating effects that surging imports had on domestic prices. In the four years between January 1968 and December 1972, while the VRA was in effect, the steel price index rose 26.7 points—or 6.67 points per year—which was twice as much as the index for all industrial products (including steel). Put differently, since the import quotas went into effect, steel prices increased at an annual rate fourteen times greater than in the preceding nine years.[23]

Another study showed that the products that had been subjected to particularly hard import pressure before the VRA evidenced greater price increases than did other steel products after the VRA became effective, again highlighting the anticompetitive effects of the quotas.[24]

Yet another study estimated that the VRA caused steel prices to increase by $26 to $39 per ton, meaning that the price of steel would have been 13 to 15 percent lower in the absence of the VRA.[25]

2. THE TRIGGER PRICE SYSTEM, 1978–1982. The lapse of the VRA in 1974, plus falling demand, caused prices to be very competitive again for a few years. But after a plethora of complaints by the domestic oligopoly before the International Trade Commission, the Carter administration granted the industry a novel form of protection in the form of the Trigger Price Mechanism (TPM). For all practical purposes, the TPM set minimum prices for all carbon steel products imported into the United States. The trigger prices (so called because undercutting by foreign suppliers was to trigger antidumping proceedings by U.S. authorities) were based on estimated Japanese production costs plus freight costs from Japan. In exchange for administrative protection, the steel industry agreed to refrain from filing antidumping, subsidy, or import-injury complaints before the International Trade Commission.

By forcing up import prices, the TPM gave domestic producers an opportunity to raise their prices as well. A sizable gap developed between U.S. and world market steel prices, and by 1981 American steel-consuming industries were paying $100 to $150 more per ton of steel than their foreign competitors were.[26] Especially hard hit were firms producing items for which the cost of steel amounted to a significant portion of total production costs, such as wire rope, fasteners, transmission towers, bridge components, oil rigs, and automotive parts. Many of them responded by purchasing these items abroad or switching production to foreign subsidiaries. American "indirect" steel trade—that is, trade in steel-containing goods—turned sharply negative during the TPM period.[27]

The TPM not only maintained American steel prices significantly above world market levels. It also distorted patterns of steel pricing and consumption in the United States. Traditionally, domestic steel prices had been low in the Great Lakes region, where the most efficient American integrated plants are located, and high on the West Coast, which lacks adequate domestic capacity. But because trigger prices were calculated on the basis of Japanese production and freight costs, they were lowest in the western states and highest in the Great Lakes region. The distribution pattern of steel imports changed accordingly.

Ultimately, the TPM system became ineffectual and irrelevant. When the value of the U.S. dollar reached inordinately high levels in the early 1980s, the TPM no longer provided protection to the domestic industry; that is, it no longer served the purpose for which it was designed.

3. THE "REAGAN/BUSH" QUOTAS, 1984–1992. The lapse of TPM came at the worst possible time for the domestic industry. The demand for steel was falling precipitously, and several newly industrializing countries, attracted by the strong dollar, were sending record volumes of low-priced steel into the U.S. market.

To remedy this situation, in September 1984, the Reagan administration began to negotiate a set of voluntary restraint agreements. By March 1985, American negotiators had concluded quota agreements with nearly every major steel-exporting nation in the world, including Brazil, Mexico, Yugoslavia, Poland, and Romania. As a result of these agreements, the total market share of imports declined from 26.6 percent in 1984 to about 21 percent in 1988. When the Bush administration decided to extend these quotas until 1992, the market share of imports declined further to 18 percent (see Table 4-3).

4. A "NEW" ERA, 1992–. When the Bush quotas were allowed to expire (over the vigorous objection of the integrated producers), the majors resumed their battle against imports by means of legal maneuvers. In June 1992, they filed eighty four complaints against companies in twenty-one countries before the International Trade Commission, alleging "material injury" or the "threat of material injury." The tenor of the complaints—a *déjà vu* of the tactics successfully used or threatened in 1977, 1982, 1984, and 1989—was familiar: Worldwide overcapacity and a soft steel market make it attractive for foreign producers to "routinely and systematically" dump their products in the U.S. market "far below fair market value and often below the cost of production." This, along with "massive foreign subsidies," Big Steel alleged, caused the U.S. industry a $2.2 billion loss in 1991 alone.

Although it costs the U.S. industry between $500,000 and $1 million to bring a trade complaint, the investment had in the past always paid off in higher prices and, ultimately, protection. The gamble, it seemed, was worth taking. But this time Big Steel's luck had run out. On July 27, 1993, the International Trade Commission administered a stunning (and generally unanticipated) defeat to the integrated majors:[28] (1) In hot-rolled sheet, a key bellwether product, the mills lost all of their cases and are now left without any import protection for the first time since 1969. (2) In cold-rolled sheets, another bellwether product, the mills suffered a sizable loss, winning only their cases against Germany, South Korea, and the Netherlands. These countries account for about one-fourth of the imports of this product; if they stop shipping to the United States altogether, other countries can easily fill the resulting "shortfall." (3) In relatively minor victories for the majors, the ITC did find "material injury or threat thereof" in the cases involving galvanized sheet, coated and electrical products, and cut-to-length plate. It ruled that import duties shall be imposed on these products.

SUMMARY

Successive systems of import restraints have facilitated pricing coordination among the dominant oligopolists, if only by making it easier for them to read one another's signals. They have abetted the efforts of the integrated majors to achieve their main objective: to stabilize and, ultimately, to raise prices. However, these schemes have not succeeded in protecting the majors against the increasingly intense competition of their domestic rivals, the minimills and the reconstituted mills. Why? Because the latter generally have lower cost structures and can, therefore, charge lower prices. Thus, Nucor's minimills in 1992 had a cost advantage of $54 per ton over the lowest-cost mill of the integrated producers. According to Wall Street analysts, the traditional integrated mills will have "to cut costs $30 per net ton over the next three years just to stay even with their peers, and $50 per net ton or more in some cases to be competitive with the new generation of thin-slab/flat-rolling units [of Nucor and other minimills]."[29] The reconstituted mills that have shed costs in the process of restructuring have also become a competitive force in the market by virtue of lower costs and a larger share of total capacity. LTV, for example, has used restructuring under Chapter 11 to achieve a cost advantage of as much as $60 per net ton over the mills of traditional integrated producers.

In short, we are now witnessing a remarkable revitalization of competition in the domestic steel industry. The dramatic decline in the market share of the major integrated companies and the concomitant rise of the reconstituted mills and the minimills is eloquent testimony to this revolutionary development.

IV. MARKET PERFORMANCE

An industry's performance is nothing more than the product of its structure and conduct. It is a measure of how well—how efficiently—it functions to serve consumer needs and contributes to the welfare of the national economy. Obviously, a precise quantitative assessment of performance is difficult, but international comparisons can yield valuable insights and guidelines for public policy.

A principal index of performance is technological progressiveness. This index measures an industry's ability to keep up with the latest advances in "best-practice" operations or to be in the forefront of such advances. Judged by this criterion, the U.S. steel industry, except for its minimill and specialty-steel segments, has lagged rather than led in the post–World War II period. Consequently, it has suffered serious erosion of its cost competitiveness, not only in world markets, but even in its domestic market.

Historically, according to the American Iron and Steel Institute, "[t]he steel industry in this country has adopted most new technologies, wherever they were developed, at least as rapidly and probably more rapidly than steel industries in other parts of the world." The industry's spokesmen were quick to label anyone who dissented from this view as "simply misinformed."

That the large steel companies have been slow to adopt state-of-the-art technology is no longer a matter of dispute. It is a fact recognized not only by the industry's academic critics but also by financial analysts and even by some steel company executives.

R&D Expenditures

According to the Congressional Office for Technology Assessment (OTA), "[t]he number of R&D scientists and engineers per 1,000 employees is smaller for steel than for any other industry except for textiles and apparel, about 15 percent of the average for all reported industries." Moreover, "foreign steel producers spend more on R&D than those in the United States."[30] Japan, for example, spends 2 percent of net sales on steel-related R&D, compared with only 0.4 percent in the United States. Furthermore, the ratio of research personnel to total steel industry employment is nearly 4 percent in Japan, compared with less than 1.0 percent in the United States.[31]

Process Innovation

Among the large number of process improvements that have been made in steelmaking during the last forty years, three undoubtedly deserve to be characterized as technological breakthroughs: (1) the basic oxygen furnace (BOF), a fast technique for converting iron to steel; (2) continuous casting, a process that bypasses both the laborious ingot-pouring process and the energy-intensive reheating of ingots and primary rolling; and (3) thin slab casting. Other major developments include methods to make steel products of higher quality by removing impurities from liquid steel and by improving the dimensional accuracy, flatness, and surface quality at the rolling end of the steelmaking process.

1. THE BASIC OXYGEN FURNACE (BOF).   The oxygen furnace was first put into commercial use in a small Austrian steel plant in 1950. It was first installed in the United States in 1954 by a small company (McLouth) but was not adopted by the steel giants until more than a decade later: U.S. Steel in December 1963, Bethlehem in 1964, and Republic in 1965. As of September 1963, several of the largest steel companies, together operating more than 50 percent of basic steel capacity, had not installed a single BOF furnace, whereas smaller companies, operating only 7 percent of the nation's steel capacity, accounted for almost half of the BOF installations in the United States.

By the late 1980s, the industry had replaced all but 5 million tons of its open hearth furnaces with BOFs. However, many of these meltshops still lag behind those of major foreign competitors with respect to such modern features as combined blowing and full process control. Steel producers in Japan, the EC, and Canada are also ahead in the installation of ancillary equipment (for example, vacuum degassing and other "ladle metallurgy" practices) that further help remove impurities and thus improve the quality of steel products.

The most likely explanation of the hesitant adoption of the Austrian converter by the large American firms is that their managements are still imbued with Andrew Carnegie's motto: "Invention don't pay." Let others first assume the cost and risk of research and development and of breaking in a new process; then we'll decide. The result was that during the 1950s, the American steel industry installed 40 million tons of melting capacity which, as *Fortune* observed, "was obsolete when it was built."[32]

2. CONTINUOUS CASTING.   The belated adoption of continuous casting by the American steel giants is a further illustration of their technological lethargy. Again, it was a small company (Roanoke Electric), with an annual capacity of 100,000 tons, that introduced this European invention into the United States in 1962. Other small steel companies followed, so that by 1968, firms with roughly 3 percent of the nation's steel capacity accounted for 90 percent of continuous-casting production in the United States.

By 1978, the U.S. steel industry (taken as a whole) was continuously casting 15.2 percent of its steel, less than one-third the 46.2 percent achieved in Japan. But these totals conceal a curious fact: American minimills, with a rate of 51.2 percent, were already ahead of the Japanese average, whereas the integrated U.S. mills were producing only 11 percent of their output by the continuous-casting method.[33] The latter managed to boost their average to 46 percent by the end of 1988, but by that time Japanese producers and U.S. minimills were achieving rates of over 90 percent. Today, the domestic industry (as a whole) is finally producing more than 80 percent of its steel by the continuous-casting method (see Table 4-4).

Over the years, continuous-casting technology has steadily improved with respect to casting speed and product quality. Indeed, the majority of steel users now specify steel made by this process.

TABLE 4-4 Continuous-Casting Adoption Rates, 1969–1992
(% of Steel Production)

Year	United States	E.C.	Japan
1969	3	3	4
1979	17	31	52
1980	21	39	60
1985	44	71	91
1989	65	88	94
1990	67	90	94
1991	75	90	94
1992	79	92	95

Source: Paine Webber, *World Steel Dynamics*, 1993, p. 122.

3. THIN SLAB CASTING. Traditionally, sheet steel was made by extruding a slab of metal 8 to 10 inches thick. The slab was repeatedly rolled and reheated to squeeze it into a thin sheet. The process required a huge capital investment, which explains why the sheet market was dominated by the integrated producers. Then, in the late 1980s, Nucor (the largest U.S. minimill) innovated the new TS/FR method which casts raw steel directly into slabs just 2 inches thick. These slabs are immediately compressed into finished steel 0.6 inches thick, at substantial savings in time, energy, and cost. (The process allows the conversion of liquid steel into hot coils in 12 to 30 minutes.) This innovation is significant because it reduces the barriers that heretofore had blocked the entry of minimills into the large and lucrative sheet market. It presents a formidable competitive challenge to both the integrated mills and foreign competitors.

4. INNOVATION GENERALLY. In 1988, the International Trade Commission released a detailed comparative analysis of the steel industry's performance in several countries. The report confirms that

> technological leadership in the integrated sheet and strip industry, in terms of both development and application, has passed to offshore competitors. . . .Major new developments in ore reduction, steel making, and casting have come from Europe (especially West Germany) and Japan, and most rolling technology originates in Japan. . . .The greatest challenge facing U.S. modernization efforts in this regard is that foreign competitors have comparatively more modern equipment than the U.S. industry. . . .Japan sets the world standards on installed technology, although West Germany, France, Korea, and Canada do have certain facilities or operating units that utilize the most advanced equipment.[34]

THE PROBLEM OF ORGANIC TECHNOLOGICAL PROGRESS

Technological progressiveness is an organic rather than a piecemeal process. It is not enough simply to add a modern continuous caster to an antiquated open-hearth furnace or to install a new BOF in a plant that is poorly located with respect to sources of raw material or markets. Efficiency, in the best-practice sense, requires a coordinated approach to modernization. Massive investment in new plant and equipment will not, by itself, produce the desired results.

In the early 1980s the large integrated steel companies did not have a technologically balanced plant structure. Noted a steel consultant at Arthur D. Little, "You've got a mishmash—100-year-old stuff fitted into two-year-old stuff." Added a company executive, "All the multi-plant mills have scattered good facilities in with bad. We've screwed things up so they'll never get untangled."[35] Said the International Trade Commission:

> Some excellent hot-end facilities are teamed with mediocre or poor rolling and finishing facilities. However, in other cases, the situation is reversed. To some extent, this has been caused by the need to retrofit existing facilities, as opposed to building complete new plants. . . .As new foreign mills are built, they incorporate state-of-the-art technologies that give those mills great advantages in terms of cost and productivity. New process technologies generally have different requirements in terms of space, flow lines, and organization than those of older technologies. Placing newer technology in an existing plant can result in non-optimal performance.[36]

By the end of the decade, drastic restructuring, including the closure of the worst integrated facilities, the streamlining of others, and the spin-off of still others, had begun to alleviate this problem.

SUMMARY

In spite of its sizable investments since World War II, the integrated U.S. steel industry never achieved a degree of modernization comparable to that of its foreign rivals. Most of the investment effort consisted of an endless "rounding out" of existing plants. Moreover, the legacy of a concentrated market structure, helter-skelter mergers, nonaggressive interfirm rivalry, and technological lethargy continued to be an impediment to modernization and international competitiveness.

In the 1980s, however, the industry started on the road to at least a partial comeback. The large producers embraced a strategy of rationalization. They shut down anachronistic facilities, and in cooperation with the United Steel Workers, they brought labor costs under control and modified unrealistic, cost-inflating work rules. At the same time, the precipitous decline in the value of the dollar—from ¥ 239 in 1985 to ¥ 125 in 1988—gave the industry a welcome respite from the intensity of foreign competition. The combined impact of these developments, according to some observers, has been to make the industry cost competitive vis-à-vis its major international competitors (see Table 4-5). It has not, however, wiped out the cost advantage of the domestic minimills over the integrated producers.

V. PUBLIC POLICY

Given the lackluster performance of the integrated steel producers, what direction should public policy take? Specifically, should antitrust enforcement be relaxed in order to facilitate mergers and encourage a "rationalization" movement among the integrated giants? Should the government, in the national interest, maintain an import-restraint policy in order to ensure the survival of an integrated steel sector in the American economy? We now turn to a brief examination of these issues.

ANTITRUST POLICY

In examining the application of the antitrust laws to the steel industry from 1890 to the present, it is fair to say that the law has had a minor impact on the structure and conduct of the industry. The Sherman Act did not block

TABLE 4-5 Pretax Steelmaking Costs, 1983–1992 (per metric ton shipped)

	United States			Japan			Germany	
Year	$	(Index)	¥	$	(Index)	DM	$	(Index)
1983	594	(100)	107,532	453	(76)	1,061	416	(70)
1984	556	(100)	95,751	404	(73)	1,060	373	(67)
1985	523	(100)	93,857	396	(76)	1,046	358	(68)
1986	511	(100)	84,806	505	(99)	929	429	(84)
1987	481	(100)	73,815	510	(106)	847	472	(98)
1988	481	(100)	68,322	533	(111)	835	476	(99)
1989	490	(100)	70,720	513	(105)	868	462	(94)
1990	494	(100)	73,449	509	(103)	875	543	(110)
1991	517	(100)	73,367	545	(105)	906	547	(106)
1992	522	(100)	74,103	586	(112)	914	586	(112)

Note: Between 1983 and 1992, Japanese and German costs (expressed in yen and deutschmarks, respectively) decreased. With the drastic decline of the dollar during this period, however, their costs (expressed in dollar terms) increased significantly. The index numbers indicate Japanese and German dollar costs vis-à-vis costs in the United States.

Source: Paine Webber, *World Steel Dynamics*, 1992, p. 71.

the formation of U.S. Steel and its achievement of market dominance in 1901; it did not result in the dissolution of U.S. Steel in 1920; it did not block the emergence of a tight oligopoly through a series of mergers among the erstwhile "independents"; it did not interdict the recent megamergers between LTV and Youngstown or LTV/Youngstown and Republic; and it did not prevent oligopolistic coordination among the leading firms of the industry, either under the conspiratorial basing-point system or other mechanisms designed to ensure collective action among ostensible competitors.

The outstanding antitrust success in the steel industry was preventing the proposed merger between Bethlehem (then the industry's second largest firm) with Youngstown Sheet & Tube (then the industry's sixth largest firm) in the mid-1950s. The net result of this action was that Bethlehem decided to build a modern greenfield plant at Burns Harbor, Indiana, in order to compete more effectively in the Chicago market. In other words, the law forced Bethlehem to expand by "building" rather than by "buying," to expand without substantially lessening competition in an already overly concentrated industry. Incidentally, the law "forced" the corporation to build a plant that even today still ranks as the most modern and efficient steel-producing unit in the Bethlehem empire.

In retrospect, one can only speculate whether a tougher antitrust stance, especially toward acquisitions and mergers, would not have helped the industry avoid some of its later difficulties in trying to remain technologically progressive and internationally competitive. One can only wonder whether a dissolution of U.S. Steel in 1920 would have created a competitive industry that was not dependent on government protection for its survival.

IMPORT POLICY

A major stimulus to competition in steel has come not from antitrust but from import competition and, to some extent, from the appearance of the minimills. As Douglas Yadon, publisher of the *Preston Pipe Report*, put it: "If we don't have imports, we have given the domestic mills a license to steal and to return to poor quality. There must be competition."[37] Not surprisingly, the major public policy battles, at least since 1960, have revolved around the industry's efforts to obtain protection from import competition. These efforts consisted of political lobbying and intensive trade litigation. As a rule, the industry argued that it needed import restraints because unfairly traded (dumped and subsidized) imports had deprived it of an opportunity to modernize as rapidly as its foreign rivals had. Such restraints were to provide the breathing space for the industry to raise its efficiency and regain its former cost competitiveness. The idea was simple: Limit imports in order to permit the industry to raise prices so that it could earn higher profits and would have more investment funds to put into new, modern facilities, which would enable it to stand on its own feet and compete effectively with best-practice firms around the world.

Critics of the industry argued that the high costs and technological backwardness of the large domestic steel firms could not be attributed to a lack of funds but to poor investment decisions and the diversion of funds into non-steel activities; that much of the imported steel was neither dumped nor subsidized; that the steel companies actually reduced investment during periods of import protection; and that protecting steel would place a heavy burden on American steel-using industries.

By and large, the critics are right. Steel-trade restrictions have been particularly harmful to "downstream" manufacturing, that is, those industries for which steel is a major input. (For the wire-drawing industry, the cost of steel amounts to 60 to 70 percent of total production costs; for heavy-equipment makers, it is about 25 percent.) As mentioned earlier, higher steel prices and other market distortions caused by protectionist interventions have forced some of these firms to abandon product lines to foreign suppliers or to relocate their businesses to countries where steel is available at lower cost. Indeed, the decline in steel imports resulting from trade restrictions has been more than offset by rising imports of steel-containing goods. Somewhat paradoxically, protection of the steel market has boomeranged against its intended beneficiaries, the steel industry and its workers. At the same time, however, it has exacted a staggering cost (estimated at between $1 billion and $7 billion annually) from American steel-using industries and ordinary consumers.[38]

CONCLUSION

Today, the steel industry is the scene of intense price competition. The stable steel oligopoly that persisted into the 1970s has disintegrated. No single cause explains this dramatic development. Technological lethargy, serious investment mistakes, the burden of inflexible and costly labor contracts, undue reliance on import restraints, the decline in U.S. steel consumption, and the startling growth of minimills all played significant roles in the demise of Big Steel.

Most important, however, was the self-delusion of the dominant oligopolists, the belief that their problems were in their stars rather than in themselves. Thomas C. Graham, the former president of U.S. Steel, makes the point bluntly:

> The American steel industry is wearing out its welcome in Washington and on Main Street. . . .For too long the traditional U.S. steelmakers as a group have laid the industry's problems at the feet of imports almost to the exclusion of other factors, when in reality the source of some of our greatest problems lies closer to home.

He deemed it irresponsible for the industry to suggest that "if only Washington would stop pandering to foreign governments, then all of our problems would be solved and our industry could get back to the business of reliving the good old days."[39]

NOTES

1. H. R. Seager and C. A. Gulick, *Trusts and Corporation Problems* (New York: Harper, 1929), p. 216. This book is an excellent source on the early history of U.S. Steel. See also Ida M. Tarbell, *The Life of Elbert H. Gary* (New York: Appleton, 1930), p. 205.
2. This chapter is confined to the carbon steel industry and does not deal with the relatively small alloy, stainless, and tool steel segment.
3. Richard Preston, *American Steel* (Englewood Cliffs, NJ: Prentice Hall, 1991), p. 11.
4. For a history of the early merger movement in the steel industry, see House, *Hearings Before the Committee on Small Business, Steel Acquisitions, Mergers, and Expansion of 12 Major Companies, 1900–1950*, 81st Cong., 2nd sess., 1950.
5. International Trade Commission, "Certain Flat-Rolled Carbon Steel Products from Argentina, Australia, Austria, Belgium, Brazil, Canada, Finland, France, Germany, Italy, Japan, Korea, Mexico, the Netherlands, New Zealand, Poland, Romania, Spain, Sweden, and the United Kingdom," vol. 1, publication 2664, August 1993, p. 309 (hereafter cited as ITC publication 2664).
6. Between 1987 and 1990, for example, U.S. Steel paid $390 million for the health care of roughly 100,000 retirees, which is more than Nucor spent building its ultramodern Crawfordsville plant.
   Wall Street analysts estimate the cost savings of mills that have gone through Chapter 11 versus the costs of typical mills as follows:

Employment	-24 percent
Material	-4 percent
Financial	-37 percent
Total	-14 percent
	(-$60 per ton)

   (Estimates by Paine Webber, cited in Japan Steel Information Center, "Steel Trade Cases Offer No Solution to Global Steel Problems," 1993, p. 7.)
7. Other reconstituted mills are Geneva Steel, Gulf States Steel, McLouth Steel, Rouge Steel, Sharon Steel, Warren Consolidated, Weirton Steel, and Wheeling–Pittsburgh. Some experts would also include LTV on this list.
8. Donald F. Barnett and Robert W. Crandall, *Up from the Ashes: The Rise of the Steel Minimill in the United States* (Washington, DC: Brookings Institution, 1986).
9. International Trade Commission, "Steel Industry Annual Report," USITC publication 2436, September 1991, p. 35.
10. *33 Metal Producing*, February 1987, p. 27.

11. USITC publication 2436, pp. 4–6.
12. Quoted in *Fortune*, July 15, 1991, p. 107.
13. Between January 1991 and March 1992, for example, Japan filled only 55.2 percent of its alloted import quota, the European Community 69.2 percent, Mexico 46.3 percent, Brazil 83.9 percent, Korea 62.6 percent, and Venezuela 28.6 percent. Overall, only 63.6 percent of the quota allowed under the voluntary restraint agreements was filled during this period. See Japan Steel Information Center, "Steel Trade Cases Offer No Solution to Global Steel Problems," 1993, p. 4.
14. U.S. Department of Commerce, "The Structure of Steel Markets in the United States" (unpublished study, 1979), p. 2.
15. ITC, *Certain Carbon Steel Products from Belgium, the Federal Republic of Germany, France, Italy, Luxembourg, the Netherlands, and the United Kingdom*, publication 1064, May 1980, pp. A-81, A-96, A-111, A-124, A-138.
16. General Accounting Office, *Report by the Comptroller General, New Strategy Required for Aiding Distressed Steel Industry*, January 8, 1981, chap. 3, pp. 4–11 (hereafter cited as GAO Steel Report).
17. See, for example, USITC publication 2664, vol. 2, p. I-161.
18. Senate Subcommittee on Antitrust and Monopoly, *Administered Prices in Steel*, 85th Cong., 2nd sess., 1958, S. Rept. 1387, pp. 17–26.
19. A. D. H. Kaplan, J. B. Dirlam, and R. F. Lanzilloti, *Pricing in Big Business* (Washington, DC: Brookings Institution, 1958), p. 175.
20. ITC, *Carbon and Certain Steel Products*, publication 1553, vol. 1, July 1984, p. 110.
21. Council on Wage and Price Stability, Staff Report, *A Study of Steel Prices* (Washington, DC: U.S. Government Printing Office, July 1975), pp. 9–10 (emphasis added).
22. Federal Trade Commission, *The United States Steel Industry and Its International Rivals* (Washington, DC: U.S. Government Printing Office, 1977), pp. 168, 240, 524.
23. Cited in Comptroller General of the United States, *Economic and Foreign Policy Effects of Voluntary Restraint Agreements on Textiles and Steel*, report B-179342 (Washington, DC: U.S. Government Printing Office 1974), p. 23.
24. See testimony by Walter Adams in ITC, "Stainless Steel and Alloy Tool Steel," investigation TA-203-3 (Washington, DC, September 1977), p. 11 (mimeo).
25. Ibid., pp. 11–12.
26. Peter Marcus and Karlis Kirsis, *Steel Strategist* 13, Paine Webber, March 30, 1987, Table 3.
27. International Iron and Steel Institute, *Indirect Trade in Steel, 1962–1979* (Brussels, 1982), pp. 17, 128–129.
28. See ITC publication 2664, vol. 1.
29. Paine Webber, *World Steel Dynamics*, 1992.

30. Office of Technology Assessment (OTA), *Technology and Steel Industry Competitiveness*, June 1980, pp. 275, 277–278 (hereafter cited as OTA steel report).
31. ITC, *Monthly Report on the Status of the Steel Industry*, USITC publication 2298, July 1990.
32. *Fortune*, October 1966, p. 135; for an analysis of the BOF innovation episode, see Sharon Oster, "The Diffusion of Innovation Among Steel Firms," *Bell Journal of Economics*, Spring 1982, pp. 45–56; and Leonard Lynn, "New Data on the Diffusion of the Basic Oxygen Furnace in the U.S. and Japan," *Journal of Industrial Economics*, December 1981, pp. 123–135.
33. OTA steel report, p. 290.
34. ITC, "U.S. Global Competitiveness: Steel Sheet and Strip Industry," Report to the Committee on Finance, U.S. Senate, January 1988, pp. 12-12, 12-13 (hereafter cited as ITC, "U.S. Global Competitiveness").
35. *Wall Street Journal*, April 4, 1983, p. 11; *Iron Age*, October 21, 1983, p. 58. In 1979 Norman Robins, research manager of Inland Steel, gave the following report after a visit to Japan:

> I was in the Fukuyama plant last year, and one of the most impressive things about it was the lack of truck and train traffic. Inland's plant, on the other hand, was begun in 1902 and undoubtedly was not conceived at that point in time to grow to the size that it has since become. Presently, there are blast furnaces in two locations and a new one being built in a third location, steelmaking at four different locations and a great deal of material handling and transportation required to move steel through the finishing facilities.

"Steel Industry Research and Technology," *American Steel Industry Economics Journal*, April 1979, pp. 49–58.
36. ITC, "U.S. Global Competitiveness," pp. 12–13.
37. *Pipeline*, March 1988, p. 12.
38. Robert W. Crandall, *Steel in Recurrent Crisis* (Washington, DC: Brookings Institution, 1981), pp. 134–139; Gary Hufbauer, "Wean the Steel Barons from Protection," *Wall Street Journal*, December 27, 1988, p. A10.
39. "Top U.S. Steelmakers Plan to Unveil Trade Lawsuits amid Great Fanfare," *New York Times*, June 30, 1992, p. A2.

SUGGESTED READINGS

Books and Pamphlets

Acs, Z. J. *The Changing Structure of the U.S. Economy: Lessons from the Steel Industry*. New York: Praeger, 1984.
Adams, W., and J. W. Brock. *The Bigness Complex*. New York: Pantheon, 1986.

Barnett, D. F., and R. W. Crandall. *Up from the Ashes*. Washington, DC: Brookings Institution, 1986.

Barnett, D. F., and L. Schorsch. *Steel—Upheaval in a Basic Industry*. Cambridge, MA: Ballinger, 1983.

Crandall, R. W. *The United States Steel Industry in Recurrent Crisis: Policy Options in a Competitive World*. Washington, DC: Brookings Institution, 1981.

Hogan, W. T. *Minimills and Integrated Mills*. Lexington, MA: Heath, 1987.

———. *Steel in the United States: Restructuring to Compete*. Lexington, MA: Heath, 1984.

———. *World Steel in the 1980s—A Case of Survival*. Lexington, MA: Heath, 1983.

Jones, K. *Politics vs. Economics in World Steel Trade*. London: Allen & Unwin, 1986.

Preston, Richard. *American Steel*. Englewood Cliffs, NJ: Prentice Hall, 1991.

Tiffany, P. A. *The Decline of American Steel—How Management, Labor, and Government Went Wrong*. New York: Oxford University Press, 1988.

Government Publications

Congressional Budget Office. *The Effects of Import Quotas on the Steel Industry*. Washington, DC: U.S. Government Printing Office, July 1984.

General Accounting Office. *Report to the Congress of the United States by the Comptroller General: New Strategy Required for Aiding Distressed U.S. Steel Industry*. Washington, DC: U.S. Government Printing Office, January 1981.

International Trade Commission. "Certain Flat-Rolled Carbon Steel Products from Argentina, Australia, Austria." Vol. 1. USITC Publication 2664, August 1993.

———. *U.S. Global Competitiveness: Steel Sheet and Strip Industry. Report to the Committee on Finance*. U.S. Senate. USITC Publication 2050, January 1988.

National Academy of Engineering, Committee on Technology and International Economic and Trade Issues. *The Competitive Status of the U.S. Steel Industry*. Washington, DC: National Academy Press, 1985.

Journal and Magazine Articles

Adams, W., and J. B. Dirlam. "Big Steel, Invention, and Innovation." *Quarterly Journal of Economics*, May 1966.

———. "Steel Imports and Vertical Oligopoly Power." *American Economic Review*, September 1964.

———. "The Trade Laws and Their Enforcement by the International Trade Commission." In *Recent Issues and Initiatives in U.S. Trade Policy*, edited by R. E. Baldwin. Washington, DC: National Bureau of Economic Research, 1984.

Aylen, T. "Privatization of the British Steel Corporation." *Fiscal Studies*, August 1988.

Crandall, R. W. "The U.S. Steel Industry in 1993." Paper sponsored by Eurofer, Korea Iron and Steel Association, and Japan Iron and Steel Exporters' Association, June 1993 (mimeo).

Dirlam, J. B., and H. Mueller. "Protectionism and Steel: The Case of the U.S. Steel Industry." *Journal of International Law*, Summer 1982.

Lynn, L. "New Data on the Diffusion of the Basic Oxygen Furnace in the U.S. and Japan." *Journal of Industrial Economics*, December 1981.

Mueller, H. "Protection and Market Power." *Challenge*, September–October 1988.

Schorsch, L. L. "Can Big Steel Change Bad Habits?" *Challenge*, July–August 1987.

5

BEER

Kenneth G. Elzinga

I. INTRODUCTION

In 1620, as every youngster knows, the Pilgrims landed at Plymouth Rock. Less commonly known is that the Pilgrims set sail for Virginia, not Massachusetts. What led them to change their destination? One voyager recorded the following entry in his diary: "Our victuals are being much spente, especially our beere." In other words, the voyage was cut short because they were running out of "beere." Historians may speculate about the effect on American history of the *Mayflower*'s beer running low. Through economics we can learn about the structure and level of competition in this important American industry.

DEFINITION OF THE INDUSTRY

Beer is a potable product with four main ingredients:

1. Malt, which is a grain (usually barley) that has been allowed to germinate in water and then dried.
2. Flavoring adjuncts, usually hops and corn or rice, which give beer its lightness and provide the starch that the enzymes in the malt convert to sugar.
3. Cultured yeast, which ferments the beverage and feeds on the sugar content of the malt to produce alcohol and carbonic acid.
4. Water, which acts as a solvent for the other ingredients.

Because the process of brewing (or boiling) is intrinsic to making beer, the industry often is called the brewing industry.

All beers are not the same. The white beverage (spiced with a little raspberry syrup) that is favored in Berlin, the warm, dark-colored drink served by the English publican, and the amber liquid kept at near-freezing temperature in the cooler of the American convenience store all are beer. Generically, the term *beer* means any beverage brewed from a starch (or farinaceous) grain. Because the grain is made into a malt, another term for beer is *malt liquor*, or *malt beverage*. In this study, the terms *beer*, *malt liquor*, and *malt beverage* are used interchangeably to include all such products as beer, ale, light beer, dry beer, ice beer, porter, stout, and malt liquor. The factor that is common to these beverages and that differentiates them from other alcoholic and nonalcoholic beverages is the brewing process of fermentation applied to a basic grain ingredient.

Beer's production process is not, however, the key to defining beer as a market for economic analysis. The concept of a market entails a group of firms (or conceivably one firm) supplying products that consumers, voting in the marketplace, perceive as close substitutes. Some avid beer drinkers may prefer light beer over malt liquor. But they would prefer either of these over a glass of skim milk. For many consumers, the cross-elasticity of demand among malt beverages is higher than the cross-elasticity of demand between beer and other beverages.[1] The fungibility among malt beverages distinguishes beer as a separate industry. This distinction is supported by the high cross-elasticity of supply among different types of malt beverages and the low cross-elasticity of supply between malt beverages and all other beverages.

EARLY HISTORY

Beer was a common beverage in England in the 1600s and among the early settlers in America. In 1625, the first recorded public brewery was established in New Amsterdam (now New York City). Other commercial brewing followed, although considerable brewing was done in homes in seventeenth-century America. All that was needed were a few vats for mashing, cooling, and fermenting. The resulting product would not be recognized (or consumed) as beer today, however, as the process was crude, the end result uncertain. Brewing was referred to as "an art and mystery."

Brewing was encouraged in early America. For example, the General Court of Massachusetts passed an act in 1789 to support the brewing of beer "as an important means of preserving the health of the citizens. . .and of preventing the pernicious effects of spiritous liquors." James Oglethorpe, trustee of the colony of Georgia, was even blunter: "Cheap beer is the only means to keep rum out."

LAGER BEER: THE JUMPING-OFF POINT

The 1840s and 1850s were important decades in the beer industry. The product beer, as it is generally known today, was introduced in the 1840s with the brewing of lager beer.[2] Before lager beers, malt beverage consumption in America resembled English tastes—heavily oriented toward ale, porter, and stout. Lager beer reflected the influence of Germany on the industry. The influx of German immigrants provided skillful brewers of, and eager customers for, this type of beer. At the start of the decade in 1850, there were 431 brewers in the United States producing 750,000 barrels of beer.[3] By the end of that decade, 1,269 brewers had produced more than a million barrels of beer, evidence of the bright future expected by many for this industry.

The latter half of the nineteenth century also saw technological advances in production and marketing. Mechanical refrigeration aided both the production and the storage of beer. Before this, beer production was partly dependent on the amount of ice that could be cut from lakes and rivers in the winter. Cities such as St. Louis, with its underground caves where beer could be kept cool while aging, lost this (truly natural) advantage with the advent of mechanical refrigeration. Pasteurization, a process originally devised to preserve wine and beer, not milk, was adopted during this period. Beer no longer had to be kept cold; it could be shipped into warm climates and stored for a longer period of time without refermenting. Once beer was pasteurized, the way was opened for wide-scale bottling and the off-premise consumption of beer. In addition, developments in transportation enabled brewers to sell their output beyond their local markets. The twentieth century saw the rise of the national brewer.

PROHIBITION

The twentieth century also saw beer sales outlawed. The temperance movement, which began by promoting voluntary moderation and abstention from hard liquors, veered toward a goal of universal compulsory abstention from all alcoholic beverages. The beer industry seemed blissfully ignorant of this. Many brewers thought (or hoped) the temperance movement would ban only liquor.

In 1919, thirty-six states ratified the Eighteenth Amendment to enact the national prohibition of alcoholic beverages. This led many brewers to close up shop; others turned to producing candy and ice cream. Anheuser-Busch and others built a profitable business selling malt syrup, which was used to make "home brew." Because a firm could not state the ultimate purpose of malt syrup, the product was marketed as an ingredient for making baked goods, such as cookies.

Prohibition lasted until April 1933, and the rapidity with which brewers reopened after its repeal was amazing. By June 1933, 31 brewers were in operation; in another year, the number had risen to 756.

THE DEMAND FOR BEER IN THE POST—WORLD WAR II PERIOD

In 1948, the demand for beer in the United States began a slow decline, from the 1947 record sale of 87.2 million barrels. During this period, the per capita consumption of beer fell from 18.5 gallons in 1947 to 15.0 gallons in 1958. It was not until 1959 that sales surpassed the 1947 total.

In the 1960s and 1970s, total demand began to grow again at an average rate of better than 3 percent per year. In 1965, for the first time, more than

100 million barrels were sold. The per capita consumption of beer increased from the 1958 level of 15 gallons to a level of 24 gallons by the end of the 1970s. The rightward shift in the demand curve for beer was due to the increasing number of young people in the United States (the result of the post–World War II baby boom), the lowering of age requirements for drinking in many populous states, and the enhanced acceptability of beer drinking by females. In addition, the number of areas in the United States that were "dry" shrank considerably.

In the early 1980s, the market demand for beer stabilized and has been rather constant since. Demographic patterns reversed as the pool of young people (18 to 34 years of age) declined. Minimum age requirements for the purchase of alcoholic beverages rose to 21 years. Other factors that have cut into demand include the pursuit of physical fitness and the increasing concern with alcohol abuse, particularly drunk driving. In some states, even laws restraining the use of one-way containers may have reduced consumption.

The market demand for beer exhibits seasonal fluctuations and also varies from region to region. The mountain and west south central states show the highest per capita consumption; the east south central and Middle Atlantic states show the lowest. By state, the demand for beer differs considerably. Utah had a per capita consumption of 13 gallons in 1991, and Nevada leads all others with a per capita consumption of 38 gallons (the figure being biased by beer-quaffing tourists). The highest per capita consumption by natives of a state probably is in Wisconsin.

No-alcohol beer is a very small but growing component of overall beer sales. Shipments in 1992 were 2.3 million barrels; by way of comparison, consumption of no-alcohol beer averaged less than 500,00 barrels per year during the 1980s.

Although economists are not able to measure price elasticity infallibly, statistical estimations indicate that the market demand for beer is inelastic— in the range of 0.7 to 0.9. Brand loyalty is not so strong as to make the demand for any particular malt beverage inelastic. Indeed, the demand for individual brands of beer appears to be quite elastic.[4] This places an important limitation on the market power of domestic brewers.

One indication of how responsive beer consumers are to price changes is provided in the records of a price discrimination case that is discussed later. Table 5-1 shows, for St. Louis on various dates, the percentage of beer sales made by Anheuser-Busch, three important rivals, and the remaining sellers. At the end of 1953, Anheuser-Busch's Budweiser brand was selling for 58 cents more per case than the three rivals and had 12.5 percent of the sales. Early in 1954, Budweiser's price was cut by 25 cents, and Anheuser-Busch's share increased. In June 1954, Anheuser-Busch reduced its price again, this time to the level of its three rivals, and became the largest selling beer in St. Louis. But after price increases by all sellers that left Budweiser at 30 cents more per case than its rivals, Anheuser-Busch's share dropped, evidence that consumers shift brands in response to price incentives.

TABLE 5-1 Percentage of the St. Louis Market Recorded by Anheuser-Busch

Brewer	December 1953	June 1954	March 1955	July 1956
Anheuser-Busch	12.5	16.6	39.3	21.0
Griesedieck Bros.	14.4	12.6	4.8	7.4
Falstaff	29.4	32.0	29.1	36.6
Griesedieck Western	38.9	33.0	23.1	27.8
Others	4.8	5.8	3.9	7.2

Source: *Federal Trade Commission* v. *Anheuser-Busch*, 363 U.S. 536 at 541. Subsequent evidence indicated that factors in addition to price accounted for the Griesedieck Bros. drop in market share.

II. MARKET STRUCTURE

According to economic theory, consumers facing a monopolist (or tightly knit oligopoly) likely will pay higher prices. For this reason, the size distribution of brewing firms arrayed before consumers is of economic interest. Is the beer industry unconcentrated, with its customers courted by many firms, or is it concentrated, leaving beer drinkers with little choice?

In the post–World War II period, two contrary trends have been at work in the industry. There has been a decline in the number of major brewers located in the United States. But there has been an increase in the size of the market area served by the existing brewers and an increase in the number of brands supplied by importers. Moreover, in recent years, specialty firms in the brewing industry (microbrewers, brew pubs, and contract brewers), have grown in number, but their total market share remains under 1 percent.

THE DECLINE IN NUMBERS

Microbrewers aside, the decline in the number of individual plants and independent companies in the brewing industry has been dramatic. In 1935, shortly after the repeal of Prohibition, 750 brewing plants were operating in the United States. Since that time, the number of brewing plants has fallen to a total of 58 in 1992 (see Table 5-2). In the period shown, the number of companies fell by over 90 percent (although beer sales doubled). Few, if any, American industries have undergone a similar structural shake-up.

Along with the decline in the number of companies, an increasing share of the market is held by the largest brewers. As shown in Table 5-3, in 1947 the top five companies accounted for only 19 percent of the industry's barrelage; in 1992 their share was 88 percent. Another way of summarizing the distribution of firm size is to compute the Herfindahl index (also shown in

TABLE 5-2 Company and Plant Concentration, 1947–1992

Year	Independent Companies[a]	Separate Plants[a]
1947	404	465
1954	262	310
1958	211	252
1963	150	211
1967	124	153
1974	57	107
1978	44	96
1983	35	73
1986	33	67
1989	29	61
1992	29	58

[a] Excludes microbreweries of less than 10,000-barrel capacity.

Source: Adapted from U.S. Department of the Treasury, Bureau of Alcohol, Tobacco, and Firearms, *Breweries Authorized to Operate*, (Washington, DC: U.S. Government Printing Office, various years); Modern Brewery Age, Blue Book (Stamford, CT: Modern Brewery Age, various years); and Staff Report of the Federal Trade Commission, Bureau of Economics, *The Brewing Industry*, (Washington, DC: U.S. Government Printing Office, 1978).

Table 5-3), the sum of the individual sellers' market shares squared (its maximum value is 10,000, with one firm in the market). The rising Herfindahl index also testifies to the industry's structural transformation. Whereas some industry observers speak of the Big Five (Anheuser-Busch, Miller, Coors, Stroh, and Heileman in 1992 order), one might also refer to the Big Two, since Anheuser-Busch and Miller in 1992 held over 65 percent of the national market.

THE WIDENING OF MARKETS

Even in the days of hundreds of brewing companies, most beer drinkers faced an actual choice of only a few brewers, because the geographic area that most brewers served was small. Beer is an expensive product to ship, relative to its value, and few brewers could afford to compete in the "home markets" of distant rivals.

Thus, at one time it was meaningful to speak of local, regional, and national brewers. Of these, the local brewer that brewed for a small market, perhaps smaller than a single state and often only a single metropolitan area, was the most common. The regional brewer was multistate but usually encompassed no more than two or three states. The national brewers, those selling in all (or almost all) states, were very few. In addition, it was uncommon for a firm to operate more than one plant.

Today, the terms local, regional, and national brewers are less meaningful. The average geographic market served by one brewer from one plant has

TABLE 5-3 Concentration of Sales by Top Brewers, 1947–1992

Year	Five Largest (%)	Ten Largest (%)	Herfindahl Index
1947	19.0	28.2	140
1954	24.9	38.3	240
1958	28.5	45.2	310
1964	39.0	58.2	440
1968	47.6	63.2	690
1974	64.0	80.8	1080
1978	74.3	92.3	1292
1981	75.9	93.9	1614
1984	87.3	94.2	1938
1987	87.9	93.9	2280
1990	88.6	94.3	2565
1992	87.6	93.4	2594

Source: Adapted from A. Horowitz and I. Horowitz, "The Beer Industry," *Business Horizons*. 10 (1967); various issues of *Modern Brewery Age*; and Beer Marketer's Insights, *1993 Beer Industry Update*.

widened because of the economies of large-scale production and, to some extent, marketing. Because the average-size brewing plant is much larger today, the brewing company may extend itself geographically to maintain capacity operations.[5] The premier example of this was the Adolph Coors Company, which once reached customers on the eastern seaboard from its single brewing plant in Colorado. But it did so at a significant transportation cost disadvantage. A second factor extending the reach of large brewers to serve new geographic regions is their propensity to operate more than one plant.

SIZE OF THE MARKET

Determining the degree of market concentration in brewing entails knowing how wide the geographic markets are for beer. If there is one national market, then concentration statistics for the entire nation are relevant. But if brewing, like cement or milk, has regional markets, delineating their boundaries is necessary before the industry's structure can be ascertained.

The federal courts have to solve this problem when deciding antimerger cases in the brewing industry. In evaluating the merger of the Schlitz Brewing Company with a California brewer and its stock control over another western brewer, the district court, noting that freight rates were important to beer marketing, singled out an area of eight western states as a separate geographic market.[6] The judge was impressed by the fact that in 1963, 80 percent of the beer sold in this area was also produced there and 94 percent of the beer produced in the area that year was sold there. The Continental Divide was seen as a transportation barrier of sorts from outside the area, as evidenced by those brewers that, having plants both in the eight-state area and outside, generally supplied the eight-state area from their western plants only.

Given these figures, many economists would agree that to understand the supply and demand for beer in this eight-state area, one might be able to ignore what was happening in the regions east of the Continental Divide. After all, in its economic sense, a market should include only those buyers and sellers that are important to explaining the supply-and-demand conditions in any one place. But Coors, located outside the eight-state area, was the leading seller of beer in the eight-state area in 1963, with 13.6 percent of total sales. Any market area that overlooks this important seller neglects an important force on the supply side. By 1980, however, 38 percent of the beer consumed in this area of eight western states was imported from outside.

In another case involving the merger of two brewers located in Wisconsin, the Justice Department's Antitrust Division asked an eminent economist at nearby Northwestern University to testify in support of the view that Wisconsin was a separate market for beer. The economics professor told the government lawyers that such a position was economically untenable. Nevertheless, the lawyers persisted in this view without him and eventually persuaded the U.S. Supreme Court that Wisconsin, by itself, is "a distinguishable and economically significant market for the sale of beer."[7]

Although Wisconsin was held to have been a separate market for legal purposes, to single it out as a market in the economic sense is to draw the market boundaries too narrowly. In 1991, brewers in the state of Wisconsin produced over 17 million barrels of beer; that year, consumers in Wisconsin consumed less than 5 million barrels of beer. Because beer also is "imported" into Wisconsin from brewers in other states, obviously more than two-thirds of Wisconsin beer is "exported" for sale outside the state. To say that Wisconsin is a separate geographic market, therefore, is to overlook the impact of most beer production in that state, not to mention the impact on the supply of beer coming into Wisconsin competing with Wisconsin brewers. In this case, the court erred by singling out Wisconsin as an economically meaningful market.

Despite disagreements about the geographic scope of the market for beer, brewing is a concentrated industry, more concentrated than the average U.S. food and tobacco industry.[8]

REASONS FOR THE DECLINING NUMBER OF BREWERS

In this section, two possible explanations for the decline in the number of brewers are considered: mergers and economies of scale.

MERGERS. A common explanation for an industry's oligopolization is a merger–acquisition trend among the industry's firms. At first glance, this seems to be the case in brewing. Between 1950 and 1983 there were about 170 horizontal mergers in the beer industry.[9] But corporate marriages between rival brewers do not explain the increase in concentration by the largest firms. It will prove instructive to review the merger track record of the top five brewers.

The first antimerger action in the beer industry was taken by the Justice Department's Antitrust Division in 1958 against the industry's leader, Anheuser-Busch, because Anheuser-Busch had purchased the Miami brewery of American Brewing Company. The government successfully argued that this merger would eliminate American Brewing as an independent brewer and end its rivalry with Anheuser-Busch in Florida. Accordingly, Anheuser-Busch had to sell this brewery and, for a period of five years, refrain from buying any others without court approval. As a result, Anheuser-Busch stopped acquiring rival brewers and instead began an extensive program of building large, efficient plants in Florida and at other locations. Anheuser-Busch deviated from its internal growth policy in 1980 when it purchased for $100 million the modern Baldwinsville, New York, brewing plant of the Schlitz Brewing Company. Schlitz's sales had dropped so much that it did not need the brewery, and because the plant's capacity was so huge, only an industry leader could absorb its output.

The second largest brewer, Miller, purchased brewing plants in Texas and California in 1966 but acquired no other breweries until 1987, when Miller acquired a small family-run brewery in Wisconsin. In 1972 Miller acquired three brand names from a bankrupt brewer. Two years later Miller bought the rights for the domestic manufacturing and marketing of Lowenbrau, a prominent German beer. And in 1993, Miller purchased the right to distribute the brands of Molson, a Canadian brewer, in the United States. The Miller Brewing Company itself was the subject of a conglomerate acquisition by the Philip Morris tobacco company in 1970. The consequences of this merger are discussed in a later section.

Third-ranked Coors has had a policy to brew its Coors brand in only one location, Golden, Colorado, but Coors has shipped beer by tankers to Elkton, Virginia, where it is bottled and canned for sale in the East. In 1990, Coors acquired the Memphis brewery of Stroh's (formerly a Schlitz plant) for $50 million. There, as in Virginia, it packages only the Coors brand (but brews the company's popularly priced Keystone brand).

The fourth-ranking firm, Stroh, acquired the F. M. Schaefer Brewing Company in 1980. This did not significantly affect its rank (it was then the seventh largest brewer). But in 1982, Stroh acquired the Joseph Schlitz Brewing Company, itself in a sales tailspin and at the time the fourth largest brewer. This acquisition catapulted Stroh to number three in the industry but also shackled the firm with debt. In consenting to the merger, the Antitrust Division required Stroh to divest a brewery in the Southeast. Stroh complied by exchanging its Tampa plant for a brewery in St. Paul, Minnesota, owned by Pabst.

The G. Heileman Brewing Company, the industry's fifth-ranking firm, is the product of over a dozen acquisitions from 1960 to the present, notably

Wiedemann, Associated Brewing, the Blatz brand, Rainier, Carling, and portions of Pabst. In 1960, Heileman was only the nation's thirty-first–ranking brewer. In 1987, Heileman itself was acquired by a large Australian brewer and holding corporation, Bond. The subsequent collapse of the Bond financial empire put Heileman into bankruptcy. In 1993, Heileman was bought by Hicks, Muse & Co., the turnaround artists who acquired and revived Dr. Pepper, Seven-Up, and A&W-brand soft drinks.

In 1981, Heileman was a rejected suitor for Schlitz (just before the Stroh–Schlitz amalgamation), and in 1982 it was a rejected suitor for Pabst. At that time, the Antitrust Division objected to both mergers. But Pabst and Heileman became substantially intertwined. During this time, Pabst purchased the Olympia Brewing Company (producer of the Olympia, Lone Star, and Hamm's brands) in a complex exchange that transferred its breweries in Georgia, Texas, and Oregon to rival Heileman, as well as the Lone Star, Red White and Blue, Blitz-Weinhard, and Burgermeister brands; it also entailed the obligation on Heileman's part to brew beer in the Southeast for Pabst at the former Pabst brewery in Perry, Georgia, that Heileman had acquired. Heileman was thwarted in a private antitrust suit from acquiring the remainder of Pabst. Most of the Pabst assets were acquired by the owner of Falstaff.

The Antitrust Division's determination to stop the Heileman–Schlitz merger indicated the limits to which expansion by merger could take place in the beer industry. Heileman–Schlitz would have resulted in a firm with 16 percent of the national market (Heileman had 7.6 percent in 1980; Schlitz, 8.5 percent). Heileman and Schlitz were the two runaway leaders in malt liquor sales (with about 67 percent of that market segment in 1980), and Schlitz's Old Milwaukee brand (one of the best-selling, popular-price beers) was in head-to-head rivalry with a stable of Heileman's popular-price beers (such as Carling Black Label, Blatz, and Wiedemann). Strong rivalry in popular-price brands benefits not only consumers favoring that segment of the market but also those purchasing premium brands, since competition at the popular-price level places a downward drag on premium prices. In what would have been the largest horizontal merger in the history of the brewing industry, the Antitrust Division said no.[10]

Most of the mergers in the beer industry did not involve firms of significant stature. Generally they represented the demise of an inefficient firm, which salvaged some remainder of its worth by selling out to another brewer. The acquiring brewer gained no market power but might have benefited by securing the barrelage to bring one plant to full capacity or by gaining access to an improved distribution network or new territory. Mergers such as these are not the cause of structural change; they are the effect as firms exit through the merger route.

Mergers have made some imprint on the structure of the brewing industry. The present stature of Stroh and Heileman is partly the result of important mergers, but these mergers have not given the firms market power. In fact, the trend toward concentration in brewing would have occurred even if all mergers had been prohibited. Much of the increase in concentration in the seventies was due to the growth of Anheuser-Busch and Miller, whose expansion was largely internal. Indeed, the enforcement of the antimerger law was partly responsible for the emphasis on internal growth by the leading brewers. The antitrust authorities recognized in the mid-1970s that the beer mergers they once would have attacked did not merit challenge, even if the merger involved sizable regional sellers.[11] One must look to factors other than mergers to explain the industry's structural shake-up.

ECONOMIES OF SCALE

When discussing economies of scale, economists generally plot a smooth, continuous average-cost curve that is the envelope of a host of similarly curvaceous short-run average-cost curves, each one representing a different-size plant. Economies of scale exist if large plants produce at lower unit costs than small ones do. What is seldom mentioned in the discussion of these curves, however, is that little confidence can be attached to the location of any point on these cost curves, notwithstanding their precise, scientific appearance in the economics literature.

With this caveat firmly in mind, Figure 5-1 is a representation of economics of scale in the brewing industry. It illustrates the fairly sharp de-

FIGURE 5-1   Economies of scale in brewing.

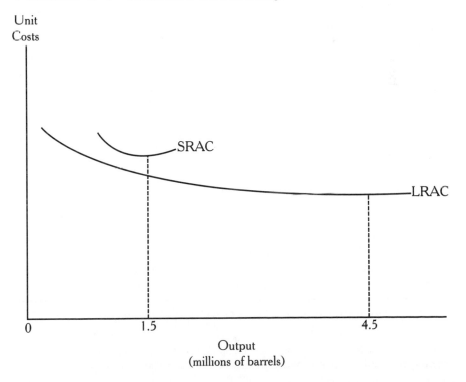

Unit
Costs

Output
(millions of barrels)

cline in long-run unit costs until a plant size of 1.25 million barrels per year of capacity is reached. Beyond this capacity, costs continue to fall, but less sharply, until a capacity of 4.5 million barrels (an enormous brewery) is attained. Here economies of large scale seem to be fully exploited.[12]

Table 5-4 shows one method (the survivor test) used by economists to estimate the extent of economies of scale. This test, like all techniques for estimating economies of scale, is not without its difficulties.[13] As its name implies, the survivor test distinguishes the sizes of those plants that have survived over time. There has been a steady decline (dramatic in some cases) for breweries with a capacity of under 2 million barrels and a large increase in breweries of 4 million barrels and above. That large brewing plants not only survived but grew in number is prima facie evidence of their lower unit costs. One can understand much better the concentration statistics in brewing after learning that the twenty-one plants of the industry leaders Anheuser-Busch and Miller in 1993 have an average capacity of 8.5 million barrels. Table 5-4 does not reflect the appearance of very small breweries—"microbreweries" or "boutique breweries," as they are called. In 1986 there were over 50 brewers of less than 10,000-barrel capacity; by 1992, their numbers had increased to more than 175. Most are new firms, and some are very small indeed, such as California's Buffalo Bill's Brewery, with a capacity of 450 barrels. Microbreweries receive much attention in the business press, in part because there has been so few new entries into the beer industry since World War II. Their owners, who often are also the managers and brewmasters, are portrayed as a new breed of entrepreneur in the beer industry. But the competitive significance of microbreweries is still small.

Figure 5-1 also has a single short-run average-cost curve above the long-run average-cost curve. A long-run cost curve represents the envelope of different-sized plants, each of which uses the latest in production techniques. The curve standing by itself better portrays the situation of many breweries

TABLE 5-4 Surviving Breweries by Capacity, 1959–1992

Listed Capacity Barrels (000's)	1959	1963	1967	1971	1975	1979	1983	1986	1989	1992
10–100	68	54	36	21	10	10	15	13	8	8
101–500	91	72	44	33	19	13	12	8	7	7
501–1,000	30	33	35	32	13	8	2	3	3	3
1,001–2,000	18	17	18	21	13	11	13	10	5	5
2001–4,000	8	10	10	12	12	13	9	10	6	5
4,001+	2	3	4	7	15	20	23	23	20	22

Source: Compiled from plant capacity figures listed in the Modern Brewery Age, *Blue Book* (Stamford, Ct: Modern Brewery Age, various years); Charles W. Parker, "The Domestic Beer Industry" (1984) and industry trade sources. These figures do not include plants listed only on a company-consolidated basis (in the case of multiplant firms) or single-plant firms not reporting their capacity, and they exclude microbreweries of less than 10,000 barrels capacity.

that met their demise. Not only were these breweries too small to exploit all the economies of scale, but their capital equipment was of such an outmoded vintage that their costs were elevated even more.

Some of the economies from larger operations come in the packaging of the beer. The newer bottling lines at the Anheuser-Busch Houston brewery have line speeds of 1,100 bottles per minute. Modern canning lines are even faster: 2,000 cans per minute. It takes a brewery of substantial size to utilize such equipment at capacity. Large plants also save on labor costs via the automation of brewing and warehousing and on capital costs as well. Construction cost per barrel is cut by about one-third for a 4.5-million-barrel plant relative to a 1.5-million-barrel-capacity plant.[14] However, no significant reduction in production costs was detected in a study of multiplant economies of scale.[15]

Cost efficiency relates not only to some finite production capacity but also to management's ability to use the capacity efficiently. Shortly after the repeal of Prohibition in 1933, there was a flood of new entrants into the brewing industry, all expecting to be met by thirsty customers. However, the demand for beer was unexpectedly low after repeal. From a high of 750 brewers operating in 1935, almost 100 were quickly eliminated in but five years. Some were second- or third-generation family-owned firms, not endowed with the brewing and/or managerial capabilities of their founders. Competitive pressures, with no respect for nepotism, eliminated such breweries.[16]

A small brewer, producing a quality product and marketing it so as to keep transportation costs at a minimum, can survive in today's industry by finding a special niche for itself. This seems to be the status, by way of examples, of Wisconsin's Stevens Point Brewery, D. G. Yuengling & Son in Pennsylvania (the nation's oldest brewing firm still in business), and the pioneer microbrewery, San Francisco's Anchor Steam Brewery. Such cases, however, are the exceptions that prove the rules. In brewing—unlike many manufacturing industries in which optimal-size plants seem to be getting smaller—large, capital-intensive plants are necessary to exploit economies of scale and survive in the industry.[17] In markets in which vigorous competitive pressures exist, firms that do not exploit economies of scale or operate with internal efficiency will not survive. This has been the fate of many brewers: They have exited from the industry because of inefficient plants, poor management, or both.

THE CONDITION OF ENTRY

The ease with which newcomers can enter an industry is a structural characteristic of great importance in ensuring competitive performance. If entry is easy, existing firms will be unable to raise prices significantly lest they encourage an outbreak of new competition. On the other hand, if entry is barred, perhaps by a patent or government license, existing firms may be able to make monopoly gains.

Entry into the beer industry is not hindered by the traditional barriers of patents and exclusive government grants. Key inputs are not controlled by existing firms. Nor are economies of scale so important that an efficient entrant would have to supply an enormous share of industry output. Nonetheless, the sheer expense of entering the beer industry is considerable, as the price of constructing a modern 4 million– to 5 million–barrel brewery is some $250 million. Marketing the new brew also is costly because entrants must introduce their products to consumers already smitten by vigorous advertising efforts.

Since World War II, none of the new entrants in the brewing industry have been top-ranking firms. New entry has come from imported beers and microbreweries. The paucity of newcomers is explained by the relatively low profitability of the industry and the ominous fate of so many exiting firms. Moreover, the industry is risky because its plant facilities have few uses other than brewing beer. New entry involves considerable sunk costs. Finally, recent demographic trends have not been encouraging to prospective entrants, especially given the abundance of existing brewing capacity. The 1980s were a decade of the market's wringing excess capacity from the industry, not inviting in new firms. One entrepreneur who recognized the beer industry's excess capacity as a new entry opportunity was Jim Koch of the rapidly growing Boston Brewing Co. This firm is primarily a contract brewer, having agreements with three regional brewers that produce, to Koch's specifications, his firm's Samuel Adams brand.

Currently, the most promising source of new competition is an established brewer moving into a new geographic area, a new imported beer, a contract brewer, or the introduction of new product lines by existing brewers. Beer imports to the United States, which come mostly from the Netherlands, Canada, Germany, and Mexico, increased tenfold between 1970 and 1987. But import consumption has tapered off since 1988. In recent years, imports (mostly in the superpremium category) have represented about 4.5 percent of domestic consumption, up from less than 1 percent in 1970. Imported beer no longer can be discounted as insignificant; in California, for example, imports in 1992 accounted for more than 7 percent of the state's consumption.

PRODUCT DIFFERENTIATION

When consumers find the product of one firm superior to the others, the favored firm can raise its price somewhat without losing these customers. This phenomenon is called *product differentiation*. Three characteristics of product differentiation in the brewing industry bear mentioning. First, several studies indicate that at least under blindfold-test conditions, most beer drinkers cannot distinguish among brands of beer.[18] Second, more expensive brands do not cost proportionately more to brew and package. Third, considerable talent and resources are devoted to publicizing real or imagined differences in beers, with the hope of producing brand preferences. Notwithstanding these efforts, product differentiation in the beer industry has not afforded individual brewers market power to a degree that would be a concern of antitrust policy.

The product differentiation of "premium beer" is important to understand. This phenomenon began years ago when a few brewers marketed their beer nationally and added a price premium to offset the additional transportation costs incurred by shipping over greater distances. To secure the higher price, the beer was promoted as superior in taste and quality, allegedly because of the brewing expertise found in their locations. At one time, this premium price was absorbed by higher shipping costs. But with the construction of efficient, regionally dispersed breweries by most of the large shippers, the transportation disadvantage was eliminated. Nevertheless, the premium image remained. With transportation costs equalized and production costs generally lower, these firms could wage vigorous advertising (and price-cutting) campaigns in areas where regional and local brewers were once the largest sellers.

The national brewers also have two other advertising advantages: (1) None of their advertising is "wasted," whereas regional brewers do not always find media markets (especially in television) that coincide with their selling territories; and (2) their advertising investment is less likely to be lost when a customer moves to another part of the country.[19]

One indicator of an industry's advertising intensity is its ratio of advertising expenditures to sales. For the malt beverage industry, this ratio is nontrivial and has fluctuated around an average of 6 percent—considerably less than the ratio for soft drinks, candy, cigarettes, soap, cosmetics, drugs, and other alcoholic beverages.[20]

The beer industry is a major buyer of television advertising time. Douglas F. Greer, Willard F. Mueller, and others have argued that advertising, particularly television advertising, is a primary cause of increasing concentration in the industry.[21] But the facts do not show a hard-and-fast relationship between the dollars spent on advertising and the market share gained.

Miller has long been a heavy advertiser. Between 1967 and 1971, it generally spent twice as much per barrel as did its rival brewers. But Miller's market share did not expand then, nor did other firms feel compelled to emulate Miller's sizable promotional outlays. Schlitz spent more on advertising in 1975 and 1976 than did either Anheuser-Busch or Miller, and yet the Schlitz brand lost sales in that very time frame. Coors experienced expanding sales with very small advertising expenditures: Between 1968 and 1974, during years of sizable growth, Coors spent only an average of $0.17 per barrel on media advertising. Coors's growing use of media correlates with its declining share position in many states. Miller was not able to secure more than a tochold in the superpremium segment of the market, notwithstanding its extraordinary advertising expenditures per barrel on Lowenbrau. In 1980 Miller High Life was the most heavily advertised brand of beer in the United States, but also in that year, its sales slowed down and began falling every year until the brand's recent repositioning at a lower price point. Anheuser-Busch significantly increased its advertising expenditures on its Budweiser brand—total and per barrel—from 1991 to 1992, and Budweiser's total sales and share of market fell. On the other hand, Natural Light has become one of Anheuser-Busch's fastest-growing brands, but the company in 1992 spent only about 20 cents per barrel on advertising (compared with a companywide average of $3.64 per barrel). Unfortunately, economic analysis does not provide a criterion for determining at what stage advertising becomes redundant and wasteful.

Rising per capita income also has contributed modestly to increasing concentration in the beer industry. Premium beer is what economists call a normal good, with a positive income elasticity. The brewers that in the 1970s came to be the major factors in the industry are, essentially, the producers of premium brands. Popular-price brands were once the leading sellers within their region. No longer. Premium brands and premium-price light beer now occupy about two-thirds of the market. Imports and superpremiums now hold about 8 to 9 percent. The loser in the decades of the 1960s and 1970s has been popular-price beer, dropping from almost 60 percent in 1970 to less than 20 percent in 1981. The traditional domestic superpremium brands (such as Michelob and Lowenbrau) also have faded, replaced by a growing emphasis on specialty beers.

Specialty brands are not just the offspring of microbrewers. Several major brewers have had modest success with specialty brands, usually priced a notch below microbrewery and import brands. Coors (Killian Red), Miller (Miller Reserve), Genesee (Michael Shea's Amber), and Stroh (Augsberger Golden) are examples of the trend. Some beverage-sector analysts speculate that the beer industry someday will have the same large array of specialty brands that one finds now in coffee and wine.

At the other end of the brand cachet spectrum, Canadian soft-drink manufacturer Cott has entered into agreements with three U.S. brewers to produce private-label beer sold directly to retailers. Without advertising and without wholesalers, Cott hopes to position its beer at a low price point and mimic the success of private-label discount cigarettes.

In 1982 and 1983, the popular-price brands (especially thoser marketed by the top four brewers) began a resurgence at the expense of premiums. For the first time, some popular-price beers now receive exposure through television advertising. Low-calorie beer, virtually nonexistent in 1970, had, by 1992, become the largest market segment, with close to 34 percent of the total market, almost all of this being brewed by the top brewers and much of it selling at premium prices.

THE PHILIP MORRIS–MILLER AFFILIATION

In the decade of the 1970s, the most dramatic single factor in the brewing industry's structural change was the rise of the Miller Brewing Company from the seventh to the second largest seller. Miller's ascent involved far more than a mere shuffling of rank among industry leaders. Rather, its production capacity and marketing methods added a new dimension to the industry that affected the industry's economic behavior throughout the 1980s. Two dates are important to understanding the Miller phenomenon: 1970, when Miller's procurement by the Philip Morris Company was completed, and 1972, when Miller purchased the brand names of Meister Brau, a defunct Chicago brewing firm.

The 1970 date initiated the management takeover of Miller by Philip Morris personnel. The new management accomplished three master strokes with the Miller High Life brand. First, it was (in marketing parlance) repositioned to appeal to the blue-collar consumer, who may drink several beers a day. As one Miller advertising executive put it, the strategy was to "take Miller High Life out of the champagne bucket and put it into the lunch bucket without spilling a drop."[22] Second, Miller improved the quality of its product: Any of its beer not sold within 120 days of its production was destroyed. Third, the company introduced a 7-ounce (pony) bottle that appealed both to infrequent beer drinkers with small capacities and drinkers who found that the beer remained colder in the small receptacle.

The virtually unnoticed Miller purchase of the three Meister Brau trademarks included one called Lite, a brand of low-calorie beer brewed by the amyloglucosidase[23] process and once marketed locally by Meister Brau to upper-middle-class weight-conscious consumers. The Miller management noticed that Lite had sold fairly well in Anderson, Indiana, a town with many blue-collar workers. In what is now a marketing classic, Miller zeroed in on "real" beer drinkers, claiming that its low-calorie beer allowed them to drink their beer with even less of a filled-up feeling. The upshot of this marketing campaign would be evident to any grocery or convenience-store shelf stocker: Lite became the most popular new product in the history of the beer industry, and other brewers now have their own brand of light beer as competition. Miller failed in its legal campaign to reserve for itself not only Lite but even the term *light*; the courts left the generic word in the public domain, thereby allowing other brewers to use this term to describe their own low-calorie emulations.

Miller's ascendancy has not been without its brickbats, which come generally in two forms. The first criticism is that Miller changed the process of beer rivalry, with the emphasis no longer on production economies but on market segmentation and brand proliferation. If "beer is beer," then it is argued that consumers do not truly benefit by the product and packaging differentiation that is now common in the industry.

Some economists are concerned that brand proliferation can erect barriers to entry to smaller firms, thereby lessening competition. The contrary argument is that Miller had to bear the financial burden of making low-calorie beer known to consumers; having done this, Miller's rivals have been able to introduce their own light beers more easily.[24]

The second criticism of Miller stems from its conglomerate ownership and the allegation that Philip Morris's "deep pocket" provides unfair advantages to Miller vis-à-vis its rivals. For example, Professor Willard F. Mueller testified before a Senate committee that "Miller's expansion after 1970 was made possible by Philip Morris's financial backing and willingness to engage in deep and sustained subsidization of Miller's operations."[25] Mueller implicitly concluded that Philip Morris's behavior would make economic sense only if the upshot of its subsidization were a monopoly position for Miller (or a shared monopoly with Anheuser-Busch) in the near future.

The matter of cross-subsidization is not simple to analyze. Because the detailed accounting records of the Miller subsidiary are not in the public domain, it remains an open question whether, for example, the depreciation (instead of expensing) of Miller's advertising expenditure would reveal any sizable loss on the Miller operations. Moreover, even if Miller were a poor investment as a beer business, it is clear now, with more than two decades of experience, that Philip Morris cannot expect to recoup any alleged losses through Miller's monopoly power in the beer industry, given the sizable rivals still remaining, their intense rivalry with Miller, and the significant lead that Anheuser-Busch enjoys over Miller. Miller High Life, the company's flagship brand, notwithstanding all the advertising critics charge as excessive, by 1992 had dropped in sales to 4.5 million barrels from its 1981 peak of 23.5 million, causing Miller to begin repositioning Miller High Life at less than a premium-price level in the hope of stemming the brand's free fall.

III. CONDUCT

PRICING

Judging from the early records of the pre-Prohibition beer industry, life in the industry was very competitive. Entry was easy; producers were many. Given these two characteristics, economic theory would predict a competitive industry, and the evidence bears this out.

In fact, the early beer industry offers a classic example of the predictions of price theory. Given the inelastic market demand, brewers saw the obvious advantages of monopolizing the industry, raising prices, and gleaning high profits. Various types of loosely and tightly knit cartels were seen as advantageous, but the difficulty of coordinating so many brewers and the lack of barriers to entry prevented any of these efforts from being successful, at least for long. The degree of competition is evidenced by this turn-of-the-century plea from Adolphus Busch to Captain Pabst:

> I hope also to be able to demonstrate to you that by the present way competition is running we are only hurting each other in a real foolish way. The traveling agents. . .always endeavor to reduce prices and send such reports to their respective home offices as are generally not correct and only tend to bring forth competition that helps to ruin the profits. . .all large manufacturing interests are now working in harmony. . .and only the brewers are behind as usual; instead of combining their efforts and securing their own

interest, they are fighting each other and running the profits down, so that the pleasures of managing a brewery have been diminished a good deal.[26]

In a free-market economy, it is best that rival managers avoid any communications about prices. But if such letters are written, this is the sort of letter that vigorous market rivalry should provoke.

The beer industry also escaped the horizontal mergers that transformed the structure of so many industries such as steel, whiskey, petroleum, tobacco products, and farm equipment during the first great merger movement. There were attempts, mostly by British businessmen, to combine the large brewers during this time. One sought the amalgamation of Pabst, Schlitz, Miller, Anheuser-Busch, and Lemp into one company, a feat that, had it been successful, may have greatly altered the structure and degree of competition in the industry. But the attempt failed, and so brewing entered Prohibition with a competitive structure that responded with competitive pricing.

THE PRICING PATTERN

Beer is generally sold on an FOB (free on board) basis. Some brewers sell on a uniform FOB mill basis, but most vary their prices at times to different customers to reflect localized competitive conditions or to test perceived changes in the marketplace. For example, in 1980, Stroh's prices differed in the contiguous states of Illinois, Kentucky, and Indiana, and its prices in Pennsylvania, a state with unusually vigorous price-cutting, were lower than in nearly all midwestern states.

The present pattern of prices dates back to the turn of the century. Premium beers generally are priced just above the price level of popular-price beers, which in turn are above those of local (or "price" or "shelf") beer. A more contemporary category is the superpremium, a beer selling at a price above premium. A number of major brewers market their own brand of superpremiums, and most imported beers and the output of microbreweries fall into this price category.

The demarcation among local, popular, premium has become blurred in recent years, not only because of the introduction of the superpremium, but also because the price differential between premium and popular-price brands has narrowed. At the same time, the distinction between local and popular beer on the basis of price has become murky because of pricing specials that regularly appear in either segment of the market. Stagnant market demand has led to discounting in the beer industry. For example, in 1992, almost 50 percent of the beer sold by Anheuser-Busch, Miller, and Coors included some form of deal: a direct price discount or an advertising/merchandising allowance. Beer price differentials cannot be attributed to some identifiable physical characteristic of the product. In the case of malt beverages, price differences are the result of customers' tastes, market competition, and history.

PRICE DISCRIMINATION

In 1955, the Federal Trade Commission (FTC) charged Anheuser-Busch with unlawful price discrimination. Anheuser-Busch had dropped the price of its premium brand to all buyers in the St. Louis area but did not make this reduction anywhere else. The FTC maintained that this would impair competition by diverting to Anheuser-Busch sales from other regional brewers serving St. Louis. The charge was brought under Section 2(a) of the Robinson–Patman Act. Proof of such a violation involves answering three questions:

1. Is there price discrimination?
2. If so, does the respondent have a defense?
3. If not, might the price discrimination lessen competition?

There was vigorous disagreement over each of these questions. After the price cut, Anheuser-Busch's Budweiser brand beer was selling for less money per case in St. Louis than anywhere in the country, and this differential could not be explained fully by the lower transportation costs from the Anheuser-Busch brewery in St. Louis. Query: Is this automatically price discrimination?

The circuit court of appeals said no, claiming that price discrimination could not exist unless different prices were charged to competing purchasers. The court put it this way:

> Anheuser-Busch did not thereby discriminate among its local competitors in the St. Louis area. By its cuts, Anheuser-Busch employed the same means of competition against all of them. Moreover, it did not discriminate among those who bought its beer in the St. Louis area; all could buy at the same price.[27]

The FTC and ultimately the Supreme Court disagreed with this interpretation. The Supreme Court stated that price discrimination is "selling the same kind of goods cheaper to one purchaser than to another."[28]

A defense of a charge of price discrimination is to show that one's lower price was offered to meet the equally low price of a rival. Before the FTC complaint, Budweiser was selling at $2.93 per case in St. Louis, and its three regional rivals were selling their beer at $2.35 per case. In two successive price cuts, Anheuser-Busch dropped its price to $2.35 per case. Query: Could Anheuser-Busch argue that it was only meeting the equally low price of its rivals?

Anheuser-Busch tried, but the FTC categorically rejected this defense. Note what this implies: Anheuser-Busch went on record that its premium beer is the same product as popular-price beer; that is, "beer is beer." The FTC, however, argued that at $2.35 a case, Anheuser-Busch "was selling more value than its competitors were. . . .[T]he consumer has proved. . .that [he or she] will pay *more* for Budweiser than. . .for many other beers."[29] The FTC claimed that because of its "superior public acceptance," Budweiser must be priced higher than regional and local beers. After the Supreme Court ruled that Anheuser-Busch had priced in a discriminatory fashion, the court of appeals heard the case a second time to decide whether competition might be lessened by the company's pricing practice. The FTC, citing the figures shown in Table 5-1, had concluded that this practice would give Anheuser-Busch market power in St. Louis by increasing its market share.

The court of appeals disagreed, ruling that diversion of business did not prove that competition in St. Louis was being lessened. The court's decision pointed out that Anheuser-Busch was not subsidizing St. Louis customers with revenues from other markets and that none of the rivals of Anheuser-Busch in St. Louis had felt so "pushed" as to lower their price in response to the Anheuser-Busch cut. The primary consequence was that consumers of beer in St. Louis could buy Budweiser for less money, which the court opined is what competition is all about.

If the FTC had won its case, companies like Anheuser-Busch would have been barred from making selective price cuts, and barring selective price cuts may be the same as barring price competition. Anheuser-Busch could respond to the loss of sales in its own backyard by cutting its prices across the board all over the country. But as one observer put it, "If a seller by law must lower all his prices or none, he will hesitate long to lower any."[30]

One unfortunate response to the pricing practices of the national brewers was the installation in a number of states of price-posting laws. Basically these laws provide that sellers must publicly post their wholesale prices, maintain them for some period of time, announce any price changes, and in some cases specify the retail price of the product as well. Although such legislation has the facade of protecting beer consumers against quick price increases, its impetus actually came from smaller brewers, who saw such laws as protection against competition from promotional price campaigns and from wholesalers and retailers, who saw these as laws providing floors under their own price structure. Price posting of alcoholic beverages has met a checkered legal response: in some states (such as Missouri) passing legal scrutiny, and in others (such as California) being struck down as violating federal antitrust law. Currently, price-posting regulations apply to the brewing industry in a minority of states. In some instances, they apply to both the brewer and the distributor; in others, only to the distributor.

QUESTIONABLE PAYMENTS

In the early 1970s, some brewers and beer wholesalers used "questionable payments" (or "blackbagging") in marketing their beer to selected retail accounts. This is a pricing practice that, depending on one's point of view, could be dubbed either a bribe or a price cut. For example, a brewer might offer, through its distributor, a thousand dollars to a restaurant chain if that company would sell only that brewer's beer on tap at its outlets. If the payment went to a restaurant employee to secure the business, it would be a bribe, but if it went into the restaurant's till, it would be (from an economic standpoint) like a price cut. One reason that brewers simply would not cut their prices openly was to avoid Robinson–Patman Act vulnerability; another was to conceal the price cut from rivals (who might match it) or other customers (who might demand it too). Price-posting laws themselves may have encouraged blackbagging. Whatever the rationale, such payments may violate certain tax and beer-marketing laws.[31]

MARKETING

Although all industries are subject to various federal and state laws that affect the marketing of the industry's product, the brewing industry faces an especially variegated pattern of laws and regulations concerning labeling, advertising, credit, container characteristics, alcoholic content, tax rates, and litter assessments.

For example, Michigan does not permit beer labels to show alcoholic content, whereas Minnesota requires an accurate statement of alcoholic content. In Indiana, advertising is strictly regulated; in Louisiana, there are no such regulations. Some states require sales from the brewer to the wholesaler to retailer to be only on a cash basis, whereas other states allow credit. Some states stipulate both the maximum and the minimum size of containers; Alabama, for example, permits no package beer containers larger than 16 ounces. States also have varying requirements on the maximum and minimum permissible alcoholic content; in some, alcoholic content is different for different types of outlets.

Government involvement in the beer industry also includes taxation. The federal tax alone on a barrel of beer is $18.00, and in 1991, the Treasury Department gathered almost $3 billion in beer taxes. The state taxes on beer vary substantially but average over $7.00 per barrel. In addition, brewers, wholesalers, and retail outlets pay federal, state, and sometimes local occupational taxes. Taxes represent the largest single cost item in a glass of beer.

BREWER–DISTRIBUTOR RELATIONS

There is little forward integration by brewing firms into the marketing of beer. In the United States, brewers are prohibited by law from owning retail outlets, leaving the wholesale distribution as the only legitimate forward vertical integration route.[32] Even wholesaling is prohibited in some states. Rather, beer is retailed through two general types of independent outlets: those for on-premise consumption and those for off-premise consumption.

Most brewers rely on independent distributors to channel their product to these retail outlets. In 1992, there were more than three thousand beer wholesalers, the vast majority being independent merchant wholesalers. Some brewers own a portion of their wholesale channel.

A brewer's keen financial interest in the distribution of its beer is self-evident: A disgruntled customer sees the brewer's name on the container, not the name of the wholesaler or retailer. Therefore, brewers negotiate contracts with wholesalers as to the marketing obligations of each party.

Not all areas of concern to the brewer are open for negotiation. For example, the determination of resale prices is not a matter for private agreement because the antitrust laws limit the contractual opportunities of a brewer in this area. As a consequence, some brewers have introduced reach-back pricing. *Reach-back pricing* is an attempt by brewers to induce price changes by distributors that are consistent with price changes by the brewer. For example, if a brewer raised its FOB price by 15 cents per case, it might expect its distributors to increase their prices to retailers by 25 cents per case (and no more), in the hopes that the price to the consumer might go up, at most, 35 cents per case. If a particular distributor increased its price to retailers by, say, 27 cents (more than a dime above the brewer's increase), under reach-back pricing the brewer would "reach back" and raise its FOB price 2 cents per case to that particular distributor, in order to encourage that distributor to drop the price to its retail accounts (whereupon the brewer would reduce its FOB price) Some distributors do not like reach-back pricing, and the practice has been challenged by certain state regulatory authorities.

Some large retail customers, notably chain stores, would prefer to purchase beer directly from the brewers, thereby eliminating the wholesale distributor. Or they would like to bargain with different distributors of the same brand of beer (possibly purchasing from a price-cutting wholesaler in another area). But brewers almost unanimously market through a three-tier distribution system.

Most brewers also support federal legislation that would immunize from antitrust attack a distributor's exclusive territorial limits. The beer industry has joined with the National Beer Wholesalers Association in championing legislation permitting exclusive territorial agreements similar to legislation secured earlier by the soft-drink industry. Both the Department of Justice and the Federal Trade Commission have opposed this antitrust exemption.

Although this legislation has not been adopted by Congress, most major brewers have proceeded with new wholesaler agreements that restrict a distributor's sales only within the boundaries of an exclusive territory. A wholesaler who makes sales directly or indirectly to customers outside the assigned territory is subject to termination. Clauses of this character earlier had been illegal under the antitrust laws, but they are now scrutinized more permissively. Exclusive territories enable brewers to offer incentives to distributors to cultivate their own territory with less fear of free riding. An example of free riding would be a distributor who transships dated beer to a territory that has been served by a distributor who, by careful stock rotation, had given that brand a reputation for freshness. Major brewers recently scored an important antitrust victory over the New York attorney general who had challenged the ability of brewers to protect against free riding by entering into exclusive territory agreements with their New York distributors.[33]

At one time, beer wholesalers distributed beer primarily in kegs for on-premise draught consumption, but now almost 90 percent of beer sales are packaged, that is, in bottles or cans suitable for on- or off-premise consumption. The beer distributor today will make many more delivery trips to grocery and convenience stores than to taverns.

This trend in beer marketing has worked to the disadvantage of the small brewer. When beer sales were primarily by the keg for on-premise consumption, the small brewer could survive by selling to taverns in the immediate area. But packaged beer sales are primarily for off-premise consumption, and the distribution of packaged beer increases the importance of product differentiation and brand emphasis.

BEER AND THE GLOBAL ECONOMY

The United States imports far more beer than it exports. In some situations, the trade asymmetry is stark. For example, in 1992, the Dutch sold about 25 million barrels of beer to the United States and imported about 9,000 barrels of U.S. beer. The Netherlands, Canada, Mexico, and Germany are the main exporters of beer to the United States (in 1992 order of volume). Japan is the number one importer of U.S. beer.

The U.S. beer industry is a latecomer to the globalization of markets. But the process is under way. For example, in 1993, Anheuser-Busch purchased an 18 percent stake in the largest Mexican brewer (Cerveceria Modelo) and entered into a joint venture with the leading Japanese brewer (Kirin). Also in 1993, Miller purchased a 20 percent interest in the Canadian brewer Molson as well as the U.S. distribution rights of Molson brands; in the same year, Miller purchased a small stake (8 percent) in the Mexican brewer FEMSA. Two years earlier, in 1991, Coors entered into a joint venture with Jinro to build a large brewery in South Korea and a license agreement with Scottish & Newcastle in Scotland to brew Coors beer for the European market. In 1992, Pabst dismantled its Fort Wayne, Indiana, brewery, shipped it to China, where it once again produces Pabst Blue Ribbon beer.

Early in the 1990s, to protect inefficiencies in its domestic beer industry, the Canadian government erected trade barriers against U.S. brewers. In response, in July 1992, the United States raised its tariff on Canadian beer. Beer drinkers on both sides of the border were the losers.

PROFITS

If an industry is effectively monopolized, one might expect to see this reflected in its profits. But this is not necessarily so, because (1) demand may not be sufficiently high to yield profits in spite of monopoly, (2) a monopolist may be inefficient, (3) accounting records often are imperfect measures of economic costs and profits and may not reflect the monopoly gains. In spite of these difficulties, economists regularly look at profit data for some insight into an industry's performance.

On the whole, in the post–World War II period, brewing firms have been less profitable than the average manufacturing firm. Profits in the industry were quite modest until 1967. During the three years 1968 to 1970, the industry's accounting rate of return on net worth after taxes averaged 9.5 percent, compared with the return for all manufacturing firms of 7.4 percent. However, between 1981 and 1985, the beer industry tallied an average return of less than 5 percent, generally below the posttax return on net worth for all manufacturing firms. In recent years, accounting profits for the industry increased, averaging 11 percent for the three years 1988 to 1990.[34]

As one might expect from our discussion of economies of scale, the largest brewers have done better than the industry average. Beginning in 1964, the top four companies began to outperform the rest of the brewing industry in terms of profits. Before that time, the profit record of the top four brewers approximated that of the rest of the industry and was usually inferior to the firms ranked five through eight.[35]

EXTERNALITIES

Externalities, or spillover effects, occur when transactions between buyers and sellers have economic consequences on persons not party to the transaction. These spillover effects can be positive or negative. To the extent that an industry generates externalities, in either the production process or the consumption of its product, the social performance of that industry is likely to be affected

The beer industry is remarkably free of two negative externalities in production commonly associated with manufacturing enterprises: air and water pollution. Brewing is a very "clean" industry (breweries must be more sanitary than hospitals, in fact), and brewing firms often are courted by areas seeking industry partly for this reason. The brewing industry performs well on this count. On the other hand, two important negative externalities affect the consumption of the product: the billions of beer cans and bottles that end up as litter and the problem of drunk drivers. These are negative externalities imposed on individuals who neither sold nor bought the beer.

Although legislation banning or restricting the sale of beer containers is frequently proposed, only a few states and localities actually have passed such laws. The most restrictive of these laws was enacted in the college town of Oberlin, Ohio, which simply outlawed the sale or possession of beer in non-returnable containers. The best known of these laws is the Oregon "bottle bill," passed in 1971, which banned all cans with detachable pull tabs and placed a compulsory 5-cent deposit on all beer and soft-drink containers. Because retail stores particularly do not want to handle returned cans, this drastically reduced the sale of beverages in such containers and offered an inducement to the use of returnable containers or on-premise draught consumption. In Oregon and Vermont, mandatory deposit legislation apparently led to reductions of 60 and 80 percent, respectively, in roadside beverage container litter. However, the statewide (or local) approach cannot solve the problem (say, in Vermont) of customers going "over the line" (to New Hampshire) to avoid the deposit requirement and higher prices.

American brewers (with the exception of Coors) oppose all taxes and bans on containers, stressing instead voluntary action and other litter-recovery programs. The latter, if generously financed, could solve the litter problem, but partially at the expense of nonproducers and nonconsumers.[36]

The costly negative externality of driving while under the influence of alcohol was responsible for raising the minimum drinking age in all states to 21 years of age. This has had little impact on drunk driving by young drivers. Economic research suggests that instead, young beer drinkers are sensitive to price increases. Indexing the federal tax on beer to the rate of inflation since 1951 would, by its impact on the retail price, have discouraged enough drunk driving by young drivers to save an estimated five thousand lives from 1982 to 1988 (more than were saved by raising the minimum legal drinking age).[37] As another strategy, the National Highway Safety Administration has endorsed a blood alcohol content (BAC) level of 0.08 (or above) as a per se driving violation. Most states currently define driving under the influence at a BAC above 0.08. Some authorities argue that the BAC minimum should be set at zero. The American Beverage Institute opposes lowering the BAC but supports stiffer penalties for those caught driving while intoxicated. The costs imposed on third parties by alcohol abuse go beyond automobile injuries and fatalities, and alcohol abuse, of course, is not limited to the consumption of malt beverages.

COMPETITION

In some industries, increasing concentration at the national level and the unlikely entry of sizable new domestic firms might pose a threat to the future level of competition. With fewer companies, given the inelastic market demand, the potential for tacit or direct collusion might be enhanced. One study argues that high two-firm concentration ratios constitute a critical measure of market power.[38] Similarly, with high concentration, the chances may be lessened that smaller firms will follow a truly independent price and production strategy.

The prospect of joint profit-maximizing behavior in the beer industry is not worrisome, however, for the foreseeable future. Thus far there is no evidence of price collusion in the industry. Even during the period of increased demand for beer in the 1960s and 1970s, competition forced the exit of marginal firms. Furthermore, competition along nonprice vectors, such as introductions of new products (no-alcohol beer, packaged draft, dry beer), promotional activities, packaging innovations, brand advertising, product freshness and availability, also is aggressive.

As the demand for beer has stabilized, excess capacity now overhangs the industry and has caused the large brewers to battle more intensely among themselves instead of competing away market share from smaller firms. For example, it was Coors whose sales dropped as a result of Anheuser-Busch's California expansion. Barring collusion or government interference, the overhang of capacity should keep brewing firms from exercising harmful market power in future years, notwithstanding the relative increase in concentration the industry has experienced.

One measure of an industry's rivalry is the extent of changes in market share or turnover in the ranking of its sellers. The beer industry exhibits high mobility in this regard. Schlitz, the nation's second-ranking firm in 1976 and the "Beer That Made Milwaukee Famous," no longer is even brewed there. Pabst was the third leading seller as recently as 1975, ahead of Miller, and the subject of Antitrust Division action. But it has now become a shell of its former self. In 1987 Schlitz and Pabst sold only 2.2 million barrels of premium beer. A decade earlier, combined, they sold 30 million barrels! Miller, number eight in 1968, rose in rank and has been number two since 1977. But Miller, the darling of the industry in the 1970s, experienced an absence of growth in the 1980s and thus far in the 1990s. Coors once "owned" Oklahoma and California, with 54 percent and 40 percent of the sales in these states. In 1992, these percentages had slipped to 22 and 13 percent.

The one constant in all this has been Anheuser-Busch: number one since 1957. Even more remarkable than its hold on number one has been its relative growth. Between 1976 and 1993, the company increased its national market share every year. Indeed, Anheuser-Busch's volume has grown every year except 1990 to 1992, when its total output was flat. Several factors contribute to Anheuser-Busch's parlaying its position into that of leader. All of its breweries are large, low-cost facilities; most of its output is sold at premium and superpremium prices; and much of that output takes the form of only one brand, produced primarily in one package format (Budweiser in 12-ounce cans). This means that Anheuser-Busch does not often incur the cost of changing brewing formulas or reconstituting packaging lines. Rather, Anheuser-Busch's pricing strategy builds on the firm's efficiencies in production. It endeavors to have prices change only in line with production-cost changes and to build overall profits through volume gains. Furthermore, Anheuser-Busch's per barrel advertising costs are significantly below those of many of its rivals because of its enormous volume. For example, in 1992, the company's expenditures were about $2.00 per barrel less than Miller's and Coors's. Currently, Anheuser-Busch is the price leader in the industry. On the other hand, the company has many chips on the Budweiser brand, whose domestic sales declined in 1991 and 1992. Should the bloom ever come off the Bud, Anheuser-Busch's position would become vulnerable.

Rivalry from foreign producers has never been a strong force in the beer industry, compared with that in markets like consumer electronic products and automobiles. Nonetheless, the amount of beer imported into the United States has been increasing and offers a modest source of rivalry among the high-priced brands. Currently, imported beer faces a relatively modest tariff.

The statistics on the structure of the beer industry, the pricing and marketing conduct of its members, and the profits it has received do not mark it as a monopolized industry. The changing fortunes of even major brewers indicate that this is no stodgy oligopoly, with firms adopting a live-and-let-live posture toward one another. The extent of exits from brewing in the last three decades demonstrates that this is hardly an industry in which the inefficient producer is protected from the chilling winds of competition.

NOTES

1. For certain individuals, this figure may be relatively high. Some Germans claim the reason that Dinkelacker is such good beer is because it must compete with the fine wines also produced in the Stuttgart area.
2. Lager beer is aged (or "stored") to mellow. Also, it is bottom fermented; that is, the yeast settles to the bottom during fermentation. The result is a lighter, more effervescent potation.
3. A barrel of beer contains 31 gallons, or 446 eight-ounce glasses (allowing for spillage), or almost 14 cases of 24 twelve-ounce bottles.
4. Thomas F. Hogarty and Kenneth G. Elzinga, "The Demand for Beer," *Review of Economics and Statistics*, May 1972, p. 197. Income elasticity is approximately 0.4.
5. Many brewers regularly ship distances of 300 to 500 miles. For example, one major brewer ships about 40 percent of its production 300 miles or more, but it ships less than 20 percent of its production more than 500 miles.
6. See *U.S.* v. *Jos. Schlitz*, 253 F.Supp. 129 (1966); aff'd. 385 U.S. 37 (1966). The states were California, Oregon, Washington, Nevada, Idaho, Montana, Utah, and Arizona.
7. *U.S.* v. *Pabst*, 384 U.S. 546 (l966) at 559.
8. John M. Connor, *The U.S. Food and Tobacco Manufacturing Industries*, U.S. Department of Agriculture Report 451, March 1980, p. 11.
9. Victor J. Tremblay and Carol Horton Tremblay, "The Determinants of Horizontal Acquisitions: Evidence from the U.S. Brewing Industry," *Journal of Industrial Economics*, September 1988, p. 22.
10. For a contrary view, see A. J. Chalk, "Competition in the Brewing Industry: Does Further Concentration Imply Collusion?" *Managerial and Decision Economics*, March 1988, 49–58.
11. Such as Carling and National, Heileman and Rainier, Olympia and Lone Star, General Brewing and Pearl, Olympia and Hamm's, Heileman and Carling, and Stroh and F. M. Schaefer.

12. See Kenneth G. Elzinga, "The Restructuring of the U.S. Brewing Industry," *Industrial Organization Review* 1 (1973): 105–109 and the sources cited therein.

13. See William C. Shepherd, "What Does the Survivor Technique Show About Economies of Scale?" *Southern Economic Journal*, July 1967, p. 113.

14. *The Brewing Industry*, Staff Report of the Federal Trade Commission, Bureau of Economics (Washington, DC: U.S. Government Printing Office, 1978), pp. 48–49.

15. F. M. Scherer et al., *The Economics of Multi-Plant Operations: An International Comparisons Study* (Cambridge, MA: Harvard University Press, 1975), pp. 334–335.

16. Alfred Marshall saw this phenomenon as one of the important factors limiting the growth and size of firms and an important determinant in the preservation of competition. See his *Principles of Economics*, ed. C. W. Guillebaud (New York: Macmillan, 1961), pp. 315–317.

17. In 1972, the total number of production workers in the beer industry was over 36,000; in 1991, the number had fallen to 23,000, yet industry production (in barrels) increased by more than 40 percent. For evidence that the economies-of-scale phenomenon is not unique to the United States, see Anthony Cockerill, "Economies of Scale, Industrial Structure and Efficiency: The Brewing Industry in Nine Nations," in *Welfare Aspects of Industrial Markets*, ed. A. T. Jacquemin and H. W. DeJong (Leiden: Nijhoff, 1977), pp. 273–301.

18. For example, see J. Douglas McConnell, "An Experimental Examination of the Price–Quantity Relationship," *Journal of Business*, October 1968, p. 439; and Ralph I. Ellison and Kenneth P. Uhl, "Influence of Beer Brand Identification Taste Perception," *Journal of Marketing Research*, August 1964, p. 36.

19. Yoram Peles, "Economies of Scale in Advertising Beer and Cigarettes," *Journal of Business*, January 1971, p 32.

20. As calculated from U.S. Department of the Treasury, *Corporation Source Book of Statistics of Income* (Washington, DC: U.S. Government Printing Office), various years; and *1987 Brewers Almanac*, Table 28. The Treasury data include figures for the malt industry as well. Because the product malt is not as extensively advertised as beer, this understates somewhat the actual ratio for brewing.

21. Douglas F. Greer, "The Causes of Concentration in the Brewing Industry," *Quarterly Review of Economics and Business*, Winter 1981, p. 100, and "Product Differentiation and Concentration in the Brewing Industry," *Journal of Industrial Economics*, July 1971, pp. 201–219; and John M. Connor, Richard T. Rogers, Bruce W. Marion, and Willard F. Mueller, *The Food Manufacturing Industries* (Lexington, MA: Lexington Books, 1985), pp. 244–259. William J. Lynk has criticized this hypothesis. See his "Information, Advertising, and the Structure of the Market," *Journal of Business*, April 1981, pp. 271–303, and "Interpreting Rising

Concentration: The Case of Beer," *Journal of Business*, January 1984, pp. 43–55. See also the response of Victor J. Tremblay, "A Reappraisal of Interpreting Rising Concentration: The Case of Beer," *Journal of Business*, October 1985, pp. 419–431.

22. "John Murphy of Miller Is Adman of the Year," *Advertising Age*, January 9, 1978, p. 86.
23. This is a natural enzyme that reduces the amount of carbohydrates (and therefore calories) in beer. The enzyme became commercially available in 1964.
24. Lutz Isslieb, a General Brewing Company executive, stated, "It's no longer necessary to sell the idea of a light beer; the issue now is which light beer." *Modern Brewery Age*, April 21, 1980, p. 12.
25. See Senate Committee on the Judiciary, Subcommittee on Antitrust & Monopoly, Mergers and Industrial Concentration, *Hearings*, 95th Cong., 2nd sess., May 12, 1978, p. 99.
26. Thomas C. Cochran, *The Pabst Brewing Company* (New York: New York University Press, 1948), p. 151 (letter of January 3, 1889).
27. *Anheuser-Busch, Inc.* v. *Federal Trade Commission*, 265 F.2d 677 (7th Cir. 1959) at 681.
28. *Federal Trade Commission* v. *Anheuser-Busch, Inc.*, 363 U.S. 536 (1960) at 549.
29. In the matter of Anheuser-Busch, FTC, Docket no. 6331, p. 19 (emphasis added).
30. F. M. Rowe, "Price Discrimination, Competition, and Confusion: Another Look at Robinson–Patman," *Yale Law Journal* 60 (1951): 959.
31. See *Brown-Forman Distillers Corp.* v. *New York State Liquor Anthority*, 476 U.S. 573 (1986).
32. Some states now permit microbreweries to sell their output on the premises, by operating brew pubs.
33. *State of New York* v. *Anheuser Busch, et al.* 811 F. Supp 848 (E.D.N.Y. 1993). For economic assessments of beer distribution, see W. Patton Culbertson and D. Bradford, "The Price of Beer: Some Evidence from Interstate Comparisons," *International Journal of Industrial Organization*, June 1991, pp. 275–289; and Tim R. Sass and David S. Saurman, "Mandated Exclusive Territories and Economic Efficiency: An Empirical Analysis of the Malt Beverage Industry," *Journal of Law & Economics*, April 1993, pp. 153–177.
34. Compiled from U.S. Department of the Treasury, *Statistics of Income, Corporation Income Tax Returns* (Washington, DC: U.S. Government Printing Office, various years). The figures include the malt industry.
35. See FTC, *Report of the Federal Trade Commission on Rates of Return in Selected Manufacturing Industries, 1960–1969* (Washington, DC: U.S. Government Printing Office, 1971); *FTC Quarterly Financial Reports for Manufacturing Corps.* (Washington, DC: U.S. Government Printing Office, various years).

36. One study done at the University of Iowa's Division of Energy Engineering suggests that mandatory deposit legislation is not the economical way to correct negative externalities in beer consumption. A cost–benefit study of one Iowa county's experience with such legislation concluded that the costs incurred by consumers and merchants of handling the returned beer containers would have allowed the purchase of road crews to clean 2,000 miles of road of all litter (the county has only a total of 750 miles of roadway). See Gustave J. Fink and Richard R. Dague, "The Iowa Beverage Containers Deposit Law," College of Engineering, University of Iowa (December 1979).
37. F. J. Chaloupka, M. Grossman, and H. Saffer, "Alcohol Control Policies and Motor Fatalities," NBER Working Paper 3831, September 1991.
38. John E. Kwoka, "The Effect of Market Share Distribution on Industry Performance," *Review of Economics and Statistics*, February 1979, p. 101. This scenario does not fit the brewing industry. See William J. Lynk, "Interpreting Rising Concentration: The Case of Beer," *Journal of Business*, January 1984, pp. 43, 55.

SUGGESTED READINGS

Books, Pamphlets, and Monographs
Baron, Stanley Wade. *Brewed in America.* Boston: Little, Brown, 1962.
Beer Marketer's Insights. West Nyack, NY (published twenty-three times a year).
Beer Marketer's Insights—Beer Industry Update. West Nyack, NY (annual).
Brewers Almanac. Washington, DC: Beer Institute (annual).
Connor, John M., Richard T. Rogers, Bruce W. Marion, and Willard F. Mueller. *The Food Manufacturing Industries.* Lexington, MA: Lexington Books, 1985.
Friedrich, Manfred, and Donald Bull. *The Register of United States Breweries 1876–1976.* 2 vols. Stamford, CT: Holly Press, 1976.
Modern Brewery Age: Blue Book. Stamford, CT: Modern Brewery Age (annual).
Norman, Donald A. "Structural Change and Performance in the U.S. Brewing Industry." Ph.D. diss., University of California at Los Angeles, 1975.
Porter, John. *All About Beer.* Garden City, NY: Doubleday, 1975.
Robertson, James D. *The Great American Beer Book.* New York: Warner Books, 1978.

Articles
Ackoff, Russell L., and James R. Emshoff. "Advertising Research at Anheuser-Busch, Inc. (1963–1968)." *Sloan Management Review* 16 (Winter 1975).

————. "Advertising Research at Anheuser-Busch, Inc. (1968–1974)." *Sloan Management Review* 16 (Spring 1975).

Burck, Charles G. "While the Big Brewers Quaff, the Little Ones Thirst." *Fortune*, November 1972.

Chaloupka, F. J., M. Grossman, and H. Saffer. "Alcohol Control Policies and Motor Fatalities." NBER Working Paper 3831, September 1991.

Clements, Kenneth W., and Lester W. Johnson. "The Demand for Beer, Wine and Spirits: A Systemwide Analysis." *Journal of Business* 56 (1983).

Cockerill, Anthony. "Economies of Scale: Industrial Structure and Efficient Brewing Industry in Nine Nations." In *Welfare Aspects of Industrial Markets*, edited by A. T. Jacquemin and H. W. Dejong. Leiden: Nijhoff, 1977.

Culbertson, W. Patton, and D. Bradford. "The Price of Beer: Some Evidence from Interstate Comparisons." *International Journal of Industrial Organization*, June 1991.

Elzinga, Kenneth G. "The Restructuring of the U.S. Brewing Industry." *Industrial Organization Review* 1 (1973).

Greer, Douglas F. "Beer: Causes of Structural Change." In *Industry Studies*, edited by Larry L. Deutsch. Englewood Cliffs, NJ: Prentice Hall, 1993.

————. "The Causes of Concentration in the Brewing Industry." *Quarterly Review Economics and Business*, Winter 1981.

————. "Product Differentiation and Concentration in the Brewing Industry." *Journal of Industrial Economics*, July 1971.

Hogarty, Thomas F., and Kenneth G. Elzinga. "The Demand for Beer." *Review of Economics and Statistics*, May 1972.

Horowitz, Ira, and Ann Horowitz. "The Beer Industry." *Business Horizons*, Spring 1967.

————. "Firms in a Declining Market: The Brewing Case." *Journal of Industrial Economics*, March 1965.

Lynk, William J. "Interpreting Rising Concentration: The Case of Beer." *Journal of Business* 57 (1984).

————. "The Price and Output of Beer Revisited." *Journal of Business* 58 (1985).

McConnell, J. Douglas. "An Experimental Examination of the Price–Quality Relationship." *Journal of Business*, October 1968.

Ornstein, Stanley I. "Antitrust Policy and Market Forces as Determinants of Industry Structure: Case Histories in Beer and Distilled Spirits." *Antitrust Bulletin*, Summer 1981.

Ornstein, Stanley I., and Dominique M. Hanssens. "Alcohol Control Laws and the Consumption of Distilled Spirits and Beer." *Journal of Consumer Research*, September 1985.

Peles, Yoram. "Economies of Scale in Advertising Beer and Cigarettes." *Journal of Business*, January 1971.

Sass, Tim R., and David S. Saurman. "Mandated Exclusive Territories and Economic Efficiency: An Empirical Analysis of the Malt Beverage Industry." *Journal of Law & Economics*, April 1993.

Tremblay, Victor J. "A Reappraisal of Interpreting Rising Concentration: The Case of Beer." *Journal of Business* 58 (1985).

Tremblay, Victor J., and Carol Horton Tremblay. "The Determinants of Horizontal Acquisitions: Evidence from the U.S. Brewing Industry." *Journal of Industrial Economics*, September 1988.

Government Publications

Connor, John M. "The U.S. Food and Tobacco Manufacturing Industries." U.S. Department of Agriculture Report 451, March 1980.

Mueller, Willard F. "Testimony in Hearings, Mergers and Industrial Concentration." U.S. Senate Committee on the Judiciary, Subcommittee on Antitrust and Monopoly. 95th Cong., 2nd sess., May 12, 1978.

Staff Report of the Federal Trade Commission Bureau of Economics. "The Brewing Industry," December 1978.

6

COMPUTERS

Neil B. Niman and Manley R. Irwin

I. INTRODUCTION

The computer industry operates in an unforgiving environment. Ten years ago, IBM manufactured over 60 percent of the mainframe computers in the United States, enjoyed revenues that exceeded $50 billion annually, increased dividend payments to its shareholders every year, and employed some 400,000 people around the globe. The company controlled virtually every facet of computer fabrication in-house, from manufacturing sophisticated chips and components to supplying software operating systems.

Today, IBM is experiencing massive and persistent losses. The company has closed some sixteen plants around the world, slashed its dividends, reduced employment to 250,000, and sold facilities. Unfortunately, a decade in the computer industry constitutes an eternity.

The computer industry is undergoing the transition from a firm-dominated to a market-dominated industry. Seismic shifts in technology competition and consumer demand have left few firms in this industry unscathed. Competitive substitutes abound; product boundaries coalesce; market demarcations overlap; and product development cycles are measured in months. A delay in component delivery risks product obsolescence. And now computer networking emerges as a new reality in the industry.

These same forces of change beset government regulation and oversight. Under a congressional mandate, the U.S. Department of State sets computer-power limits beyond which a firm must apply for an export license in order to sell abroad. Such oversight is promulgated on the premise that computers have years, if not decades, of longevity. Yet personal computer power grows at 30 percent per year, and computers from Taiwan, Hong Kong, and Malaysia sell throughout the world at speeds three times the U.S. export limits. Can government, as an organization, keep pace with an environment driven by competition, torrential innovation, and its counterpart—rapid obsolescence?

Clearly, the computer industry is precipitating a crisis in organizational response. True, firms in the industry must adapt or die. But it is disconcerting to witness the industry's preeminent corporation reassessing every aspect of its illustrious past. The computer is assaulting the monopoly of information once possessed by top management that gave order and coherence to a vertical organizational hierarchy. Today, information power is accessible at the lowest corporate tier, thus deconstructing long-held assumptions of scale economies, scope economies, and the efficiency of market size. Indeed, it is more fashionable to speak of the nonhierarchical, "horizontal" corporation.

Similarly, the information revolution has dissipated the presumed exclusivity of technology long exercised by government entities. If corporations find their decision-making responses driven by the threat of insolvency, can government entities that embody due process, political trade-offs, and legislative deliberation hope to control the economic incentives of global capital markets connected by 200,000 computer terminals? Are we not, therefore, witnessing the erosion of state sovereignty as we have understood government power in the past?

This dual crisis—private sector, public sector—is unprecedented. The United States, indeed the world, is on the threshold of an information revolution. At the base of that revolution stands the computer, the microchip, and software.

What is the nature of this turbulent industry?

II. HISTORY

The electronic digital computer originated during World War II. The British developed a machine named the Colossus that was capable of breaking German naval codes in 1943. The U.S. computer effort was spurred by the need for military artillery tables. The resulting ENIAC machine incorporated 18,000 vacuum tubes, weighed 30 tons, and, at that time (the 1950s), was the fastest machine in the world[1] (see Figure 6-1, which sketches the evolution of the industry from 1930 to the present).

The cold war expedited computer research and development in the 1950s. Funded by the Department of Defense, computers were designed as part of an early warning system—SAGE and BMEWS—to counter the Soviet system's development of ballistic missiles and thermonuclear weapons. (These computers were the first to incorporate transistors.) IBM participated in these programs; indeed, government spending accounted for about half the company's revenues.[2]

IBM's prewar business had concentrated on electric business calculators and punch-card machines. In the era that followed World War II, the company identified and seized the opportunity presented by the electronic computer. IBM's ability to move from the world of mechanical accounting machines to electronic computers rewarded the company handsomely. Its first machines (the "701" and the "702") were commercial successes. IBM soon distanced its rivals in the mainframe computer industry, partly because its "fanatical devotion to deadlines and delivery schedules was legendary."[3]

FIGURE 6-1 Evolution of the computer industry.

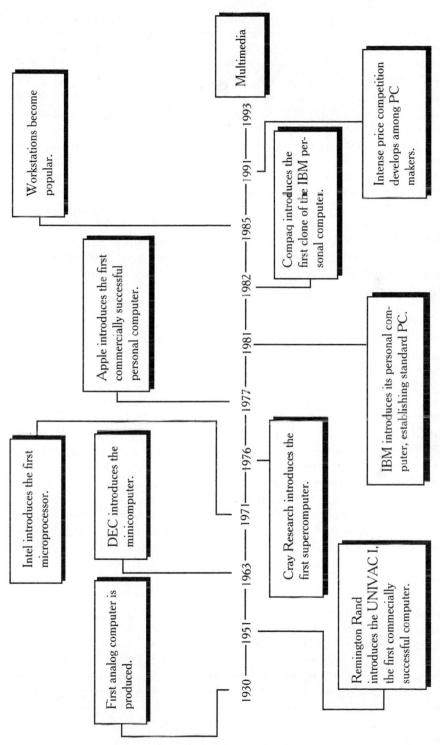

Source: U.S. International Trade Commission, *Global Competitiveness of U.S. Advanced Technology Industries: Computers,* Investigation 332-339, ITC Publication 2705, December 1993, p. 22.

By the 1960s, IBM not only dominated the mainframe industry, but the company also allocated more resources to computer research than did the federal government. The company's "360" large and small machines, however expensive and arduous the innovation process, clearly set IBM apart from its rivals. Although these rivals—GE, RCA, UNIVAC, Control Data, NCR, and so on—were massive in size and resources, they became known in the industry as the "Seven Dwarfs." By the mid-1960s, IBM accounted for some 60 percent of all computers delivered in the United States.

Despite IBM's market preeminence, two submarkets did invite new entry. The first, the minicomputer market, traced its origins to government R&D of the cold war. Minicomputers, smaller and cheaper than the million-dollar mainframes, took off when such firms as Digital Equipment Corporation (DEC), Data General, Hewlett-Packard, and others entered the fray. IBM also offered a minicomputer series, but market growth diluted IBM's share of this market. Every computer firm adopted separate or proprietary operating software in an attempt to differentiate its products from its rivals'. Product differentiation, however, led to product incompatibility. IBM machines could not run on DEC machines and vice versa.

A second computer submarket, the so-called plug-compatible peripherals, opened up in the late 1960s. These memory attachments, disk drives, and tape drives were supplied by independent manufacturers. The devices, plugged into sockets tied to the company's mainframe processors, simulated IBM tape drives.

Independent peripheral suppliers challenged IBM's concept of total system control. A user could mix and match IBM's mainframe to a tape drive or disk manufactured by Telex, Memorex, or CalComp. Indeed, large IBM users, such as DuPont, encouraged the growth of the plug-compatible industry, which permitted independents to field products that were often superior in performance and lower in price than comparable IBM equipment. Apparently, the heavy capital investment in IBM's 360 computer system had exacted a development cost in terms of the company's peripheral products.

IBM's response to entry in this submarket was nothing if not spirited. First, the company extended the length of its system leases and thus reduced its customers' opportunities to switch to another supplier. Moreover, once under contract, the customers were assured of getting an IBM peripheral upgrade whenever the new product became available.

Second, IBM reengineered the interface standard linking peripherals to its mainframe. In fact, the controller and disk units were repackaged into one unit. Independent suppliers found their equipment was now incompatible and hence obsolete.

Third, IBM quickened the development and introduction of new peripherals. Independent suppliers asserted, however, that the development and marketing costs of new tape drives were spread across several product lines, so that the price of the products did not reflect the assigned costs.

IBM's strategy of leasing, interface changes, and introduction of new products devastated the independents in the plug-compatible market. Revenues dried up and costs increased as firms redesigned their products to fit IBM's new interface standards. Moreover, as profits plummeted, the market shares of Telex and Memorex also fell, thus reducing their capital for future product innovation. Indeed, an IBM memo noted that one rival, Memorex, would experience an economic squeeze that would leave the company less than "viable."[4] Yet when IBM's practices were challenged in court, federal judges upheld the company's response to competition as a model worthy of emulation by any American firm.[5]

In 1969, the Department of Justice filed an antitrust complaint against IBM, charging that the company had monopolized the mainframe computer industry and thus had run afoul of the Sherman Act. The complaint was later amended to include the charge that IBM's response to plug-compatible competition was predatory in nature.

IBM vigorously defended its conduct and practices. Some estimated that the company spent up to $1 billion in its defense against the government complaint. Reportedly, the IBM president "told his legal staff to spend whatever was necessary and they still went over budget."[6] One MIT consultant used his fees to buy a yacht and named it *Section 3*.

In 1982, the government dropped its antitrust suit on grounds that the complaint was "without merit." This action was highly controversial. Given the industry's dynamic change, one school of thought held that prosecuting a company in the 1980s for monopoly in the 1960s constituted a policy anachronism. Another insisted that the "seeds of self destruction lie within all monopolies." A third held that if IBM had gone through divestiture, "it would have had to develop new entrepreneurial opportunities."[7] In any event, the fact was that IBM dominated the industry in terms of market share, profits, pricing, and industry standards. So formidable was the company's control of the software market that when Japanese companies filched IBM's operating manuals, IBM sued and collected $833 million from Fujitsu and $250 million from Hitachi.[8]

Two developments occurred in the mid-1970s that injected unparalleled changes into the industry. First, the microprocessor chip—the computer on a chip—was invented by Intel. Then, Steve Jobs and Steve Wozniak plugged a circuit board of these chips into a display, thus giving birth to the Apple computer. Other firms—Atari, Osborne, Radio Shack—entered a segment of the personal computer market destined for video games and home entertainment.

IBM responded in the early 1980s to the personal computer challenge. In an attempt to break out of its mainframe computer culture, IBM set up an independent computer subsidiary in Boca Raton, Florida. The subsidiary violated virtually every tradition of IBM's mainframe culture: It bought components on the open market and purchased power supplies, disks, circuit boards, and printers offshore. The affiliate engaged a firm (Microsoft Inc.) to write its PC operating system (MS-DOS) and offered its computer for sale by Computerland and Sears. IBM's Boca Raton operation was indeed run by "wild ducks."

The result was spectacular. From zero revenues in 1981, IBM's PC captured 30 percent of the market and sales grew to $4 billion by 1984. By the mid-1980s it would not be too much to say that IBM's strategy of open architecture had become the PC industry's standard.[9]

Today, the company's fortunes have reversed dramatically. A $60 billion company is drenched in red ink, and it is downsizing, shutting down plants, laying off employees, and slashing dividends. The firm's CEO has taken early retirement. The reason for this reversal of fortune will become apparent when we examine the current structure of the industry.

III. MARKET STRUCTURE

Figure 6-2 shows the structural components of the computer industry and indicates the relative importance of each segment, as measured in dollar revenues.

In terms of market size, the mainframe computer has traditionally dominated the computer industry. These large machines, capable of filling an entire room, cost millions of dollars, approach a total population of some 50,000 units worldwide, and represent an investment value of $1 billion.[10] It is in this segment that IBM has maintained its traditional market presence over the past forty years.

Table 6-1 lists the current participants in the mainframe market segment and gives their respective market share as of 1992. Note that IBM's share of

FIGURE 6-2   Structure of the computer industry and size of the global market by segment, 1992.

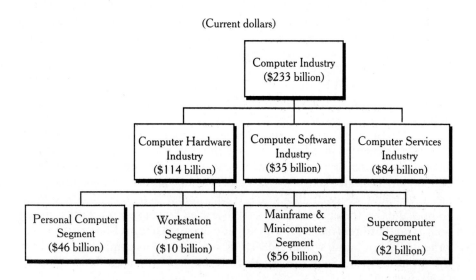

(Current dollars)

Source: U.S. International Trade Commission, *Global Competitiveness of U.S. Advanced Technology Industries, Computers,* Investigation 332-339, ITC Publication 2705, December 1993, p. 22.

TABLE 6-1 Large-Scale Systems

	Company	Revenues ($ million)	Market Share[a]
1.	IBM	8,190.0	29.1
2.	Fujitsu	4,431.3	15.8
3.	Hitachi	4,043.0	14.4
4.	NEC	3,063.5	11.0
5.	Unisys	1,966.0	7.0
6.	Amdahl	1,489.6	5.3
7.	Nihon Unisys	1,029.3	3.7
8.	Siemens Nixdorf	962.2	3.4
9.	Groupe Bull	857.3	3.0
10.	Cray Research	550.5	2.0

[a] Percentage of Datamation 100 revenues.

Source: Datamation, June 15, 1993, p. 22.

29.1 percent is almost twice as large as that of its next two rivals, Hitachi and Fujitsu.

The personal computer market is depicted by Table 6-2. Again, IBM's share of 17.2 percent heads the list, followed by Apple, Compaq, and NEC. Figure 6-3 captures the dynamic change in the computer industry, that is, the shift away from mainframe units to smaller personal computers. What factors account for that transition? The short answer is that the traditional barriers to market entry have eroded precipitously over time.

The first barrier consisted of the capital cost of manufacturing large computer systems. The cost of research, component fabrication, assembly, and manufacture embodied the essence of a capital-intensive industry. IBM's 360 computer, for example, cost $750 million in engineering outlays and required

TABLE 6-2 Personal Computers

	Company	Revenues ($ million)	Market Share[a]
1.	IBM	7,654.5	17.2
2.	Apple	5,412.0	12.1
3.	Compaq	4,100.0	9.2
4.	NEC	3,986.8	8.9
5.	Fujitsu	2,618.5	5.9
6.	Toshiba	1,949.4	4.4
7.	Dell	1,812.5	4.1
8.	Olivetti	1,348.7	3.0
9.	AST	1,140.4	2.6
10.	Gateway 2000	1,107.1	2.5

[a] Percentage of Datamation 100 revenues.

Source: Datamation, June 15, 1993, p. 22.

FIGURE 6-3 The mighty micro.

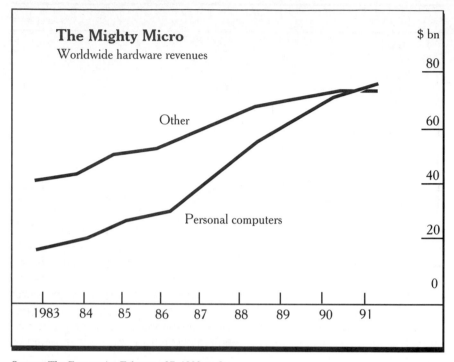

Source: *The Economist*, February 27, 1993, p. 8.

another $4.5 billion investment in plant facilities.[11] The need for huge invest-ment outlays rationed and effectively controlled market entry.

A second barrier was the cost of developing compatible software. Such costs amounted to approximately 50 percent of the total computer expenses borne by mainframe suppliers. Each manufacturer pursued its own proprietary oper-ating systems. As we have seen, IBM set the software standard for the industry.

Proprietary software served as a further barrier to potential competition. Once a firm adopted, say an IBM computer system, the cost of switching to a rival system was formidable. Proprietary software thus locked a user into a particular firm and its hardware configuration. And software, though critical, was not the only lock-in variable. If a user required a computer system up-grade or expansion, the price of software, peripherals maintenance, services, and tracing reflected the user's price inelasticity. Software lock-in translated into higher revenues to the computer vendor.

Vertical integration added further disincentives to the market entry process. Computer firms manufactured their own components, fabricated subsystems internally, supplied their own software, assembled computers in-house, and leased hardware and software systems as a total package. Poten-tial entrants to the industry had to contemplate replicating vertical integration if they sought to compete directly with the industry leader: IBM.

Today, these traditional constraints to market access have eroded with a vengeance. The multiplication of firms supplying computer components and parts now reaches into the thousands. Independent specialized firms can achieve unit-cost economies that can exceed the efficiency of in-house computer operations.

The microprocessor has clearly provided an entrée for potential rivals. The microprocessor can be purchased off the shelf and is available to all comers. Falling costs and each succeeding chip generation confer more power to the computer fabricator. Lower chip costs further ease access to computer manufacturing.

Customers seek computer compatibility. Many computer firms now accept and embrace open systems architecture. As a result, competitors now collaborate in developing a set of specifications that allow machines to run on nonproprietary software, and these operating software packages can be purchased off the shelf.

More recent software developments permit personal computers and workstations to cluster around data networks. PCs operating on LANs or WANs (local-area networks, wide-area networks) enable users to tap into dozens of other PCs, linked by cable, fiberoptics, or satellite. By 1994, some 50 percent of all office PCs will be tied to a network.[12] As personal computer prices continue their inexorable drop and as PCs incorporate more processing and storage power, that power can be leveraged by the computer network, and it is this network development that has caught the mainframe industry by surprise.

Originally, the PC was viewed as a product complementing the use and demand for the mainframe computer. But clusters of microprocessors and workstations on data networks multiply the computer power available to the user at a fraction of the cost of a mainframe, and so PC networks now loom as a competitive substitute. As a rearguard action to prevent further market erosion, computer mainframe suppliers must reduce their prices. But as Figure 6 1 suggests, the tide currently runs with the PC. Indeed, the value of PCs shipped nearly equals the combined value of minicomputers and mainframes.

The eroding entry barriers and the availability of competitive substitutes have caused revenue losses not only in the mainframe computer segment of the market but in the minicomputer segment as well. The turmoil experienced by IBM testifies to both the extent and the velocity of these structural forces. In a word, the microprocessor and networking software have decentralized computer power and access. Seldom have developments in the factor input market fueled such monumental repercussions in the final product market.

IV. CONDUCT

As the industry has embraced open operating standards, computer firms have adopted pricing, cost, product differentiation, and cooperative alliances as strategies to compete in this market environment.

PRICING STRATEGIES

Before the 1990s, pricing in the computer industry followed a predictable pattern. Name-brand computer firms set the established price for the industry, and clones or computer copies sold at discounts of up to 30 percent. To retain market share, therefore, the larger players were forced to cut prices. Almost inevitably, one round of cuts would be followed by another.

Ease of entry into the PC market has accelerated this pricing dynamic. Cheap "clones" have nibbled away at the market shares held by IBM, Compaq, and Apple. Indeed, some two hundred personal computer firms supply IBM clones in Hong Kong alone.[13] In 1992, a price war, started by Compaq, instigated price reductions of up to 50 percent. The net result was not only lower prices for the whole industry but the elimination of an entire tier of rivals.

Workstation products, beginning at $5,000, have not been immune from this frenzied pricing strategy. To preserve an aging market, mainframe and minicomputer companies are offering discounts to their customers. Yet the explosion of competitive substitutes shows little sign of abating, and mainframe pricing is vulnerable. Indeed, in the past twelve months, mainframe prices have fallen 40 percent.[14]

LOW-COST STRATEGIES

Price reductions inevitably invite computer firms to engage in rigorous cost reduction. Traditionally, market size was thought to yield purchasing discounts and hence reduced unit costs for parts and related components. With high product development costs and large fixed investments in plant and equipment, it seemed reasonable to conclude that the major computer firms would enjoy the advantage of large-scale production.

Today, however, computer firms no longer require high volume to achieve manufacturing efficiencies. As noted, computer firms can purchase standard parts and components from hundreds of third-party suppliers. These suppliers produce generic components that appeal to a broad spectrum of buyers, and the outsourcing suppliers themselves combine specialization and scale economies and pass forward those efficiencies to computer assemblers in the form of lower prices.

Some firms elect to retain their in-house or upstream operations, on the premise that vertical affiliates yield cost savings. Although many of these firms have chosen to remain in the United States, others have moved offshore to Southeast Asia in an effort to achieve labor-cost economies. Not only are wage rates and plant investment attractive there, but many circuit board suppliers can circumvent U.S. environmental laws governing the disposal of hazardous wastes such as PCBs.

The computer industry has experienced a transformation on the demand side as well. Firms such as Dell and Gateway have pioneered the selling of products through direct mail order service, in which firms purchase components, computers, and peripherals and provide customers with access to their products via an "800" telephone number. Low-cost production and marketing mean lower prices to users through direct purchase. Moreover, Dell's telephone network is in contact with its customers and therefore can constantly monitor its customers' changing needs. (Dell reputedly started his company in a college dormitory.)

IBM has not been unresponsive to these new channels of distribution. In 1990, the company introduced the PS/1 computer line, targeted to mass merchants and discount stores. Two years later, IBM moved into the direct mail order business. Under the name Ambra, IBM suppressed its own brand name in order to compete with Dell and other direct-marketing companies.

PRODUCT DIFFERENTIATION

To avoid the ravages of commoditylike price discrimination that accompanies product homogeneity, most firms in the industry embed in their hardware distinctive software features and higher performance capability. A laptop computer, for example, now contains the processing power of a mainframe computer of twenty years ago. At the same time, microminiaturization further reduces product size, resulting in portable computers that fuel the growth of wireless communication networks.[15]

Nonprice competition assumes many forms in today's market. Some firms offer longer warranties, on-site service, bundled software, and submarket differentiations. Other firms offer personal computers capable of multitasking, word processing, data processing, fax transmission, wireless telecommunications, and teleconferencing.

COOPERATIVE ALLIANCES

Some firms elect a third competitive strategy by forming joint ventures with industry rivals (see Table 6-3). Corporate alliances constitute a relatively new phenomenon in the industry. Again, the driving force behind this strategy appears to be the rapid pace of technology in the industry. As one industry participant put it, "Technologies are changing so fast that nobody can do it all alone anymore."[16]

The semiconductor industry is a case in point. The cost of developing new chips is now estimated to be $1 billion. Not surprisingly, corporations are finding that the risk of doing this alone now exceeds the potential gain of cooperation. Therefore, companies like Toshiba, IBM, and Siemens have gotten together to develop a new generation of advanced memory chips.

Sometimes an alliance is driven not so much by a desire to reduce capital outlays as to secure the special expertise enjoyed by another company. Motorola, IBM, and Apple, for example, have formed an alliance to develop and produce a new microprocessor called the PowerPC. Although IBM created the initial design, the company needed Motorola's help to modify the design so that it could be manufactured on a single piece of silicon and produced at low cost.

TABLE 6-3 Selected Manufacturing and R&D Joint Ventures and Alliances

U.S. Firm	Allied Firm	Headquarters	Product	Year
Apple	General Magic	U.S.	Networking	1993
	IBM	U.S.	Software	1991
	Motorola & IBM	U.S.	PowerPC chip	1991
	Sharp	Japan	Palmtop computers	1992
	Sony	Japan	Notebook computers	1991
DEC	Alcatel	France	Display terminals	n.a.
	Apple	U.S.	Network interfaces	n.a.
	Cray Research	U.S.	Supercomputer/ minicomputer interfaces	1992
	Escom	Germany	Network services	1992
	Fluent	U.S.	Video networking hardware and software	1992
	MasPar	U.S.	MPP computers	1991
	Mitsubishi	Japan	Alpha AXP procesors	1993
	Motorola	U.S.	Data interface chip sets	n.a.
	Olivetti	Italy	Network interfaces	1992
	Siemens Nixdorf	Germany	Semiconductors	n.a.
IBM	Apple	U.S.	Software	1991
	Canon	Japan	Printers	1992
	Digital	U.S.	Disaster recovery	1992
	Groupe Bull	France	Workstations	1992
	Hewlett-Packard	U.S.	Fiber optic components	n.a.
	Intel	U.S.	Microprocessors	1991
	Motorola	U.S.	Phoneless modems	1990
	Motorola & Apple	U.S.	Power PC Chip	1991
	Motorola & Nat'l Semiconductor	U.S.	LAN products	1992
	Picturetel	U.S.	Video conferencing	1991
	Siemens Nixdorf	Germany	Semiconductors	1991
	Thinking Machines	U.S.	Supercomputers	1991
	Toshiba	Japan	Flat panel displays	1991
			Memory chips	1992
Sun Microsystems	Fujitsu	Japan	SPARC chips	1986
	Intergraph	U.S.	64-bit microprocessor	n.a.
	Kalpana	U.S.	LAN technology	1992
	Moscow Center of SPARC Technology	Russia	Workstation software	1992
	Texas Instruments	U.S.	Super-SPARC chip	n.a.
	Toshiba	Japan	RISC technology	n.a.

n.a. = not available.

Source: U.S. International Trade Commission, *Global Competitiveness of U.S. Advanced Technology Industries: Computers,* Investigation 332-339, ITC publication 2705, December 1993, p. 2-12, 2-13.

A different type of arrangement is for companies to agree on technical standards that they hope will be adopted by the entire industry. For example, Unix, invented by Bell Laboratories, has more than thirty-five different Unix versions in the marketplace. Without a common standard, independent software vendors would be reluctant to develop applications that could run on only a small percentage of the Unix machines. To rescue Unix from obscurity, Novell, IBM, Sun Microsystems, and Hewlett-Packard have formed an alliance to establish COSE (common open software environment) as a step toward unifying the Unix standard.

Alliances, of course, do not guarantee that the partners will not use this shared knowledge to develop new products that compete with their existing product offerings. Under these circumstances, two companies may pool assets to form a third, independent company, in other words, a joint venture. Joint ventures tend to bind both companies to formally commit resources to the new entity, thereby ensuring all the partners' continued participation and cooperation. Following the breakup of the Microsoft/IBM alliance for the development of OS/2 (an operating system designed to replace MS-DOS), IBM formed a joint venture with Apple (Taligent) to design an even more advanced operating system.

SUMMARY

In sum, few computer firms can escape the turbulence of a market of eroding entry barriers. In that environment, a number of competitive strategies are possible. Some firms may elect to cut costs and play the price elasticity game. Other firms may choose to differentiate their products via incremental feature innovation. Still others may pursue corporate alliances in order to remain technologically current or to spread out research expenditures.

Market pressures necessitate a constant reexamination of corporate response and organization. Mainframe firms, as noted, traditionally supplied the bulk of their chips, boards, software, and peripherals via in-house affiliates. Internal economies were presumed to supersede the transaction costs associated with buying components on the open market.

Today's competitive pressure has transformed this make/buy decision. Some fifty thousand component suppliers now populate the world, including Pacific Rim firms that manufacture chips, computers, and assembled systems.[17] These firms set the benchmark for the performance of in-house suppliers. IBM, for example, has been forced to restructure into thirteen autonomous divisions that, for the first time, are allowed to buy and sell on the open market. Indeed, the company is beginning to sell components to OEMs (original equipment manufacturers) that had previously been restricted to internal consumption only. Possessing the only license to produce the 486 chip (as long as it is sold as part of a circuit board), IBM has aggressively moved into the OEM market with a line of speed chips for the same price that Intel sells a single chip.

V. PERFORMANCE

To assess the performance of the computer industry, it is useful to examine the indices of price, cost, marketing, research, and innovation.

PRICES

Few industries approach, much less approximate, the breathless performance of the computer industry. Competition has spurred price reductions across an entire spectrum of computer products. Over the past thirteen years, personal computer prices have plunged 32.7 percent annually.[18] Similarly, workstation prices have dropped between 10 to 25 percent annually in the past decade. Even mainframe and minicomputer prices have not been immune from price discounts. In short, this industry has come to experience the ravages of deflation.

COSTS

Competition has inspired quantum leaps in computer productivity. Over the past decade, computer power has grown by 30 percent annually. Every eighteen months, advances in microprocessors have doubled the computing power that a dollar will buy. An MIT study comparing the cost to obtain 4.5 MIPS of computing power in 1980, 1990, and 2000 estimates that the cost of 4.5 MIPS in computing power was $4.5 million in 1980 (or the equivalent of 210 employees with a certain skill level).[19] By 1990, the cost of the same 4.5 MIPS had fallen to $100,000, or the equivalent of 2 employees of the same skill level. By the year 2000, the cost of 4.5 MIPS of computing power is projected to be just $10,000, or the equivalent of 0.125 workers. Figure 6-4 illustrates the falling cost of PCs over the past several years.

In the search for operating efficiency, market competition has forced many firms to reappraise the premise of their organizational integration. IBM, of course, set the industry standard by owning all stages of the production–distribution–marketing process as a means of achieving internal economies. Today that model has been inverted. Transaction-cost efficiencies often supersede the economies of backward integration. Corporations are outsourcing components and selling off in-house affiliates. A key to corporate growth is fast response time and accelerated decision cycles. Today, computer firms are organized by collaborative teams from within the corporate organization.

Finally, the transition from a mainframe to a personal computer environment has forced management to reappraise marketing costs. A million-dollar final product can carry a sales expense much more easily than can a $3,000 personal computer. Indeed, mail order distribution systems have come into their own as a consequence of product miniaturization, and increasingly, sophisticated computer buyers now order computer components via "800"

FIGURE 6-4 Less is more.

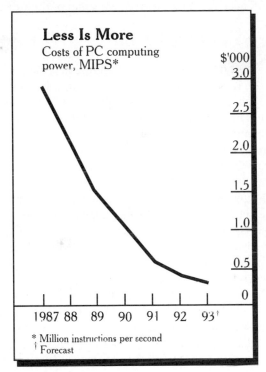

Less Is More
Costs of PC computing
power, MIPS*

$'000

* Million instructions per second
† Forecast

Source: The Economist, February 27, 1993, p. 6.

telephone numbers. Dell computer, for example, generates 85 percent of its revenues by telephone under the promise of four-day delivery service.[20] Recently, a new PC firm announced a policy of one-day service.

PROFITS

Corporate profits are nothing if not volatile in this industry. They rise or collapse on the ability of a firm to anticipate demand and control expenditures. A decade ago, mainframes and minicomputers dominated the industry's profits, but now both market segments are awash in red ink. IBM has lost close to $18 billion in the last two years, and the firm's attempt to cut employment from 400,000 to 225,000 (1994) is well known.

But IBM is not alone. DEC, a major minicomputer supplier, wrote off $3.2 billion in two years, vacated 165 plant facilities, and cut employment by 18,000. UNISYS lost $436 million in 1992, and Data General lost over $59 million, despite eliminating half its workforce. Groupe Bull (French) recently closed 8 out of 13 plants and announced 8,000 layoffs. Olivetti (Italian) is downsizing by 20 percent. Siemens–Nixdorf (German) announced layoffs of

6,000. Even Japanese mainframe computer suppliers (NEC, Hitachi, and Fujitsu) have experienced revenue shortfalls in some of their operations.

RESEARCH AND INNOVATION

Research and product innovation, nonetheless, remain a hallmark of the computer industry. If patents provide one proxy for industry performance, the industry ranks high relative to other industries.[21] When measuring R&D spending of the top nine hundred companies, a *Business Week* study concluded that with total spending of $11.9 billion, computer hardware manufacturers as a group outspent every other industrial group.[22] What makes this expenditure somewhat startling is that it occurred while the industry experienced $8.5 billion (1992) in losses.

R&D expenditure in and of itself is no guarantee of marketing success. To wit, over the past ten years, IBM has allocated some $55 billion to research.[23] Yet the company has been outmaneuvered and outinnovated by small entrepreneurial firms. Perhaps IBM has been reluctant to compete against its own product. More and more, the name of the game is to cannibalize your own products or be cannibalized by rivals.

The converse side, product obsolescence, is indicative of the frantic pace of the computer industry. Product design cycles are shrinking, and product life cycles are now measured in months. Product components cycles are even more contracted. But the competitive experience of the U.S. mainframe industry is by no means unique. Mainframe systems in Europe and Asia are being similarly assaulted by waves of product substitutes and product innovation.

The innovation track record of the U.S. computer industry—particularly microprocessor networking—is viewed with less than equanimity from abroad. The number of personal computers in Japan, for example, is about a quarter of that in the United States. Over 50 percent of U.S. personal computers are tied to data networks (local-area networks), whereas in Japan 14 percent are so linked. Indeed, the State of California has six times more computers tied to Internet (a domestic and international data network) than all the computers in Japan. As one Japanese official observed, "We are almost ten years behind the U.S. in networking."[24]

To sum up, a market structure that encourages and solicits market entry confers the gift of inspired economic performance on seller and buyer alike.

VI. PUBLIC POLICY

The past two decades have seen the computer industry make the transition from a firm-dominated to a market-dominated industry. We have already alluded to the forces of entry—technology, standardization, competition, user change—that have contributed to this transformation. But what about the role of government policy during this era? What issues remain on the public policy agenda, and how will they be resolved? Today, four such issues can be discerned: market dominance, intellectual property rights, U.S. import policy, and U.S. export policy.

MARKET DOMINANCE

Ironically, market concentration questions focus not on computer makers directly but, rather, on their factor input suppliers—microprocessor chips (Intel) and operating software (Microsoft).

Intel supplies the "brain" of the personal computer and has been credited with undermining the oligopoly structure of the industry in the 1980s. But today the question is whether Intel will continue to dominate the microprocessor industry as it has since the mid-1980s. This is unlikely. In a world of rapid technological change, today's victor may well find itself tomorrow's vanquished. Product innovation is an exogenous force. New microprocessors are introduced every eighteen months; parallel processing appears to be the next computer wave; networking and system integration are explosive; IBM, Apple, and Motorola, in concert, are researching the next generation of microprocessors.[25] The forces of technology, entry, innovation, and obsolescence show little sign of moderating. Under these circumstances, some policymakers perceive any market concentration as a transitory phenomenon.

Concentrated control over another critical input to the industry, computer software, is another matter, however. Microsoft supplies 90 percent of the operating software for personal computers (MS-DOS): The firm's share of the PC software market in the 1990s is equivalent to IBM'S mainframe dominance in the 1970s. As one buyer put it, "Doing business with Microsoft is like water-skiing behind the Queen Mary."[26]

One software rival, Novell, the market leader in the networking of personal computers (Netware and Unix), has alleged that Microsoft accords price discounts to PC makers on condition that Microsoft receive a license fee for every personal computer sold. Novell insists that Microsoft's policy restricts entry, concentrates the software market, forestalls industrywide standardization of computer operating systems, and has prevented its (Novell's) clone of the DOS operating system from gaining a foothold in the industry.

That these allegations are not trivial can be seen in the fact that the Federal Trade Commission, the Department of Justice, and the EC Commission of Europe all have investigated Microsoft's marketing conduct.[27] The Federal Trade Commission recently ended its investigation without taking action, but the Department of Justice is looking into the anticompetitive effects of Microsoft's pricing and marketing policies. Whatever the outcome, some industry observers insist that Microsoft's predominant market position is subject to potential erosion by the merging of products—PCs, telephone, fax, and teleconferencing; the digital integration of voice, data, text, and video; and object-oriented technology, computers incorporating parallel-processing technology and software, and interactive and multimedia computers. In a world of shifting technology, some industry observers insist that market concentration is but a temporary aberration.

PROPERTY RIGHTS

A second issue, not unrelated to software, turns on the question of intellectual property rights. What exactly constitutes an original idea, and who should have a legitimate claim to its creation? Novell, a software firm, is currently facing a legal challenge from an inventor who asserts that a 1986 patent covering the concept of networking entitles him to infringement damages. Another inventor successfully obtained a basic patent covering the microprocessor and now wants to collect royalties from the semiconductor industry. The industry views these developments with some apprehension, fearing that patent awards may be so broad or generic that the industry may be held for ransom.

Any government policy that attempts to strike a balance between the protection of intellectual property rights and the industry's interest in open standards is bound to be controversial. Litigation between Borland International and Lotus Development Corporation typified this policy tension. Lotus's spreadsheet 1-2-3 has for years dominated the market in DOS-based computing. As a new entrant, Borland developed Quattro Pro, a competitive product that permits macros (complex computer instructions to be executed with a single keystroke) in a language compatible with the Lotus product. Although Borland did not directly copy Lotus's source code (the equivalent of plagiarizing), the question before the court is whether Lotus can exclude products compatible with its own technology. Recently, the court ruled that Borland not only had to pay damages to Lotus but also must remove its products from retailers' shelves.[28]

This judicial ruling has sent ripple effects throughout the software industry. Virtually every company in the industry acknowledges that directly copying another company's source code violates a firm's intellectual property rights. But by not allowing competitors to create compatible products, a patent can have the unintended effect of eliminating the benefits of open industry standards.

Some industry members argue that if a firm abuses its position as the holder of the industry standard, then rivals will seek to replace that standard with something different. Others maintain that if companies are not able to establish a strong competitive position by employing existing standards, their ability to launch an offensive against the current standard can be nullified. User switching costs are expensive, and thus industry standards, once adopted, tend to take on the status of permanence. As noted, users of computer technology often make a substantial investment in hardware, software applications, training, and support services that precludes a painless move from one standard to another. Without the resources to provide a compelling reason for customers to make the switch, these impediments can remain a substantial barrier to market entry.

If the courts adopt a narrow interpretation of intellectual property rights, the industry may find itself undergoing a period of consolidation and the establishment of fewer standards. On the other hand, by lowering barriers to entry, open standards enable small companies to compete with and challenge incumbent rivals. How the courts will balance open standards against U.S. private property rights will be watched closely by industry participants in the years ahead.

IMPORT TARIFFS

A third set of government policies focuses on import tariffs. Should the U.S. government pursue programs designed to protect domestic submarkets from foreign competition? The temptation to do so is often irresistible. Yet the result of such policies can disappoint their sponsors or advocates. Flat-screen technology illustrates the unintended results of import tariffs. Employed in computer notebooks, laptops, PCs, auto instrument panels, and the like, flat-screen technology originated in the United States. Today, however, some 95 percent of this $3.2 billion market is supplied by Japanese firms.[29]

In 1991, the U.S. government imposed a 62.7 percent tariff in response to a complaint that Japanese producers were dumping active-matrix displays on the U.S. market. The Commerce Department argued that such a tariff would encourage companies to invest and build plants in the United States. But the tariff applied only to the display component rather than to displays embedded in assembled computers. To circumvent the tariff, U.S. computer companies moved the assembly of notebook computers offshore—a transfer that carried with it a concomitant loss of U.S. jobs.

Obviously, higher input costs handicap a domestic computer firm's ability to sell its products in overseas markets. Whatever its form, the law of unintended effects generally haunts policies of protectionism.

EXPORT LICENSES

A fourth facet of government oversight centers on computer regulations destined for overseas customers. Computer power, measured as MTOPS (Millions of Theoretical Operations Per Second), is subject to national security oversight. Exports by U.S. firms are restricted to a speed of 12.5 MTOPs, and so firms that exceed that processing power must secure an export license from the Department of State. License application is not a free good, however. Today a 13-MTOP computer falls within the Department of State's classification of "munitions."

The United States, of course, is not the sole supplier of powerful processing computers. India, Taiwan, China, and South Korea manufacture machines capable of up to 67 MTOPS.[30] Predictably, U.S. firms contend that the export-licensing requirement handicaps their ability to compete in overseas markets, since their rivals are not subject to comparable export restrictions.

Recently, the State Department relaxed some of its strictures, but the standoff between computer power and national security remains an ongoing issue. As computers experience 30 percent annual growth in productivity, it does not take long for technology to encounter the ceiling of government regulation. Indeed, clusters of personal computer/workstations hooked together by data networks now approximate the processing power of supercomputers. Thus, on-line terminals located in the United States are candidates for export regulation if they can be connected to international telephone lines.

National security regulation also monitors the export of software programs by U.S. firms. In the past, the National Security Agency (NSA) has regulated encryptographic (scrambling) codes as part of its surveillance of international data networks. The NSA naturally wants to preserve its ability to tap into criminal or terrorist network activities.

On the other hand, privacy is of critical concern to U.S. financial institutions operating in global markets. U.S. banks, for example, send 350,000 overseas messages daily over encrypted wire transfer lines, and not surprisingly, the industry has been in the forefront of developing codes to ensure the privacy of financial transactions that total $350 trillion annually.[31]

Government policy insists that computer encryptographic technology be cleared by the NSA, but U.S. corporations counter that government regulation of private encryption codes stifles and dampens scrambling technology in the United States.[32] Some computer firms claim that government regulation handicaps the ability of U.S. firms to develop new technology while overseas rivals, unfettered by comparable oversight, gain a distinct competitive advantage.[33] Still others observe that networks of microprocessors that approximate the power and speed of supercomputers have made government surveillance "a near impossible task."[34]

In sum, technology, innovation, entry, and substitutes impose their own imperative on both the public and private sectors of the economy. Whether public policy can resolve the issues of an industry in the throes of turbulent technological innovation remains an open question. As the government attempts to impose regulation on the industry, the tension between policy oversight and innovation change will likely remain a running dialogue for the foreseeable future.

VII. CONCLUSION

The computer industry serves as a metaphor for fundamental crosscurrents besetting an economy in rapid transition. Barriers to entry continue to fall; market access is an unrelenting process; design schedules are calculated in months; and product life cycles are measured in quarters. To survive, firms are now coerced into institutionalizing both product and process innovation.

Products, services, and industry sectors now intersect and overlap. A computer no longer stands alone. Today's products incorporate a telephone, a fax, a video conference terminal, and a word processor. The current interest in "multimedia" and "interactivity" suggests that the computer industry can no longer be easily distinguished from such industries as cable, broadcasting, entertainment, software, and telephones. The rash of interindustry mergers—for example, cable and telephone—underscores the importance of cross-boundary technology.

As industry demarcations fall, corporations find themselves competing not only with domestic and international rivals in traditional markets but also with competitors occupying adjacent or even distant industries. The computer industry is both a cause and an effect of unparalleled changes in the marketplace.

How will corporations adjust to the imperatives of an information-intensive economy? Clearly, a rapidly changing environment is both promising and treacherous. Some firms will flourish and grow; others, beset by losses, will downsize, restructure, or disappear. In a sense, IBM's experience illustrates how a model of U.S. organizational performance can be blindsided by technology, competition, and shifting user demand. (It is noteworthy that the stock market assigns a higher value to Intel and Microsoft—two firms nonexistent thirty years ago—than it does to IBM.)[35] Yet who is to say whether tomorrow IBM may make a comeback. Parallel processing, interactive multimedia terminals, and wireless data networks offer inviting market opportunities.

In a real sense, however, computer technology has diluted the monopoly of information once held and exercised by corporate CEOs. Networks, for example, now give "the rank-and-file new access: the ability to join in on-line discussions with senior executives. In these interactions, people are judged more by what they say than by their rank on the corporate ladder."[36] Such developments are creating the "horizontal corporation." Hierarchy is dying, says *Business Week*: "In the new corporate model, you manage across—not up and down."[37] Certainly something profound is occurring within the structure of today's business organization. Firms are reducing overhead, contracting size, abandoning vertical integration, and pushing decision making to lower and lower levels of the corporate ladder, forming clusters of development teams and engaging in industry alliances. As one observer noted, "Today with the average industry cycle at 6 to 8 months, the boundaries between development and manufacturing and distribution must disappear."[38]

All of this is an attempt by today's firm to adjust, if not master, enormous changes in an environment of entry and rivalry. Indeed, the firm of tomorrow may well be "a new institution that holds a portfolio of smaller faster moving enterprises."[39]

If environmental turmoil is besetting the private sector, what about decision making in the public sector? Can domestic policies—rules, regulations, oversight, quotas, subsidies—possibly be relevant to a world whose capital markets are monitored vigilantly by 200,000 global computer terminals?[40] If corporations find that they no longer can exercise total control over technology, products, or customers, can government policy—plagued by deliberation, political trade-offs, and interminable due process—possibly be germane to a knowledge-based world?[41] Has the information explosion dissipated the monopoly power of the state? Do the networks of microprocessors serve as a subversive force that undermines state sovereignty and control, with respect to monetary affairs, taxes, regulation, technology, or even employment policy?[42]

Presumably, institutions in both the private and public sector are searching for some answers to a world of information transparency. In the meantime, the environment of change shows no signs of attenuating. If this trend continues, it is not clear whether many entities, public or private, will be able to withstand the assault of "creative destruction." What is clear is that the future will place a premium on those organizations endowed with a spirit of adventure, entrepreneurship, and risk taking. More important, the future may well be seized by those institutions humble enough to master the art of learning how to learn.

NOTES

1. Charles H. Ferguson and Charles R. Morris, *Computer Wars* (New York: Random House, 1993), p. 4.
2. Ibid., p. 5.
3. Ibid., p. 4.
4. Richard T. Delamarter, *Big Blue* (New York: Dodd Mead, 1986), p. 174.
5. Gerald W. Brock, "The Computer Industry," in *The Structure of American Industry*, ed. Walter Adams, 8th ed. (New York: Macmillan, 1986), p. 257.
6. Ferguson and Morris, *Computer Wars*, p. 11.
7. James B. Stewart, "Whales and Sharks," *New Yorker*, February 15, 1993, p. 38.
8. Ibid., p. 13.
9. Daniel Ichbiah and Susan Knepper, *The Making of Microsoft* (Rocklin, CA: Prima, 1991), p. 93.
10. Laurie Hays, "IBM Tries to Keep Mainframes Afloat Against Tide of Cheap, Agile Machines," *Wall Street Journal*, April 12, 1993, p. B7.
11. Ferguson and Morris, *Computer Wars*, p. 8.
12. "Rethinking the Computer," *Business Week*, November 26, 1990, p. 117. Also see John R. Wilke, "Shop Talk: Computer Links Erode Hierarchial Nature of Workplace Culture," *Wall Street Journal*, December 9, 1993, p. 1.
13. Thomas Burdick and Charlene Mitchell, "Downsizing to Continue for Several Years," *Washington Times*, March 11, 1993, p. C2.
14. Alan Cane, "Information and Communications Technology," *Financial Times*, March 23, 1993, sec. 3, p. 3.
15. Peter Nulty, "When to Murder Your Mainframe?" *Fortune*, November 1, 1993, pp. 110–114.

16. "The Virtual Corporation," *Business Week*, February 8, 1993, p. 100.
17. David Manasian, "The Computer Industry," *The Economist*, February 27, 1993, p. 5.
18. Cane, "Information and Communication Technology," p. 3.
19. Michael Scott Morton, ed., "Introduction," *Corporation of the 1990s: Information Technology and Organizational Transformation* (New York: Oxford University Press, 1991), p. 9.
20. "Deconstructing the Computer Industry," *Business Week*, November 23, 1992, p. 90. Also see "Duel," *The Economist*, January 30, 1993, p. 57. Finally, see Jim Carlton, "Popularity of Some Computers Means Buyers Must Wait," *Wall Street Journal*, October 21, 1993, p. B12.
21. "Global Patent Race Picks up Speed," *Business Week*, August 9, 1993, p. 60.
22. "In the Labs: The Fight to Spend Less, Get More," *Business Week*, June 28, 1993, p. 120.
23. "What Went Wrong at IBM," *The Economist*, January 16, 1993, p. 24. Also see "To Save Big Blue," *The Economist*, January 16, 1993, p. 18.
24. Andrew Pollack, "Now It's Japan's Turn to Play Catch-Up," *New York Times*, November 21, 1993, p. F6.
25. "Deconstructing the Computer Industry," p. 90. See also "A Spanner in the Work," *The Economist*, October 23, 1993, p. 75. (A PC's life can be as short as six months.)
26. Louise Kehoe, "The Hottest Act in Town," *Financial Times*, March 8, 1993, p. 13.
27. "Novel vs. Microsoft: What's Behind the Hate?" *Business Week*, September 27, 1993, p. 128.
28. Shawn Willett and Doug Barney, "Borland to Rush Quattro Pro 5 to Market: Omits 1-2-3 Macroreader," *Infoworld*, August 23, 1993, p. 1.
29. "Did Commerce Pull the Plug on Flat Screen Makers?" *Business Week*, July 5, 1993, p. 32.
30. John Markoff, "Shift Expected on Computer Export," *New York Times*, August 27, 1993, p. D2. Also see John Burgess, "Encryption Plan Prompts Question," *Washington Post*, April 7, 1993, p. 40: "The White House plan grew from federal concern that private scrambling technology is advancing so rapidly that government could find itself unable to tap the communications of criminals or terrorists."
31. John Markoff, "A Battle over Secret Codes," *New York Times*, May 7, 1992, p. D2.
32. Robert Palmer, "Freeing the U.S. Computer Industry," *Boston Globe*, October 16, 1993, p. 13. The president of DEC stated,

Computers based on Alpha AXp platforms run so fast that today's export regulations would classify even our desktop PCs and low end stations as super computers and subject them to stringent restriction even for sale to friendly countries like Switzerland and Ireland. We are being penalized for too much innovation.

33. John Markoff, "Group to Set Rules for Computer Encoding," *New York Times*, July 13, 1993, p. 17; also see Ellen Messmer, "Encryption Restriction Policy Hurts Users, Vendors," *Network World*, August 23, 1993, p. 34.
34. "Duking It out for the Decoder Ring," *Business Week*, November 22, 1993, p. 12; also see Markoff, "Shift Expected on Computer Export," p. D2.
35. Carol J. Loomis, "Dinosaurs?" *Fortune*, May 3, 1993, p. 42; also see Louise Kehoe, "Who's Got the Right Stuff?" *Financial Times*, January 29, 1993, p. 13.
36. Wilke, "Shop Talk," p. 1.
37. *Business Week*, December 20, 1993, p. 76. See also Margie Wylie, "Will Networks Kill the Corporation," *Network World*, January 11, 1993, p. 51; and Wilke, "Shop Talk," p. 1.
38. Louise Kehoe, "Big Blue Decides Small Is Beautiful," *Financial Times*, October 23, 1992, p. 14.
39. Ani Hadjiian, "Welcome to the Revolution," *Fortune*, December 13, 1993, p. 70.
40. John Markoff, "Rethinking the National Chip Policy," *New York Times*, July 14, 1992, p. D6.
41. Walter B. Wriston, *The Twilight of Sovereignty* (New York: Scribner, 1992).
42. Walter B. Wriston, "The Decline of the Central Banks," *New York Times*, September 20, 1992, p. F11.

SUGGESTED READINGS

Carroll, Paul. *Big Blues—The Unmaking of IBM*. Chicago: Crown, 1993.
Ferguson, Charles H., and Charles R. Morris. *Computer Wars*. New York: Random House, 1993.
Ichbiah, Daniel, and Susan Knepper. *The Making of Microsoft*. Rocklin, CA: Prima, 1991.
Morton, Michael Scott. *Information Technology and Organizational Transformation*. New York: Oxford University Press, 1991.
Stewart, James. "Whales and Sharks." *New Yorker*, February 19, 1993.

7

PHARMACEUTICALS

William S. Comanor and Stuart O. Schweitzer

I. INTRODUCTION

The pharmaceutical industry is an American success story: Its products dominate the world market. It leads in the development of new technology and for decades has achieved a rapid pace of innovation. It pays high wages to its employees and offers advanced products to its customers.

Despite these facts, however, the pharmaceutical industry has been subject to continual criticism from American political leaders. Politicians from both parties have asserted that its products are overpriced and its profits are excessive. As recently as 1993, Senators David Pryor (Arkansas) and William Cohen (Maine), Democratic chairman and ranking Republican member of the Senate Committee on Aging, respectively, wrote: "This pattern of excessive inflation by drug manufacturers has made it extremely difficult for millions of Americans to afford life-saving medications,...[and it is time therefore] for pharmaceutical cost containment."[1]

The obvious question is, Why should a well-performing industry be subject to such attack? Are these senators merely responding to political pressures and failing to acknowledge other industry dimensions in which its performance is exemplary? Are pharmaceutical prices truly high and increasing, and if they are, is this the expected cost of a rapid pace of pharmaceutical innovation? Or does this industry function as a traditional monopolist so that its claims of good performance are overstated? In that case, should we view the pharmaceutical industry's performance as relatively poor rather than relatively good?

A striking feature of this debate is its constancy. It originated in lengthy hearings before the Senate Subcommittee on Antitrust and Monopoly during the 1950s and early 1960s. Although the debate has become more informed and more knowledgeable, it has stuck on how to characterize the performance of this industry and on the subsequent issue of what policy actions should be taken, if any, to improve its performance.

That the debate has continued for so long has several implications. One is that the industry's performance has not changed very much. In 1962 and again in 1984, new laws were enacted to reform the regulatory process, and yet controversy surrounding the pharmaceutical industry has not diminished. Whatever problems exist were apparently not solved by past legislative efforts.

In this chapter, we explore and evaluate the economic issues raised in the continuing policy debate. To do so, we first look at the structure of the pharmaceutical industry, which deals with how firms are organized and what functions are carried out and also how the various firms together comprise the larger industry. Next, we look at typical patterns of firm conduct among both leading firms and smaller rivals. In particular, we examine how pharmaceutical prices are set. Finally, we consider how these factors interact to determine the industry's performance.

II. STRUCTURE

FUNCTIONAL CHARACTERISTICS

All firms in the United States economy are vertically integrated to some degree, meaning that they carry out activities that could be done elsewhere. That is, specific inputs could be purchased rather than produced, and the firm's output could be limited to a single product. Although it is difficult to find a firm that is so fully nonintegrated that it purchases all possible inputs and produces only a single output, still the extent of integration is a distinguishing characteristic of firms and industries.

In the pharmaceutical industry, the leading companies are engaged in three distinct activities, which in principle could be carried out separately and which together characterize this industry. These activities are manufacturing and production, research and development, and selling and promotion. Together they determine the products that are sold and the prices and quantities that are set. They also fix the total costs of member firms.

Although production costs comprise a substantial proportion of total costs, they still account for less than half. As indicated in Table 7-1, they represent less than 30 percent of sales revenues for a sample of large, U.S.-based companies and less than 40 percent of total costs. The next largest category is advertising and promotion, although the actual amounts are not reported by most companies. And third are research and development outlays, which in 1991 were approximately 12 percent of revenues. This percentage allocation is nearly double that reported for earlier years.

1. THE MANUFACTURING FUNCTION. In their manufacturing functions, pharmaceutical firms engage in two fairly distinct activities. The first is similar to that carried out in other segments of the broader chemical industry: the generation and production of basic chemicals that serve as the active and inert ingredients in their final products. In most cases, these substances are generated from basic materials by means of chemical reactions, although some products, such as antibiotics, are produced through biological processes.

TABLE 7-1 Revenue Allocation (%) for Leading Pharmaceutical Companies

	1958	1966	1991
Production costs	32.1	35.0	27.7
Research and development	6.3	6.5	12.1
General and administrative	10.9	35.0[a]	35.0[a]
Advertising and promotion	24.8	– –	– –
Income taxes	12.8	10.0	7.1
Net profits	13.0	13.5	16.0
Number of companies sampled	22	17	10

[a] These figures include advertising and promotion as well as general administrative expense.

Source: 1958: Senate Subcommittee on Antitrust and Monopoly, *Report on Administered Prices: Drugs* (Washington, DC: U.S. Government Printing Office, 1961), p. 31; 1966: Task Force on Prescription Drugs, "The Drug Manufacturers and the Drug Distributors," *Background Papers* (Washington, DC: U.S. Government Printing Office 1969), p.14; 1991: Survey of annual reports for 1991.

In the second stage of the manufacturing process, the resulting chemicals are combined with other chemicals and with inert materials to make the final product. For the most part, this involves purifying and mixing the materials and then encapsulating the resulting substance. In many instances, the basic materials are purchased from others so that manufacturing activities are limited to the second-stage processes.

2. RESEARCH AND DEVELOPMENT. In sharp contrast with its manufacturing functions, the research and development functions of the pharmaceutical industry provide its special characteristic. Before World War II, this function was almost entirely absent. The leading firms of that time produced a limited number of well known products that did not change much from year to year. These firms typically sold active drug ingredients through wholesalers to retail pharmacies. In many cases, pharmacists made their own pills, filled capsules, and prepared liquid suspensions and tinctures. Because the pharmacist's skill was critical to making the products, the brand name of the manufacturer of the ingredients was less important.

Then, at the end of the war when industry leaders realized that their future depended on research and development, the pharmaceutical industry underwent a veritable revolution. Indeed, the introduction of the first antibiotics demonstrated that more effective drugs could be discovered that would generate high consumer demand. The pharmaceutical companies saw that new and improved products could be highly profitable, and so they began to invest large and increasing sums on creating them.

What occurred in these years was a major technological breakthrough that had important ramifications for the future of the industry. The advance was not in the development of a single product or in the manner by which pharmaceuticals are produced. Rather, it was an advance in the process by which new pharmaceuticals were discovered. As Peter Temin noted, the technological revolution that created the modern pharmaceutical industry "was a method of research rather than a method of production."[2] The firms that grew and prospered were those that rapidly adopted the new technology.

In the years that followed, expenditures on research and development continued to increase. By 1991, the member firms of the Pharmaceutical Manufacturers Association (PMA) spent $8.9 billion for research and development on drugs designed for humans, of which 18 percent was spent abroad and the rest in the United States.[3]

This research effort led to a vast array of new pharmaceuticals. Between 1946 and 1991, as reported in Table 7-2, the industry introduced over 1,200 new entities, or 27 per year on average. This figure, however, disguises some important differences. In the seventeen years between 1946 and 1962, before new regulatory requirements were imposed, 684 entities were introduced, which is more than half the total number introduced during the entire forty-five-year period. In the seventeen years after 1962, however, only 275 entities were introduced, which represents a decline of about 60 percent. A new regulatory regime, described later, had changed the ease with which new drugs could be developed and introduced. Still, the introduction of new products remained a major feature of this industry, and the rate of drug development has even accelerated in more recent years.

Although research spending rose throughout the postwar era, the number of new products dropped from its early heights, and accordingly, the research costs per new product exploded. In the years before 1962 these costs are estimated to have been $6.5 million per product[4] but then jumped to between $65 million and $75 million per successful new product by the end of the decade. When capitalized to the date of the Food and Drug Administration's (FDA) approval, these costs ranged from $108 million to $124 million per new chemical entity.[5] And they have continued to rise. For new drugs first entering human testing in the 1970s, capitalized after-tax R&D costs were reported at between $140 million and $194 million per product.[6] The development and introduction of new pharmaceuticals had become a very costly enterprise.

The industry's response, however, was not to withdraw from the research enterprise but to expand its efforts. The rapid pace of new product introduction has continued. New drugs are introduced to replace older ones, a process that has given the pharmaceutical industry very different characteristics than it would have if products changed very slowly or even remained the same from year to year. This pattern has determined the structure, behavior, and performance of the industry.

TABLE 7-2 New Single-Entity Drug Introductions to U.S. Markets, 1946–1991

Year	Number of Entities	Year	Number of Entities
1946	19	1970	16
1947	26	1971	14
1948	29	1972	10
1949	38	1973	17
1950	32	1974	18
1951	38	1975	15
1952	40	1976	14
1953	53	1977	16
1954	42	1978	23
1955	36	1979	15
1956	48	1980	13
1957	52	1981	19
1958	47	1982	26
1959	65	1983	22
1960	50	1984	15
1961	45	1985	20
1962	24	1986	24
1963	16	1987	20
1964	17	1988	19
1965	25	1989	24
1966	13	1990	24
1967	25	1991	30[a]
1968	12		
1969	9	Total	1,215

[a] This figure is the number of new entities approved by the FDA in 1991 rather than the number of actual introductions.

Source: Pharmaceutical Manufacturers Association analysis of Paul De Haen data.

3. ADVERTISING AND PROMOTION. The marketing function is the third essential task performed by leading firms in the pharmaceutical industry. Since new products are introduced at a rapid pace and older ones are quickly discarded, physicians need information about them. Indeed, for physicians even a few years beyond training, many, if not most, of the leading products were introduced after they started to practice, so the pharmaceutical companies must have some means of telling them about the new drugs.

The leading pharmaceutical companies have long accepted this role and accordingly have spent large sums on it. In 1958, the principal companies spent approximately one-quarter of their total revenues on advertising and promotion, and most likely spend similar proportions today. Of the 185,900 people employed by PMA-member firms in 1989, fully 55,500, or nearly 30 percent, were in marketing.[7]

Since pharmaceuticals are prescribed only by health professionals, the industry's products do not have a broad distribution but, rather, are directed specifically at medical personnel. These efforts have traditionally taken two forms: First, the companies recruit a class of specialized sales personnel, called detailmen and -women, who call on physicians individually to present the products offered by their firms; and second, the companies spend substantial sums on advertising in professional journals and presentations at medical conferences.

To a large extent, these advertising and promotional efforts are directed toward new products. These expenditures, for both firms and therapeutic categories, typically are greatest in a product's early years but fall steadily after that.[8] Since physicians and other health professionals decide whether to use a new product in their practice soon after it has been introduced, the pharmaceutical companies have a strong incentive to promote their products heavily at that time. Subsequently, when physicians' prescribing patterns have become established, their marketing expenses often decline.

Although sales personnel usually emphasize new products, some older products also carry large detailing expenditures. For example, in 1971, Abbott Laboratories spent over $2 million detailing Erythromycin, which had been introduced in 1952.[9] Such efforts suggest that pharmaceutical advertising and promotion are designed to gain physicians' loyalty as well as to provide information.[10]

Although most firms continue to push products still protected by patent, their efforts fall sharply when that protection is removed. Branded advertising and promotion decrease by roughly 10 percent per year for the two years before the patent expires but drop even more when the patent actually does expire: 20 percent with the entry of the first generic rival and another 40 percent when the number of generic rivals reaches five. Apparently these efforts are designed to expand the market for the product, and the original supplier no longer finds it profitable to do so when it must share the product with others.[11]

Whatever the impact of these expenditures, the accuracy of the information provided along with the advertising and promotion remains in dispute. Critics argue that a substantial share of the industry's advertising and promotion is misleading. One study found that most of the pharmaceutical advertising in medical journals did not meet the FDA's criteria for scientific quality and, moreover, did not have much educational value.[12] In effect, the authors found that journal advertising did not give physicians full information about the characteristics and uses of new drugs.

Leffler, however, disagreed with this view, arguing that the study overstated the purpose and function of advertising. He has "serious doubts whether the purpose of medical journal advertisements is to inform the reader about the details of the product rather than simply to announce the availability and/or remind the physician of the product's name." Also, "one would certainly hope that physicians are reasonably knowledgeable about pharmaceuticals, that they are reasonably skeptical about the self-serving claims of advertisers and that journal ads are not the main source of their information about the specific details of the drugs."[13]

ORGANIZATIONAL CHARACTERISTICS

Just as the pharmaceutical industry has a variety of functions, it also consists of a variety of firms that together make up the industry. Although before World War II, there were several traditional drug manufacturers, the revolutionary changes that took place after the war changed this structure. It still includes many of these old-line firms, but a steady process of consolidation has joined many of them together. In addition, many of the major European companies have entered the U.S. market and established subsidiaries that have become important parts of the American industry. The latter include the American subsidiaries of Ciba–Geigy and Hoffman–LaRoche, which are major Swiss companies; Glaxo, a British company; and Beecham Laboratories of London, which entered the U.S. market through a merger with SmithKline to form SmithKline Beecham. The leading firms are reported in Table 7-3.

An important factor leading to the consolidation process was the greatly increased cost of introducing a new product. A firm had to be big to support an extensive research establishment that could undertake many projects at the same time, so that the research successes and failures could balance out to reduce the firm's overall risk. For small firms and small laboratories, therefore, research can be exceedingly risky, but that risk is diminished substantially through diversification across a number of projects.

TABLE 7-3 Largest Pharmaceutical Firms (1992 Sales in $ Billions)

	Wholesale Pharmaceutical Sales	
Firm	*U.S.*	*Worldwide*
Merck	$ 3.7	$ 8.2
Bristol-Myers Squibb	2.9	6.1
Glaxo	2.7	7.5
American Home Products	2.7	n.a.
Eli Lilly	2.6	4.4
Johnson & Johnson	2.2	4.2
Pfizer	2.2	4.5
SmithKline Beecham	2.1	5.2
Marion Merrell Dow	1.9	n.a.
Upjohn	1.7	n.a.

n.a. = not available.

Source: Booz-Allen & Hamilton; Bernstein research.

Not only have new pharmaceutical firms from abroad entered the industry, but also a new technology has given birth to new firms. During the latter part of the 1980s and into the 1990s, a number of biotechnology firms have entered the industry. As before, the new technology refers not to the production of individual products but instead to the process by which new products are discovered. In effect, this type of technological change is analogous to what occurred in the 1950s, in that it deals with the research rather than the manufacturing.

This new technology applies genetic engineering to pharmaceuticals, in which a gene from one organism is inserted into another and the result is made into a usable product. Nonetheless, the pharmaceuticals discovered and produced in this manner still must compete with those developed and produced using more conventional methods; the products are sold in the same markets.

Finally, during the latter part of the 1980s and early 1990s, many small generic pharmaceutical manufacturers entered the industry. Ever since regulatory barriers were removed in 1984, generic producers have poured into the industry. Unlike the leading manufacturers, these firms do little or no research and have only minimal marketing expenditures. Rather, they rely on the research and marketing efforts of others, with their own activities generally limited to manufacturing. But for many of these firms, even this is abbreviated. Instead, their activities are limited to the second stage, encapsulating the basic chemicals or otherwise producing the final products. Such firms often buy the basic chemicals from large pharmaceutical companies or other fine-chemical manufacturers. Usually, however, they produce products that meet the standards set by the FDA and do so at a cost that is not much higher than those of the major companies.

The presence of generic drugs in the marketplace has expanded quickly in recent years. In 1980, before the new legislation, they accounted for only a small share of pharmaceutical sales in the United States, but by 1992 that share had risen to approximately 37 percent.[14]

Throughout the 1980s, this process of consolidation continued, as did the entry of foreign and U.S. firms. For the most part, these trends have offset each other, but at the same time, the share of generics has grown. The market concentration of prescription drug sales is shown in Table 7-4. Although the aggregate share of the largest companies fell somewhat, the share of the fifty largest companies, including many generic manufacturers, did not. (To be sure, these aggregate concentration ratios are not indicative, one way or another, of competitive conditions in this industry.)

THE GLOBAL INDUSTRY

Although the United States has the largest national market for pharmaceuticals and accounts for 27 percent of the total sales in the developed world,[15] it is nonetheless an international industry. Precisely because research has become so expensive, it is essential for firms to recoup their investments

TABLE 7-4   Aggregate Share of U.S. Prescriptions, 1982–1990 (%)

Company Group	1982	1986	1990
Leading company	8.4	6.7	7.6
Four largest companies	29.5	25.0	24.5
Eight Largest companies	48.9	43.4	41.6
Twenty largest companies	79.4	74.9	74.7
Fifty largest companies	96.7	97.5	95.1

Note: These data report new and refilled prescription dollars as dispensed at the pharmacy-cost level.

Source: National Prescription Audit, IMS America, Ltd., as reported in Pharmaceutical Manufacturers Association, *Statistical Fact Book*, 1991, Table 1-12.

over as large a territory as possible. Because most countries have their own drug regulations, the largest companies have formed subsidiaries throughout the world, and those without such subsidiaries often sign cross-licensing agreements with those that do. Indeed, the major companies have become so visible in each important market that it is often difficult to tell where a company's "home" is.

Because the marginal production costs of most drugs are fairly low, it is profitable for a firm to sell its products even where national regulations or practices force relatively low prices. As long as prices exceed marginal costs, some contribution is made to the research and development overhead, and as a result, there can be large differences among the prices charged for the same drugs in different countries.

Using drug sales to pharmacies in nine industrial countries and applying U.S. quantity weights, U.S. prices per standard dose are higher on average than in four countries but lower on average than in four others. On the other hand, U.S. prices per gram of active ingredient are higher on average than in six countries but lower than in two.[16] Such comparisons depend greatly, however, on the sample, price measure, and weighting scheme employed, and so there are no simple conclusions to be drawn.

Not only do prices differ among countries, but there also are substantial differences in the availability of products. Because the government agencies responsible for approving pharmaceuticals function differently in different countries, not all drugs are approved at the same time. Although some countries are more likely to approve drugs developed at home, this is not true for the United States, which instead is more likely to approve drugs after they have been introduced elsewhere.[17]

Another question is whether the United States lags behind other countries in approving new drugs. It is slow in approving some new drug approvals, but then again, all countries are slow in approving some new products.[18] No country is consistently ahead in approving all new drugs.

III. MARKET CONDUCT

This section deals with the actions of pharmaceutical firms in regard to both their price-setting behavior and their investment decisions on research and development.

PRICE BEHAVIOR

Although the research costs required to introduce a new drug are substantial, they are usually incurred before the product is sold to even a single consumer. As a result, research costs are fixed costs in that they are required to sell the product but do not vary with the amount sold. Similarly, most marketing costs are incurred in the early years of a product's life cycle and are designed to introduce it to the medical community. Like research costs, they do not vary with output and also are fixed costs. For the most part, therefore, the only variable costs in this industry are at the manufacturing stage. For large research-intensive companies, variable costs represent only about 30 percent of the product's value. The marginal costs for these products are quite low and reveal little about the prices charged for pharmaceuticals.

Research, marketing, and manufacturing costs all reflect conditions on the supply side of the market. None of them has a major impact on pharmaceutical pricing behavior. In contrast, prices depend on demand-side considerations. That is, the prices charged for pharmaceuticals are determined largely by how valuable or therapeutically useful they are and what consumers are willing to pay for them. Also important are the prices charged for rival products: The seller of a new or old pharmaceutical cannot sell its product for more than the price charged by its rivals unless it is therapeutically superior. If it is not, the seller usually prices its product no higher than those of its rivals and indeed sometimes must offer a discount.

When a new product is introduced, whether it offers a small or great therapeutic advance, usually there are existing products that are used for the same or similar indications. These alternative products are what physicians would prescribe in the absence of the new product and thus are the rivals with which the new one must compete. Note that this concept of relevant market, resting on specific therapeutic indications, is far narrower than the conventional standard of a therapeutic category, whereas classifications such as antibiotics or hypertensives are so broad that they include pharmaceuticals with very different indications.

Why should the prices charged for rival products be relevant, since the physicians only prescribe the pharmaceuticals, whereas their patients must actually pay for them? The answer is that many physicians do pay attention to the prices charged for the drugs they prescribe, particularly physicians in rapidly growing managed care settings. Especially when insurance companies or health maintenance organizations (HMOs) pay for pharmaceuticals, the relative prices of rival products influence prescribing decisions. Price behavior in this industry cannot be explained, therefore, by assuming that prescribing physicians pay little attention to actual prices. In such a case, demand would not be important. But this is not the case.

The demand-side factor most important in determining the price charged for a new pharmaceutical is its therapeutic advance compared with older products already on the market. To explore the importance of this factor, one study examined the amount by which new products are priced above their existing substitutes.[19] The results are given in Table 7-5 for new products, divided between those used for acute and for chronic ailments. As indicated, the average price premiums for more therapeutically advanced products are substantially greater than those offering only modest gains, and largely imitative products are generally priced at or below the levels set for existing products. A second factor is the number of existing substitutes. The more substitutes that are available, the lower the introductory price will be.

Another dimension of pharmaceutical price behavior is the rate of price advance following a product's introduction. For the most part, largely imitative products follow a penetration strategy, in which the introductory price is fairly low but then increases steadily over time. On the other hand, more innovative products are introduced at a higher price but then have few price increases and sometimes even small price declines over time. Moderately innovative products follow an intermediate path. The price levels for highly innovative products are fairly stable after they have been introduced, although this is not true for more imitative products.

A third milestone in the picture of pharmaceutical prices occurs when patents expire, enabling generic producers to enter the market, generally offering the products at much lower prices. Even if their manufacturing costs are higher than those of the original producers, they still set their prices much below those of the original seller.

The prices set by generic producers are greatly affected by the number of entrants that sell the product. As their number increases, the price competition becomes more vigorous, and prices fall below the level found when there is only a single entrant. In the case of anti-infectives, the largest price effect came when the fourth and fifth firms entered. The average price per prescription

TABLE 7-5 Relative Prices at Time of Introduction

Primary Use	Important Therapeutic Gain[a]	Modest Therapeutic Gain[a]	Little or No Therapeutic Gain[a]
Acute	2.97/3.22	1.72/ 3.09	1.22/1.37
Chronic	2.29/3.12	1.19/1.58	0.94/1.07

[a] These categories are determined by the Food and Drug Administration at the time of introduction. These price relatives report both median and mean values in each class of products from a sample of 135 new products introduced between 1978 and 1987.

Source: S. John Lu and William S. Comanor, "Strategic Pricing of New Pharmaceuticals," unpublished manuscript, January 26, 1994.

dropped from nearly $30 with two or three sellers to roughly $17 with the presence of a fourth rival, and then to $9.25 when a fifth firm entered.[20]

Average prices fall even though the prices charged for the original branded products are *increased* and not *reduced* when another firm enters. The original manufacturers do not usually compete with the new generic entrants on the basis of price but, rather, find it more profitable to concentrate on the segment of the market that includes brand-loyal customers. Such buyers are physicians and patients who prefer a particular brand and so continue to use it despite the presence of a lower-priced substitute. When generic manufacturers enter production, the price differential widens as the prices charged for the original branded products increase.

Another feature of pharmaceutical pricing is generally found in markets where demand rather than cost factors dominate. Since demand conditions can vary greatly among different sets of buyers, with each having its own price elasticity, one would expect firms to charge different prices to different buyers. Although pharmaceutical companies establish a list price for each drug, many sales are made by discounting that price, and these discounts can be substantial. A survey of prices in a particular area found that the average price charged for a selection of well-known products sold to hospitals was only 19 percent of that charged to a local pharmacy.[21] Since hospitals' demands for specific products are likely to be more elastic than those of individual pharmacies, which must stock a large number of products in order to fill individual prescriptions, the hospitals' prices should be much lower than those charged to pharmacies. When prices are driven by demand, the differences in demand elasticities are reflected in the differences in the actual prices.

These discounts may differ between individual and chain-store pharmacies and between hospitals and HMOs. A critical fact about this market is that there is no single price for an individual product even at a specific time; instead, prices depend on the demand conditions presented by particular buyers.[22]

Pharmaceutical prices are not determined by costs but by demand and competitive factors. Since marginal costs are only a small share of total costs, it is not surprising that prices exceed marginal costs and also that the extent of these differences varies greatly among buyers. The most important demand-side factor affecting price is the therapeutic advance embodied in the product, which is precisely what competitive processes should enforce. In addition, the number of substitute products lead to lower prices at the time of introduction and afterward. Both results indicate that competitive forces play important roles in determining pharmaceutical prices.

RESEARCH BEHAVIOR

As noted earlier, the leading pharmaceutical companies are continuing to increase their investment in research and development. Between 1971 and 1991, members of the Pharmaceutical Manufacturers Association raised their spending from $362 million per year to $8,908 million, with both amounts expressed in 1991 dollars.[23] Like other investments, these expenditures are made because the net present value of future returns is positive, even with the enormous costs associated with bringing a new product to market. However, unlike other investments, competitive factors can be important.

Some economists have suggested that pharmaceutical companies use research as a competitive weapon to gain markets ahead of their rivals and in effect to "race" to develop new products. However, despite the plausibility of this idea, it seems not to be the case. Instead, the research programs of major companies are determined mainly by three criteria: "the size of unmet medical need, the scientific potential of the field, and the idiosyncratic capabilities of their researchers."[24] Research efforts are not so easily shifted among therapeutic areas that research directors can readily redirect their programs to compete with others.

The leading pharmaceutical companies are not mirror images of one another but have developed special expertise in certain research areas. Despite the role of existing medical needs in determining research directions, such capabilities are very important. The leading pharmaceutical companies are heterogenous organizations with their own strengths and capacities, thereby making competitive research and development difficult.

IV. GOVERNMENT REGULATION

The technological revolution that created the modern pharmaceutical industry led also to a new regulatory structure, which eventually became full government supervision of the research process. In 1938, Congress established the Food and Drug Administration, whose mission was to ensure the safety of drugs sold in the United States. Before a new drug could be marketed, its seller was required to submit both animal and clinical studies that demonstrated its safety for human consumption. The FDA was given the power to reject this application, but had to do so within six months.

The FDA's mission was broadened in 1962 to include assurance of effectiveness as well. The FDA now was required to rule on therapeutic efficacy as well as safety before permitting a new product to be marketed, and the time constraint previously imposed was removed.

Although the new requirement to demonstrate efficacy seemed at first like a harmless and sensible regulatory provision, its consequences were enormous. The FDA now specified what tests were required and what records must be kept. To rule on the effectiveness of a new product, the regulators required a considerable amount of new information from both animal and clinical trials. Furthermore, these trials had to meet rigorous scientific standards, for otherwise how could it be shown that a new product was truly efficacious? Demonstrating efficacy is far more complex than demonstrating safety.

The process starts when a manufacturer files an Investigational New Drug (IND) application that is based on preclinical studies. The FDA's approval of these studies is then signaled when it grants a New Drug Approval (NDA) application. Clinical trials are then conducted in a series of phases. Phase I trials assess the drug's safety among a relatively small number of normal, healthy subjects. Phase II trials are the first opportunity to test the new drug on patients having the medical condition that the product is designed to treat. Because safety has not been firmly established, Phase II trials are limited in size. Phase III trials build on past experience and are larger and of longer duration, often lasting several years. Phase IV and V trials follow and are used to determine the drug's long-term effects and to compare the new drug with alternatives already available.

The review process begins when a manufacturer's files an IND application for a promising new entity, continues when it receives permission to start human testing through an NDA application, and concludes when it receives an NDA approval to market a drug for five to seven years. Compared with the 1970s, there has recently been a drop in approval rates and also in the average time required by the Food and Drug Administration to permit the introduction of new drugs.[25]

A direct result of this regulatory delay has been a substantial decline in effective life of a patent. The great majority of new pharmaceuticals are patented, which gives their developers a period of exclusivity. To ensure their position as inventor, companies apply for a patent early in the development process, even though the introduction of the product must await the FDA's approval. As a result, effective patent lives are much shorter than the statutory period of seventeen years.

As can be seen in Figure 7-1, the average life of a patent shrank until the early 1980s. Then, in part to correct for the regulatory delay, the Drug Price Competition and Patent Term Restoration Act of 1984 was enacted, which extended the term for up to five years but no more than fourteen years. It also granted three-year periods of exclusivity, regardless of patent status, for new or supplementary NDAs on which new clinical trials were required. In recent years, therefore, these periods of exclusive marketing have recovered somewhat.

The regulatory delays have attracted considerable attention and many complaints. In response, the FDA established in 1987 the Treatment Investigational New Drug program to deal with life-threatening diseases for which no alternative therapy exists. In such circumstances, the FDA permits physicians to use, on a case-by-case basis, drugs still being tested. Although manufacturers are permitted to charge for these drugs, they rarely have done so. As of 1993, twenty-three such drugs had been supplied under this program, but in only five cases were patients charged for them.[26]

The 1984 law granted a period of exclusivity for all new products, but it also facilitated generic competition when that period ended. Although previously the FDA had required that generic producers duplicate the studies and

FIGURE 7-1 Effective patent life for drugs approved, 1968–1989.

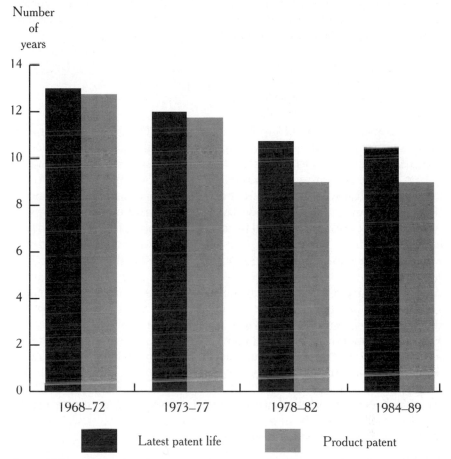

Number
of
years

Latest patent life   Product patent

Source: Office of Technology Assessment, *Pharmaceutical R&D Costs, Risks, and Rewards* (Washington, DC: U.S. Government Printing Office, February 1993), p. 83.

tests carried out by the original developer, the 1984 act permitted these firms to file an Abbreviated New Drug Application (ANDA), which requires them only to demonstrate bioequivalency of their product and the original drug. On average, an ANDA requires only eighteen months for approval. As a direct result of this law, there has been a major expansion in the number and sales of generic producers.

Finally, in 1992 Congress passed the Orphan Drug Act, whose purpose was to stimulate the development of drugs with limited market potential, because either the condition was so uncommon or the drug's development and manufacturing costs were too high. It authorized the FDA to assist manufacturers in designing research protocols, to subsidize clinical and preclinical drug studies, and, finally, to grant seven years of marketing exclusivity to the first firm to receive NDA approval for a particular drug. By September 1992, seventy-nine new orphan drugs had been approved. Furthermore, approximately two-thirds of the firms gaining NDA approval under the act were small firms rather than major pharmaceutical manufacturers.

V. INDUSTRY PERFORMANCE

Since market power is conventionally measured by differences between prices and marginal costs and since drug prices typically exceed marginal costs, the pharmaceutical industry does not function as a standard competitive industry. Its leading firms enjoy substantial degrees of market power. At the same time, industry performance is not adequately measured by the conventional criteria of a competitive industry. If the pharmaceutical industry were typically competitive, there would be little investment in research and development; few if any new products would be introduced; and the companies would have little need to spend large sums on marketing their products. That was the status of the industry before World War II. In the modern pharmaceutical industry, however, the appropriate criteria for industry performance are more broadly stated, since few industry watchers believe that returning the industry to its prewar status would improve its performance.

The dominant characteristic of the modern industry is the substantial allocation of resources to research and development and the extensive array of new therapeutic agents that have resulted. To be sure, not every research project is successful; there are more "wrong turns" than "right ones." Many research dollars are spent on projects for which no new products result, and the average research costs of new products includes this work as well. Consequently, as we noted earlier, millions of dollars are spent on each new product introduced.

In addition, all new products are not equally successful or innovative; instead, the revenues received are highly skewed. The relevant data on revenue generation per product are shown in Figure 7-2. As can be seen, only the top three deciles of new products introduced between 1980 and 1984 generated sufficient revenues to exceed average research costs, estimated at $202 million per product introduced. The fourth decile contains those products whose average present value of net revenues are $177 million, less than the average research costs, and for the rest, the net revenues are even smaller.[27] Of course, this analysis presumes that all research costs are the same, which is unlikely. Still, prospective revenues are highly skewed, with a minority of new products (largely those embodying major therapeutic advances) generating most of the industry total.

To be sure, aggregate pharmaceutical industry profits depend on both revenues and costs. For this comparison, we need to evaluate the average returns per new product after all costs are deducted. A government report on phar-

FIGURE 7-2   Present values by decile: new chemical entities, 1980–1984.

After-tax PV in millions
of 1990 dollars

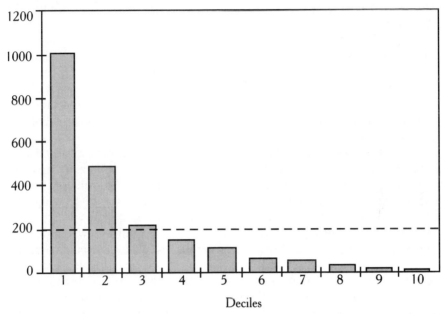

Deciles

Source: Henry G. Grabowski and John M. Vernon, "Returns to R&D on New Introductions in the 1980s," *Journal of Health Economics*, 1994.

maceutical research found that for new pharmaceuticals introduced between 1981 and 1983, the average returns after deducting research as well as all other costs were $36 million per product. This surplus is 4.3 percent of the average product price over its life cycle.[28] The Grabowski–Vernon study noted earlier offers an alternative estimate of $22 million for new chemical entities introduced between 1980 and 1984.[29]

These estimates are similar, and together they describe a profitable enterprise at the start of the 1980s. At that time, the industry could expect a substantial increase in its stock market valuations, as in fact was the case. Furthermore, these returns promoted a major increase in spending on pharmaceutical research and development that was 10 percent per annum throughout the 1980s.[30] These profits served to attract new resources into the industry, which is the principal function of profits in a market economy.

The critical dilemma for evaluating the performance of the pharmaceutical industry has long been the conflict between static and dynamic efficiency. Low prices are desired not only because they benefit consumers but also because higher prices limit patients' ability to buy and use essential drugs. Although the theoretical standard for the maximum economic welfare of marginal cost pricing is too stringent for this industry, society still prefers the lowest possible prices that are consistent with intensive research and the rapid introduction of beneficial new drugs. Unfortunately, this criterion is not precise because high prices and profits stimulate research spending and because therapeutically better products are priced above their current substitutes. In effect, there is a trade-off between static and dynamic efficiency.

If public policy actions are taken that reduce prices too severely, the prospective returns from investing in research and development will decline; there will be less investment in product introduction; and fewer new products will be developed. On the other hand, if the rewards from pharmaceutical innovation expand too much, then consumers will suffer, and important new products will not be as accessible as they might be. The main issue is balancing these competing objectives.

Ideally, society would gain most from a mechanism that permitted it to achieve both objectives at the same time, through an institutional change that permitted intensive research along with low prices for the fruits of that research. But that would require an industry substantially different from the one that now exists, and such a change would raise many new issues.

With the existing industrial structure, there is a conflict between society's two objectives. At the same time, it is apparent that we have decided that relatively high prices are worth the benefits that flow from major new investments in pharmaceutical research and development. We have made our choice, but through our political representatives, we nonetheless complain about it. The political statements reported at the start of this chapter represent our ambivalence regarding the decision we have made. Yet it is surely true that we could make none other.

NOTES

1. Senate, *Staff Report to the Special Committee on Aging* (Washington, DC: U.S. Government Printing Office, February 1993), p. v.
2. Peter Temin, *Taking Your Medicine: Drug Regulation in the United States* (Cambridge, MA: Harvard University Press, 1980), p. 87.
3. Pharmaceutical Manufacturers Association, *Statistical Fact Book* (Washington, DC: Pharmaceutical Manufacturers Association, September 1991), Figure 2-1.
4. These studies are reported in Peter Borton Hutt, "The Importance of Patent Term Restoration to Pharmaceutical Innovation," *Health Affairs*, Spring 1982, p. 9.
5. U.S. Office of Technology Assessment. *Pharmaceutical R & D: Costs, Risks, and Reward* (Washington, DC: U.S. Government Printing Office, February 1993), pp. 47–72. These estimates are reported in 1990 dollars.
6. Ibid, p. 72.

7. Pharmaceutical Manufacturers Association, *Statistical Fact Book* (Washington, DC: Pharmaceutical Manufacturers Association, September 1991), Table 1-19.
8. These findings are reviewed in William S. Comanor, "The Political Economy of the Pharmaceutical Industry," *Journal of Economic Literature*, September 1986, pp. 1196–1199.
9. Reported in Keith B. Leffler, "Persuasion or Information? The Economics of Prescription Drug Advertising," *Journal of Law and Economics*, April 1981, p. 62.
10. Leffler finds that "pharmaceutical promotion serves at least two functions. The informational function is indicated by the promotional intensity of new products and reliance of new products on detailing. Yet the variance of advertising across therapeutic categories and the longevity of intense advertising for some products supports the empirical relevance of a 'habit formation,' persuasion theory of pharmaceutical promotion" (ibid., p. 64).
11. Richard E. Caves, Michael D. Whinston, and Mark A. Hurwitz, "Patent Expiration, Entry, and Competition in the U.S. Pharmaceutical Industry," *Brookings Paper on Economic Activity: Microeconomics*, 1991, pp. 39–40.
12. Michael S. Wilkes et al., "Pharmaceutical Advertisements in Leading Medical Journals: Experts' Assessments," *Annals of Internal Medicine*, June 1, 1992, pp. 912–919.
13. Keith B. Leffler, "Assessing Prescription Drug Advertising: Is Information the Proper Criterion?" Comments for the UCLA Seminar on Pharmaceutical Economics and Policy, December 10, 1992, pp. 4–5.
14. This figure was provided by the Generic Pharmaceuticals Industry Association and rests on IMS data. It includes generic prescriptions for single-source drugs and is not limited to those sold by generic manufacturers.
15. This is the current share of U.S. sales in the twenty-four-nation Organization for Economic Cooperation and Development (OECD).
16. Patricia M. Danzon, "International Drug Price Comparisons: Uses and Abuses," unpublished study financed by Pfizer Inc., November 1993, p. 15 and Table 2. She concludes that "a cross-national comparison reveals at most a confused and partial picture of manufacturer pricing policies, constrained by. . .diverse price regulatory regimes and health care reimbursement systems" (p. 6).
17. E. Reis-Arndt, "A Quarter of a Century of Pharmaceutical Research— New Drug Entities 1961–1985," in *Drugs Made in Germany*, ed. Viktor Schramm (Aulendorf: Editio Cantor, 1987), vol. 3, pp. 105–112.
18. William M. Wardell, "Introduction of New Therapeutic Drugs in the United States and Great Britain: An International Comparison," *Clinical Pharmacology and Therapeutics* 14 (1973): 773–790; and Stuart O.

Schweitzer, "Comparaison France–États Unis de l'autorisation de mise sur le marche de nouveaux produits pharmaceutiques," *Journal d'économie médicale* 11 (1993): 33–44.

19. Z. John Lu and William S. Comanor, "Strategic Pricing of New Pharmaceuticals," unpublished manuscript, February 2, 1994.
20. Steven N. Wiggins and Robert Maness, "Price Competition in Pharmaceutical Markets," unpublished manuscript, October 1993, p. 5.
21. *Los Angeles Times*, January 30, 1994, sec. A, p. 1.
22. Such price differences are not unusual in American industry. For example, airline prices differ greatly between trips that do and do not extend over a Saturday night. The airlines apparently believe that this is the principal factor distinguishing business from vacation travel, and they set their prices accordingly.
23. Quoted in Rebecca Henderson and Iain Cockburn, "Racing or Spilling? The Determinants of Research Productivity in Ethical Drug Discovery," unpublished manuscript, October 1993, p. 2.
24. Ibid, p. 9.
25. U.S. Office of Technology Assessment, *Pharmeceutical R&D*, pp. 162–163.
26. Ibid, p. 156.
27. Henry G. Grabowski and John M. Vernon, "Returns to R&D on New Drug Introductions in the 1980s," unpublished manuscript, September 1993, p. 20.
28. U.S. Office of Technology Assessment, *Pharmaceutical R&D*, p. 1.
29. Grabowski and Vernon, "Returns to R&D," p. 15.
30. U.S. Office of Technology Assessment, *Pharmaceutical R&D*, p. 2.

SUGGESTED READINGS

Caves, Richard E., Michael D. Whinston, and Mark A. Hurwitz. "Patent Expiration, Entry and Competition in the U.S. Pharmaceutical Industry." *Brookings Papers on Economic Activity: Microeconomics*, 1991.

Comanor, William S. "The Political Economy of the Pharmaceutical Industry." *Journal of Economic Literature*, September 1986.

Congressional Record, March 11, 1992, pp. S 3181–S 3257.

Office of Technology Assessment. *Pharmaceutical R&D: Costs, Risks and Rewards*. Washington, DC: U.S. Government Printing Office, 1993.

Scherer, F. M. "Pricing, Profits and Technological Progress in the Pharmaceutical Industry." *Journal of Economic Perspectives*, Summer 1993.

Temin, Peter. *Taking Your Medicine: Drug Regulation in the United States*. Cambridge, MA: Harvard University Press, 1980.

8

MOTION PICTURE ENTERTAINMENT

Barry R. Litman*

The motion picture entertainment industry will be one of the most fascinating and dynamically changing industries of the twenty-first century. Awash in a whirlwind of merger activity involving domestic and international communication conglomerates and intricately linked to new telecommunication and computer technologies, even now it hardly resembles that quaint, turn-of-the-century invention of Thomas Edison that would fulfill the master inventor's dream of "doing for the eye what the phonograph does for the ear." Rather, it is at the center of the information and communication revolution that has transformed modern society over the last quarter century and that will shortly bring hundreds of channels of entertainment and information into the household through a fiberoptic cable.

Indeed, the motion picture is one of America's most treasured art forms, more popular than ever, here and abroad. As an industry, it has met the test of time, enduring and adapting in the midst of continual competitive threats from new media. The major studios no longer fear new competition from any source; rather, they embrace it and use it to expand their overall business enterprise, both domestically and internationally. No longer reliant on theatrical attendance, motion pictures can be viewed by consumers through an expanding panoply of exhibition "windows," including television, cable, and the videocassette recorder. The traditional order has been completely upset as historic business relationships and lines of demarcation have become intertwined and blurred. It is now proper to refer to this broadened landscape as encompassing a motion picture "megaentertainment" industry.

To gain insight into the breathtaking developments that have transformed the industry, it is important to begin with an historical perspective. The past is indeed prologue to the present, since attempts to control content and exhibition windows are rooted in a hundred years of industry history.

* With Scott Sochay.

I. HISTORY

Shortly after receiving his patent for the motion picture camera and its viewing machine, in 1894 Edison introduced motion pictures to an unbelieving public.[1] These early peephole-viewing machines (kinetoscopes), with their continually rotating 50-foot strips of celluloid film, proved to be an instantaneous success.

The arrival of the nickelodeon theater in 1905 provided the nascent industry with its first opportunity to stand on its own as an entertainment medium. Later on, one-reelers gave way to two-reelers and then to full-length feature films, and the nickelodeon was replaced by first-class and then deluxe movie houses. Thus motion pictures developed as a mass medium in a series of cycles, each with some driving force that further cultivated a mass audience with the moviegoing habit. Each new invention and quality improvement propelled the industry through a ratchet-type effect to a new, higher equilibrium.

THE SEARCH FOR MARKET POWER

Concomitant with the development of motion pictures as a cultural medium was the perennial attempt by a small cadre of companies to acquire monopoly power. In the early years, Edison and a few other inventors tried to dominate the market by exploiting their patents on cameras, projectors, film, and other integrated components. After a cycle of costly patent infringement suits threatened the stability of the industry, a truce was called in 1909 among the sixteen key patent holders, including Edison, Biograph, Armat, and Eastman-Kodak. They created a collusive patent-pooling arrangement that was implemented through the Motion Picture Patents Company. By cornering the market on equipment and offering top-quality productions, the trust believed that it could monopolize the industry and exclude independents at all access points. The trust ran into trouble, however, when it was unable to prevent maverick exchanges and nickelodeons from dealing with independents for additional needed product. This instability was alleviated through integration into the distribution stage. The General Film subsidiary of the trust acquired almost every licensed exchange, thereby creating monopoly control over every stage of production, save exhibition, during the first few years of the industry's existence.

With the rise of feature films, a new distribution organization was needed to handle this different species of film. Under the leadership of Adolph Zukor, the head of Famous Players, an exclusive alliance was formed in 1914 with an association of state's-rights distributors (known as Paramount) to distribute feature films. Within two years, Zukor assumed control and merged his Famous Players company with Paramount, thus beginning a trend toward the vertical integration of the production/distribution sectors that continues to this day.

When Paramount's terms became too expensive, the largest first-run exhibitors integrated backward into production to supply their own needs and thereby circumvent the power of the Zukor organization. A merger race ensued as the large companies in each stage of the industry sought partners to guarantee either an assured supply of films or access at reasonable terms. By 1925, there were only a handful of giant vertically integrated firms left in the industry.

The arrival of sound occurred in the mid-1920s at precisely the time when something new and exciting was once again needed to stimulate movie attendance and compete with radio. During the slump of the Depression years, the industry tried to solicit new business by appealing to more salacious themes as well as by introducing the double feature. Most important, it crafted the Code of Fair Competition under the National Industrial Recovery Act of 1933 to protect itself against the ravages of unbridled competition. The code regulated various trade practices and was administered by the Motion Picture Producers and Distributors of America, the industry's trade association. Such trade practices as block booking and blind selling, which severely disadvantaged independent theater owners, were now legally sanctioned and enforceable in a court of law.[2] Within two years, the National Industrial Recovery Act was declared illegal and the code was discontinued, but the industry had learned an important lesson: the beneficence of mutual cooperation. From now on, it would follow a policy of tacit collusion—a form of shared monopoly[3]—that yielded handsome rewards over the years.

THE PARAMOUNT CASE

During the late 1930s, the five fully integrated firms ("the majors") deliberately sought to eliminate the remaining independents. Since none of the majors individually possessed enough of the highly desirable first-run theaters or produced enough A quality feature films to be self-sufficient, they needed one another. Acting in concert, they had enough first-run theaters to provide a nationwide exhibition showcase for their films, and their combined production efforts (in association with three minor distributors) were sufficient to supply an entire year's schedule of films.[4] Achieving near 100 percent self-sufficiency severely restricted the freedom of the independent producers, distributors, and exhibitors from gaining access at any one stage in the vertical chain. At the exhibition level, the Big Five collectively operated 70 percent of the all-important first-run theaters in the ninety-two cities with a population of 100,000 or more and 60 percent in those cities with a population of between 25,000 and 100,000.

Beginning in 1938, this pattern of vertical organization was attacked by the Antitrust Division of the Justice Department. The ultimate outcome of the decade-long case was a court order mandating vertical divestiture of the exhibition level from the production/distribution level and furthermore requiring that those theaters illegally acquired or used as part of the conspiracy be sold off by the newly reconfigured exhibition circuits. For the most part, the divested chains of theaters could not acquire new theaters without specific court approval, and none of the affected parties could reintegrate (either forward or backward). In addition, competitive bidding was suggested (but not mandated) as a way of further ensuring that decentralization of power and open access would prevail in the exhibition level.

THE ADVENT OF TELEVISION

The motion picture industry initially ridiculed the early television programming of its new rival, comparing it with the B films of the 1940 era. Nevertheless, as TV penetration increased and Americans stayed at home rather than going to movie theaters, the industry soon understood the enormity of the situation. Its initial strategy was to boycott television by refusing to permit its creative personnel (primarily actors) to appear on television programs, produce television series, or license films for television exhibition. It also sought to counter the inroads of television by introducing a number of product innovations, including Cinerama, three-dimension movies, and big-budget films with lavish production values.[5]

After recognizing the revenue-generating capability of television production, Warner Brothers became the first major to break the boycott in 1955 by agreeing to produce a weekly series. Shortly thereafter, the rest of the majors hitched their wagons to television's rising star rather than retain their purity in films. They also realized that the television networks (and, later, the television stations) could become subsidiary markets for licensing theatrical films once those films had played out their theatrical run. The temporal price-discrimination pattern of multiple theatrical runs that had worked so well when only theaters were involved could now accommodate the new television technology. Therefore, what had begun as a major confrontation between two entertainment media gradually evolved into a pattern of stability, mutual interdependence, and economic symbiosis.[6] This stable business relationship continues and has been extended to the newest technologies of cable, videocassettes, and fiber-optic telephone systems—all of which have been folded into the distribution price-discrimination process.

II. MARKET STRUCTURE

Because of the various subsidiary markets and the historical pattern of vertical integration, the structure of the motion picture entertainment industry is extremely complex.

DISTRIBUTION

The distribution stage remains under control of a handful of major companies, although a few names have changed since the Paramount divestiture.[7] In terms of the majors, RKO has left the scene; in 1983, MGM took over distribution for United Artists and eight years later acquired the company completely. Monogram became Allied Artists and was later purchased by Lorimar, which in turn was acquired by Warner Brothers.[8] Another independent, Embassy, was sold in 1985 to Columbia and then resold that same year to Dino DeLaurentis and eventually reacquired by Columbia; it is currently in financial reorganization. American International was sold to Filmways in 1979, and its name was changed to Orion in 1980. Another key player is Tri-Star. Formed in 1983 as a joint venture of Columbia, HBO, and CBS, it was purchased late in 1987 (including its theater circuit) by Columbia and folded into the new Columbia Entertainment Division.[9] Finally, the latent Disney studio was reorganized in the mid-1980s, shed its exclusive family orientation, and established itself as an industry leader with four separate labels.

The market shares of the leading distributors are given in Table 8-1. The familiar majors still collectively account for three-fourths of the total industry, even though their stranglehold on industry power has slipped somewhat over the last decade[10] owing to the entry of significant newcomers like Tri-Star and the intensified competition provided by Disney, Orion, and independents, such as Cannon and New Line. The four-firm concentration level declined from 68 percent in 1982 to 60 percent by 1991 but has since increased to 73 percent in 1992. (The Herfindahl–Hirschman index fell correspondingly from 0.16 to 0.13 until it rebounded to 0.16 in 1992.) Even discounting recession-laden 1992, these numbers still indicate considerable oligopoly power remaining in this sector of the industry. Interestingly, market shares and industry leadership seem very volatile, often influenced by having one or two box office smashes in any given year.

BARRIERS TO ENTRY

It is not surprising that the distribution level is highly concentrated. This is a characteristic common to all the mass media and is due to economies of scale that accompany the national distribution of the entertainment product. To service the 25,000 North American movie screens and exploit the foreign market potential, a distributor must have a vast worldwide network of offices. Furthermore, if the other subsidiary markets are to be efficiently tapped, additional bureaucratic layers must be created. Focusing on the domestic distribution of motion pictures, a major company needs twenty to thirty regional offices and a sizable sales and marketing force. It also needs a steady stream of features released throughout the year so that these offices can work at full capacity.

Although some critics argue that ownership of studios and back lots is an albatross around the necks of the major distributors (because it drives up overhead costs), there still may be some strategic advantage that accrues to those so situated, especially during a production boom. There also must be a corresponding level of demand to justify such a large enterprise and to use the capacity most efficiently. Here is where product ideas, differentiation, and

TABLE 8-1 Market Shares for North American Film Distributors,[a] 1982–1992

Company	1992	1990	1988	1986	1984	1982
Warner Bros.	20	13	11	12	19	10
Buena Vista (Disney)	19	16	20	10	4	4
Twentieth-Century Fox	14	13	12	8	10	14
Columbia	13	5	4	9	16	10
Universal	12	13	10	9	8	30
Paramount	10	15	15	22	21	14
TriStar	7	9	6	7	5	–
New Line	2	4	–	–	–	–
MGM/UA	1	3	10	4	7	11
Miramax	1	1	–	–	–	–
Orion	<.5	6	7	7	5	3
Embassy	–	–	–	–	<.5	1
Atlantic	–	–	–	1	1	<.5
Cannon	–	–	–	2	1	<.5
New World	–	–	–	3	1	1
DeLaurentis	–	–	–	3	–	–
All Others	1	2	5	3	2	2
CR$_4$	73%	58%	58%	53%	66%	68%
CR$_8$	89%	94%	95%	84%	91%	96%
H-H	.16	.12	.12	.11	.13	.16
CR$_6$–Majors	77%	71%	68%	64%	81%	89%
Instability[b]		28%	25%	28%	49%	40%

[a] Market shares = percentage of annual film rentals to distributors.

[b] Instability = $\sum_{i=1}^{N} S_i^t - S_i^{t-1}$; Tri-Star and Columbia combined after 1987.

Source: *Variety*, annual editions.

managerial skill play such an important role. The studio must consistently make correct decisions concerning the composition of its product. Unlike a modern industrial enterprise, in which automobiles are turned out in large quantities over long manufacturing runs, each film is handcrafted, with a very short product life and only occasional opportunities for reusing the creative inputs. The successful studio requires a continual stream of new ideas.

Given the escalating production and advertising costs of recent years, the average release by a major distributor runs more than $20 million plus at least half that amount to give it a national day-and-date launch. Understanding these economic risks and the fact that they have a portfolio of such projects at different stages in the two-year cycle, the major distributors and their corporate parents tend to act rather conservatively, seeking to minimize their financial risks by searching for some formula like that of their television counterparts. This formula may take the form of employing bankable movie stars, top directors, and writers who at some point have touched the magic of a box office smash, or else spreading the risk through coproduction with foreign investors.[11] Alternatively, they may repeat ideas, themes, or characters that worked well in the past, thus generating the ever present sequel.[12]

VERTICAL INTEGRATION AND CONGLOMERATION

Part of the risk can also be lessened if distributors have vertical integration or deep corporate pockets. Even though the Paramount decrees forbade reintegration between the distribution and exhibition levels, as time has passed, the original companies have been resold or absorbed into larger conglomerate corporations. Furthermore, entry has occurred at both levels, thus introducing new companies that are not covered by the original decrees. Most important, new subsidiary exhibition markets have opened up, thus reducing the fears of an earlier era, and the antitrust-tolerant Department of Justice has not opposed reintegration by the original Paramount defendants.

There currently exists a significant degree of forward integration by the leading distributors into theatrical exhibition. The advantages of vertical integration are numerous—greater distributor control over admission prices, release patterns, and the avoidance of anti-blind-bidding statutes. Most important, it permits all of the box office dollars to remain with the distributor rather than sharing them with exhibitors.

The biggest players have been MCA (Universal) and Paramount; Disney, Fox, and MGM-UA have diversified into other sectors. In 1986, MCA acquired a 50 percent interest in Cineplex–Odeon, the second largest theater circuit in North America. Similarly, Paramount has roughly 470 domestic and 433 Canadian theaters, making it a large North American circuit. It also is a partner with MCA in the 76-screen circuit in Europe, known as Cinema International Corporation.[13] In late 1987, Columbia Entertainment acquired Tri-Star, including its circuit of 310 theaters (the former Loews circuit). In early 1988, the Loews subsidiary acquired the USA circuit, thereby adding another 317 screens to Columbia's holdings and making it the seventh largest circuit in North America. Meanwhile, Cannon, which had a 525-screen circuit in Europe, bought the domestic Commonwealth circuit in 1986, with its 432 screens.

Most interesting is the position of the industry leader, United Artists Theater Circuit. Not only has it aggressively sought merger partners and embraced new theater construction programs, but it is controlled by Telecommunications, Inc. (TCI), the dominant multiple-system cable television operator. Given this strong position across multiple exhibition windows and with vertical integration into the distribution arena through such cable networks as Turner Broadcasting Systems (CNN, WTBS, TNT), Black Entertainment Network, and American Movie Classics, we may be witnessing a vertical chain of control never dreamed of by the Paramount conspirators. TCI's attempted blockbuster, $26 billion merger with Bell Atlantic suggests even greater size across many different communication technologies and services.

In terms of conglomeration, the major motion picture companies are subsidiaries of large communication corporations or other conglomerates. As these major distributors rely on the resources of their parents for production loans or capital expansion, this further enhances the barriers to entry and widens the historic gap between the majors and independents.

Table 8-2 gives a brief synopsis of each parent company and the percentage of its 1990 revenues that came from filmed entertainment. Columbia, Disney, Universal, Twentieth-Century Fox, and Warner Brothers are subsidiaries of very large corporate giants and consequently account for a rather small percentage of the parent company's operations. These parents have their businesses based in either electronic equipment (Sony, Matsushita) or other mass media or entertainment properties (Time Warner, News Corporation, Viacom and Walt Disney). Smaller production companies, such as Carolco, New Line, or Republic, have no significant family ties and hence have low diversification percentages. These may be termed *pure motion picture companies*.

EXHIBITION

The days of the stand-alone theater owner or other media enterprise are fast diminishing. Chain ownership yields efficiencies in spreading managerial (such as motion picture booking) and legal expertise, as well as advertising and marketing costs, across a large number of outlets; yet it has the potential for raising entrance barriers and enhancing market power.

The indices given in Table 8-3 indicate only a moderate degree of national concentration in theater ownership, but the recent trend is definitely on the rise.[14] Since December 1983, the four-firm concentration index has increased by 11 points while the eight-firm index has risen by 20 points. The current industry leaders—United Artists, Cineplex–Odeon, and American Multi-Cinema—have more than doubled their theater holdings over the last few years. United Artists has risen from second place to first by adding some 1,374

TABLE 8-2 1990 Diversification Ratios for Selected Film Distributors

	(A) Total Corporate Revenues[a]	(B) Total Communication Revenues[a]	(C)=B/A Comm. Total(%)	(D) Filmed Ent. Revenues[a]	(E)=D/A Diversification (%)
Paramount Communications[b]	$ 3.87	$ 3.87	100.0	$2.45	63.3
Sony (Columbia)	25.65	5.19	20.2	1.83	7.1
Disney (Buena Vista)	5.84	2.25	38.5	2.25	38.5
News Corp. (Twentieth-Century Fox)	6.72	3.04	45.2	.99	14.7
Time Warner (Warner Bros.)	11.78	11.78	100.0	2.90	24.6
Matsushita (Universal)	n.a.	n.a.	n.a.	n.a.	n.a.
MGM-Pathé	1.23	0.69	56.2	0.35	28.4
ORION	0.58	0.58	100.0	0.58	100.0
Republic	0.57	0.57	100.0	0.57	100.0
New Line	.13	.13	100.0	0.13	100.0
Carolco	0.26	0.259	96.2	0.259	96.2

[a] Revenues are stated in billions of dollars.
[b] Acquired by Viacom, Inc. in 1993.
n.a. = not available.

Source: Veronis-Suhler & Associates, *Communications Industry Report,* 1990.

TABLE 8-3 Number of Screens for Top U.S. Theater Circuits, 1983–1992

	May 1992				December 1983		
Rank	Company	Number of Screens	(%)	Rank	Company	Number of Screens	(%)
1	United Artists	2,379	9.7	1	General Cinema	1,050	5.6
2	Cineplex-Odeon	1,715	7.0	2	United Artists	1,005	5.3
3	AMC	1,623	6.6	3	AMC	736	3.9
4	Carmike Cinemas	1,512	6.1	4	Plitt	605	3.2
5	Harcourt General	1,405	5.7	5	Martin	431	2.3
6	Cinemark USA	956	3.9	6	Commonwealth	362	1.9
7	Loews	870	3.5	7	Mann	315	1.7
8	Nat'l Amusements/Redstone	725	2.9	8	Redstone	302	1.6
	Total U.S. Screens	24,639				18,884	
	CR_4	29.4%				18.0%	
	CR_8	45.4%				25.5%	
	H-H	0.029				0.011	

Source: Data from *Variety*, annual editions; *Encyclopedia of Exhibition*, annual editions; company annual reports.

screens, and AMC leaped over long-time industry leader General Cinema by adding nearly 900 screens. It should be noted that the expansion of circuit size is not limited to the acquisition route; a net of nearly 6,000 new screens have also been built. The industry leaders have clearly initiated a major merger wave but have also built new theaters as well. Interestingly, the theater circuits originally divested by Paramount have been absorbed by other equally large circuits.

Nonetheless, it would be wrong to examine only national concentration, since movie theaters are local retail outlets. In local market after market, the trend toward concentration of ownership repeats itself with a rotation of leadership among the top national chains and the occasional appearance of a smaller regional chain. In most localities, therefore, the consumer is faced with a handful of oligopolists.

NEW DEVELOPMENTS

Between 1985 and 1991, five of the seven Hollywood majors either were acquisition targets or were involved in mergers, signaling the beginning of a new conglomerate firm era for Hollywood. These changes included News Corporation buying Twentieth-Century Fox, Sony acquiring Columbia, Time and Warner Brothers merging, Pathé buying MGM/UA, and Matsushita acquiring MCA. All were predicated on the belief that to survive in the new global media environment, companies must become large and take advantage of the potential for hardware/software efficiencies and synergies between in-house units. Such efficiencies include the ability to produce and distribute major motion pictures through all the new video windows that new technology is creating. After these initial consolidations, only Paramount and Disney remained unattached majors, and both were acquisition minded themselves. The only other alternative was to build a major Hollywood studio from scratch, a path followed by Penta (Italy) and Polygram (Netherlands).

On a parallel path, media firms began to realize that no one firm had all the financial and technological resources to fully exploit a media environment consisting of digital compression, computers, and interactive services. With such uncertainty and huge capital risks involved, a mind-set developed that it was better to engage in joint ventures and alliances with other firms than to shoulder the whole risk and cost alone.

Both trends—globalization through the acquisition of Hollywood studios and entrance into the media technologies of the future through alliances and joint ventures—continue. Examples include Bell Atlantic's attempted buyout of TCI (and indirectly QVC's attempt to acquire Paramount), the recent Viacom–Paramount merger, Turner's acquisition of Castle Rock and New Line, and Disney's buyout of Miramax. Warner Brothers and Paramount also are exploring the possibilities of setting up a fifth television network.

The most recent of these developments was the effort by Bell Atlantic to take over TCI. The significance of the transaction for the motion picture industry is that both are leading "exhibition" firms in their respective industries. In a deal valued at $26 billion, Bell Atlantic would have obtained access to 42 percent of American homes and the capability of developing a superhighway for informational and entertainment digital services. Among the purported rationales for the merger were the potential economies of scale in digital technologies and the synergies that may come from marrying the fiber optics, programming, and interactive expertise of TCI with the switching and computer technology of Bell Atlantic.[15] The development of these superhighways to the home would offer revolutionary new approaches to consuming motion pictures, no longer confined to the theater setting; they would create almost an instant supermarket for consumer choice.

A second key development is the recent merger of Viacom and Paramount. Viacom, valued at $8 billion, and Paramount, valued at $10 billion, came under the control of the former, making the new entity a conglomerate giant on a par with Time Warner. This vertical merger brings together Paramount's movie and television studios with Viacom's cable networks (for example, MTV and Showtime), local television stations, and cable systems. In addition, Viacom's presence in overseas markets (MTV is strong in Europe and Latin America) will create additional distribution outlets for Paramount products. Thus there is a strong programming/distribution fit in the merger that will give Viacom an assured flow of programming to accompany its move into interactive media, in association with AT&T and Sony.[16]

In fact, the long run raises the specter of the pre-Paramount days, especially if several telephone–cable–film–broadcast giants are formed in the next few years: Although no single company would be strong enough individually to monopolize the mass media industries, it could do so in concert with others and realize the benefits of cooperation and reciprocity. There could evolve a system of arrangements with the power to regiment or control competition at virtually all access points in the media industry. Such a scenario necessitates a watchful eye by government regulators as they monitor for potential antitrust violations.

Finally, this possibility of monopoly further heightens the importance of the Hollywood studios. Each of these new distribution channels will need programming, and those media conglomerates that have established some measure of control over film production and distribution will find themselves in a better long-run competitive position than those that have not. As long as this remains true, Hollywood will continue to remain at the core of the new global media environment.

Only weeks before the Viacom–Paramount announcement, Ted Turner made another foray into Hollywood. Previously, Turner had acquired the rights to the MGM film library by buying the studio and then selling it back without the film library. With a vast library of film classics in hand, TNT was created, which was followed by the later acquisition of the Hanna Barbara cartoon library and the founding of the Cartoon channel. In his most recent Hollywood dealings, Turner acquired the rights to three hundred Paramount films to use as the backbone of his new Turner Classic Movies (TCM) channel. Not content with film libraries, Turner Broadcasting has now taken the plunge into Hollywood studio ownership by acquiring New Line Cinema and Castle Rock Entertainment, moderately successful independent production companies. Such purchases should put Turner Broadcasting in a better position to ensure a steady flow of theatrical products and stabilize its future.[17]

Until recently, only Paramount and Disney remained as unacquired Hollywood majors, yet each is active in its own realm. Disney's expansion plans (beyond film-related growth such as EuroDisney) involved Miramax, a successful independent film producer. This acquisition gives Disney a larger presence in film production, creating a fourth movie label to go with Walt Disney Pictures, Touchstone, and Hollywood Pictures. In 1993, Paramount was acquired by Viacom.

In other diversification areas, Warner Brothers has been exploring the possibility of creating a fifth network through possible alliances with Chris-Craft or Tribune Broadcasting, which own local broadcast television stations. Such an alliance, fleshed out by the availability of Time Warner cable systems, would provide a guaranteed national outlet for the growing first-run syndication programming being produced by Warner Brothers.[18]

Further complicating the race for a fifth network is the potential entrance of Paramount, the current leader in first-run syndication with such series as *Star Trek: The Next Generation*, *Deep Space Nine*, and *The Untouchables*. Paramount recently acquired WKBD-TV in Detroit with an eye toward further acquisitions that could increase its local television station holdings to the FCC limit of 25 percent of the country. Paramount has also talked with Chris-Craft about a possible alliance.

These developments in the mass media industries are a sign of increased competition across multiple venues, at least in the short run until positions of dominance can be established. At the same time, however, the competitive viability of unaffiliated, nonintegrated independents may once again be in jeopardy.

III. CONDUCT

The process of product pricing and revenue sharing has changed very little since the Paramount restructuring. At the exhibition level, theaters will either bid for upcoming features or individually negotiate the contract. The Paramount decrees require that motion pictures be contracted picture by picture at the local rather than chain level so that all theaters will have an equal opportunity. The contract generally has standard provisions with respect to admission prices, beginning date, length of run, minimum guarantee, dollar advance, terms for extended runs and early cancellation, and, most critical, rental terms. *Rental terms* refer to the percentage split of box office dollars between the exhibitor and the distributor. The industry standard is a 90–10 split in favor of the distributor after the house expenses are deducted. This small profit margin for theater owners demonstrates the relative bargaining power of the distributors vis-à-vis the exhibitors.

PRICE DISCRIMINATION

With the arrival of television and other subsidiary markets, the theatrical tiering system of runs and clearances was replaced by the sequencing of exhibition "windows." Both systems represent forms of second-degree price discrimination: charging different prices to classes of customers who place different utility valuations on the product. With motion pictures, these groups cluster according to the time dimension. Avid moviegoers who absolutely must be the first to see newly released films are willing to pay the highest price per ticket.[19] Those willing to wait until the second theatrical run or until the movies appear in the video stores or on cable pay lower prices, and those with even less interest may wait several years until their appearance on network or local television.

Formulating the optimal time sequencing of exhibition windows maximizes profits for those distributors who have obtained rights across all these windows, provided that everyone follows the same pattern and there is no significant leakage or resale between the windows.

EXHIBITION CONDUCT

The process of downsizing theater size and building multiple-screen auditoriums illustrates a different profit-maximization approach by the theater owners. This permits a more cost-efficient utilization of seat capacity than was possible in the old single-auditorium deluxe theaters. With six to eight screens per site, a theater owner can manage a portfolio of different pictures with confidence that at least one or two of the screens will have a hit and that the others will do moderate business until their run is over. In this way, average load factors can be increased, in contrast with the hit-and-miss strategy associated with having a single screen.

The multiple-screen concept is also tied to the shopping-mall phenomenon.[20] The movie theater complex is an integral part of the modern shopping mall, since it generates a lot of foot traffic for other store owners. For this reason, the recent construction boom in movie theaters is more a shopping-mall phenomenon than a by-product of fundamental industry economics. If one looked only at theaters in isolation, given the stagnant demand for admissions due to the VCR and pay-cable alternatives, one could not explain the addition of five hundred to one thousand new screens per year. Theater construction can be justified only as part of the profit-maximizing calculus of building a successful shopping mall.

The theater concession stand also fits into this traffic-flow analysis and multiple-screen concept. Since the theater retains 100 percent of all concession revenues (which have an extremely high markup) yet only makes about

10 percent profits from box office gross, any strategy that maintains a high load factor can be very profitable.

The theaters can further minimize their excess seat capacity by negotiating escape clauses in their contracts that permit shortened runs for unpopular films and lengthened runs, additional screens, and move-overs (to a larger auditorium) for films that prove unexpectedly popular. Although theaters seldom charge differential prices based on the perceived quality or popularity of the individual film, they do practice discounting for off-peak time periods (such as matinees, twilight, and, occasionally, certain weekdays) and for customers with elastic demands (such as children and senior citizens). Once again, this brings more customers into the theaters and sells more popcorn.

THE TELEVISION NETWORK MARKET

The major movie distributors have maintained an extraordinary presence in the market for regularly scheduled prime-time programs, collectively averaging in excess of 40 percent of regular series since the mid-1960s. The networks themselves produce a small percentage of their programming needs, concentrating primarily on news, sports, soap operas, and specials rather than entertainment series.[21] Prime-time series are also produced by independent production houses, many of whom have market shares comparable to those of the major studios. The majors have no economic advantage in this programming area because of a well-developed rental market for inputs, minimal economies of scale, and easy entry.[22]

Given such competitive conditions, the program supply industry has been unable to countervail the coordinated buying power of the Big Three networks. The networks understand that their bargaining advantage is greatest in the initial developmental stages, when the quality of the scripts and pilot is unknown and the future success of the program is uncertain. At this stage, venture capital is scarce because of the financial risk. Once a show becomes a hit, its true value is known, and it can command a high price on the open market.

Motion pictures, by contrast, represent a different species of network programming because there is prior knowledge about their value in the theatrical market and each movie is a uniquely crafted entity. The known quality of the movies reduces the risk to the networks and enables the movie companies to demand prices reflecting the marginal worth of their movies.

The movie majors initially received high and increasing prices for their films. Prices for theatrical movies have always been two to three times higher (per hour) than the prices of regularly scheduled programs of comparable quality. To countervail the studios' bargaining power in this market and stabilize licensing prices, the networks took a number of steps. In the mid-1960s, they began commissioning made-for-television movies that were a substitute for theatrical movies and could be produced at significant cost savings. Second, ABC and CBS entered directly into motion picture production. This vertical integration into production resulted in some eighty theatrical movies between 1967 and 1971 and 40 to 50 percent of their annual requirements of made-for-television movies. Private and public antitrust litigation ensued.[23] Distributors charged that the television networks were violating the Paramount decrees, since they had set up a fully integrated market chain through network distribution, ownership, and affiliation with broadcast exhibition outlets and now production of theatrical movies. The case ended with a consent decree that basically ratified the FCC's rules and limited the networks' ability to produce in-house television series. (Paradoxically, the decrees did not forbid the networks from producing movies or from broadcasting them on their own network after the theatrical run.[24])

THE PAY-TELEVISION MARKET

The development of the pay-television market filled a void created by the scarcity of space in the electromagnetic spectrum that limited the number of available very high frequency (VHF) stations in any local market. Given the powerful economic incentive to share programming expenses through networking, and the necessity for networks to have local affiliates to transmit their programs, there was room for only three national TV networks. With only three network signals and the incentive to maximize ratings by seeking the lowest common denominator of programming, many program forms were not used, even though public broadcasting sought to bridge the cultural programming gap.

Proponents of pay TV argued that direct consumer payment would make programming more sensitive to viewers' preferences than would the advertiser-supported system of "free television."[25] With the development of cable television during the 1960s, a collection mechanism was now in place for excluding free riders—the basic market-failure problem associated with over-the-air broadcasting.

After the broadcast interests lobbied to forbid pay TV from gaining a foothold, the issue was resolved with a series of FCC rules that permitted pay TV to exist but restricted its programming to those types that were not available from commercial broadcasting (for example, recent theatrical movies that had not yet appeared on network television). In 1972, Home Box Office initiated a pay-cable program service consisting of recent uncut, uninterrupted movies, Las Vegas night club acts, and special sporting events. This channel of programming was sold on a monthly subscription basis to those areas already wired for cable. Yet its nationwide expansion would not come until a new cost-effective satellite-to-dish transmission system replaced microwave and until the U.S. Court of Appeals, in 1977, declared null and void all programming limitations imposed by the FCC on pay-cable networks. This provided a lift for pay-cable networks, since they could now become true competitors of the three commercial networks. (Cable television now penetrates nearly two-thirds of all U.S. households).

Such movie-driven premium networks as the Movie Channel and Showtime joined HBO, the industry leader, to exploit this new market.[26] In recent years, other specialty networks have entered, differentiating their product in other dimensions (for example, Disney, Playboy, and American Film Classics) rather than offering a full range of motion pictures and entertainment specials.

Ironically, in these pre-VCR days, consumers appeared willing to subscribe to several redundant networks to gain more viewing flexibility, and soon most cable systems began offering multiple networks. For example, HBO set up a second service, Cinemax, to try to capture the extra business, but the genie could not be put back in the bottle, and competition through differentiation became the new reality. At first the networks continued licensing box office hits on a nonexclusive basis and licensed exclusively the less successful movies (sports and specials) and all-time classics. By 1981, exclusive rights were obtained for all movies (just as the commercial TV networks did). Throughout this period, HBO/Cinemax collectively dominated this market, with a combined market share of nearly 60 percent, over twice that of Showtime. HBO used its monopsony power to reduce the license prices to the motion picture distributors, paying on a flat rental, take-it-or-leave-it basis rather than the customary per subscriber method. Nevertheless, this represented additional revenue for the motion picture distributors.

Various attempts by the major motion picture studios to establish their own distribution networks for this important subsidiary market, and thereby circumvent HBO, did not withstand antitrust scrutiny or were short lived in nature. Nevertheless, HBO had learned an important lesson: Lacking an assured source of films, it was vulnerable to such an "end around." It thus decided to become vertically integrated through ownership and long-term contracts to avoid future problems. It established a production subsidiary called Silver Screens, joined CBS and Columbia in launching a new minimajor company called Tri-Star, and signed long-term exclusive contracts for the full line of theatrical films of Columbia, Orion, and CBS.

To counteract this move, in 1983 the owners of Showtime and the Movie Channel (number two and three, respectively) in conjunction with Warner Brothers, Universal, and Paramount, announced a new joint venture. Even though the principals claimed that these networks would continue to be run separately, the Justice Department believed otherwise.[27] Upon the withdrawal of the major distributors, Showtime and the Movie Channel were permitted to merge. The premium pay-cable industry would now be restructured as a duopoly under the control of subsidiaries of Time Warner (HBO) and Viacom (Showtime–The Movie Channel).

Although some large studios have not signed long-term agreements and prefer to negotiate cable rights picture by picture, a significant share of the supply industry has been committed to a form of vertical tying arrangement that severely limits the ability of a potential pay-cable distributor from obtaining hit movies to compete with HBO or Showtime.

THE VIDEOCASSETTE MARKET

Of even greater importance to the motion picture industry than cable has been the videocassette revolution of the past decade. VCR penetration, at nearly 80 percent, now exceeds cable penetration by nearly 15 percent and has even surpassed the theatrical box office in revenues generated.[28] It soon may be considered the main reason for a universal service on a par with telephones and over-the-air broadcasting. The popularity of the VCR is its versatility; it permits viewers to time-shift programming; it allows playback of family home video photography; and finally, it allows consumers to access, through purchase or lease, a wide array of prerecorded videocassettes.

Unlike the television and cable markets, the major studios act as VCR distributors themselves rather than relying on specialized distributors. In fact, VCR distribution is very like that of magazines, paperback books, and records, with retail establishments acting as the point of sale to consumers. The market shares for VCR distributors are similar to those given in Table 8-2. It is clear that the major studio distributors have transplanted their power into this market, although their dominance in VCR software is much smaller than in the traditional theatrical market.[29] The VCR market is consequently a substantial source of revenue for the motion picture distributors and has moved toward the front of the line of exhibition windows.

THE INTERNATIONAL MARKET

Historically, the second most important exhibition window has been the international market, an aggregation of some eighty or more trading partners of the United States. As far back as World War I, the American film distributors have dominated this world market for film, later extended to television and most recently to videocassettes.[30] In fact, American control has been so pervasive that charges of media imperialism have frequently been leveled.

The reason for this dominance rests primarily on the enormous size and strength of the American market compared with those of other countries (see Table 8-4). American producer-distributors can recoup a greater percentage of their production costs from the domestic market alone, and given that the greatest expense is the first-copy production cost, distribution prices to foreign lands need only cover the incremental expenses. This pricing practice is often mislabeled as dumping. Since prices are based mainly on the strength of a country's demand, the richer and more populous countries pay higher prices for the same video product. For example, a theatrical movie currently distributed in France would yield $30,000 to $150,000 in rentals; $60,000 to $200,000 in Japan, and only $3,500 to $7,000 in Norway or Denmark.[31]

The major American distributors have bolstered their economic advantage by developing far-flung distribution networks throughout the world, including significant ownership of foreign theaters!

The protectionist response of foreign countries to American dominance has been the erection of trade barriers, including import quotas, tariffs, strict

TABLE 8-4 The Top Fifteen Export Markets for Major U.S. Distributors
(Theatrical Film Rentals to Members of the MPEA, $ Millions)

Country	1992	1988	1983
Japan	165.1	141.9	114.0
Germany	162.3	100.8	77.1
France	141.0	98.8	84.2
Canada	130.4	124.2	94.2
U.K./Ireland	127.4	90.3	59.6
Spain	122.5	67.5	40.8
Australia	67.4	45.3	51.6
Italy	65.2	73.4	51.0
S. Korea	39.6	–	–
Mexico	36.9	14.7	17.8
Sweden	31.8	27.6	15.2
Belgium	26.1	16.8	11.9
Switzerland	24.3	18.3	15.9
Argentina	24.2	–	–
Brazil	23.1	18.7	17.9
Taiwan	–	–	–
Netherlands	–	15.9	12.8
S. Africa	–	12.4	17.7

Totals	1992	1988	1993
Top 15 mkts.	1187.3	867.3	681.6
Other mkts.	252.7	153.0	157.2
Total	1440.0	1020.3	838.8
Exports			
USA only	2000.0	1413.6	1297.4
Total World	3440.0	2433.9	2136.2
Domestic (%)	58	58	61

Sources: Compiled from Variety, June 17, 1987, pp. 1, 36; May 18, 1988, pp. 1, 30; June 14, 1988, pp. 11, 14; June 13, 1990, pp. 7, 10; June 17, 1991, p. 10; and June 28, 1993, p.11. Revenues are film rentals from theatrical release, not the box office revenues from which the rentals were paid as exhibitor license fees. These rentals include theatrical license fees from films released in all or some areas by the MPEA members, including films that were released only in certain export areas. Furthermore, these rental figures include some modest revenues from foreign sales of advertising-publicity accessories for the promotion of the feature.

licensing procedures, limitations on the percentage of screen time for imported films (and TV programs), and measures preventing local currencies from leaving the country. In addition, foreign governments have encouraged indigenous producers through subsidies, prizes, tax breaks, or loans at favorable (or zero) interest rates.

In general, one should expect that the foreign markets for U.S. films will increase in importance in the foreseeable future, owing to developments such as "Europe 1992," the opening of Eastern Europe, cataclysmic changes in the

Soviet Union, the international multiplexing phenomenon, and increased scrutiny of piracy in Asia.

IV. PERFORMANCE

If an industry has a concentrated market structure, tries to coordinate pricing behavior, and has significant barriers to entry and an inelastic demand, the end result should be high prices for the consumer and excess profits for the companies.

According to a study by Veronis and Suhler that separated motion picture entertainment from the other product lines, the average operating-income margin (on sales) for 38 companies from 1986 to 1990 was 8.06 percent, and the cash-flow margin was 11.30 percent. For the 344 companies occupying all ten communication industry segments, the corresponding margins were 14.06 percent and 20.12 percent, respectively.[32]

It seems that the motion picture distributors have not been able to exert as significant a degree of market control in their own industry segment as have some other communication distribution companies; yet compared with a broader grouping of leisure and service companies throughout the economy, they have done relatively well. This internal failure is undoubtedly due to the vast degree of differentiation accompanying the motion picture product, the unpredictability of consumer tastes, and the instability of market shares from one year to the next. This is compounded by the fact that industry leadership rotates among the top firms and that market shares are more evenly dispersed than in other industries with comparable concentration ratios. It may also reflect the fact that there is now a plethora of different media through which consumers can obtain their entertainment.

What about the impact of this market structure on the public? One way to answer this question is to examine inflationary trends in motion picture admission prices compared with other products and services in the economy at large. Using 1982–1984 as a common base period, indoor admission prices rose by 65.2 percent during the 1980s; the CPI category of entertainment services and products increased by 51.3 percent; and the category of "all goods and services" climbed by 50.5 percent. Hence, admission price inflation has risen at roughly the same rate as these other categories. Thus the evidence is consistent with the profitability data and implies that the full force of the concentrated market structure is not inflicting unusual pain on consumer pocketbooks.

DIVERSITY

Because the motion picture industry produces cultural products, a critical question is whether concentration of control affects product diversity for the American people. Various studies of other mass media all have reached the same conclusion: Without the spur of competition, oligopoly firms with diversified parents tend to lead a quiet, imitative, unimaginative life rather than engage in costly product experimentation and new ideas.

Hence the incentive for the studios may be to stress the bottom line rather than worry about the impact on motion pictures as an art or cultural form. This would lead the studios on a constant search for formulaic content that reduces risks for a motion picture as it winds through all the subsidiary markets. This might mean a reliance on sequels that have a built-in recognition factor or extravagant spectacles with universal themes and international stars who appeal to international audiences. In either case, the end result is the homogenization of content. In recent studies, Dominick, Litman and Kohl, and Sochay documented the strong correlation between the concentration of control and the lack of content diversity.[33]

V. CONCLUSIONS

The American people's love affair with motion pictures has lasted for a century. Although production budgets have skyrocketed since the days of the one-reelers and the space-age and computer technologies have opened up many new points of competitive access, the motion picture industry remains a cherished institution on the American landscape, as venerable as the automobile. The basic structure of this industry was forged in the 1920s and continues largely intact except for the growth of subsidiary markets to replace the traditional system of theatrical runs. The concentration of market power, which had its roots in economies of scale and a vertically interlocking, self-sufficient system of arrangements, was decimated by the Paramount consent decrees but has regrouped and refocused at the distribution stage. Throughout all of the technological innovations, the resilience and adaptive capability of the industry have been truly remarkable. As distributors and circuits seek new merger partners and form strategic alliances with hardware and other diversified conglomerates under the laissez-faire attitude of the antitrust authorities, the historic dangers of expanding market control may resurface and even take on global proportions. In this regard, most dangerous and hence most important is the kind of multimedia acquisition strategy of companies like TCI and Time Warner which seek dominance of multiple exhibition markets. Constant antitrust vigilance is required lest the type of vertical market control—originally confined to the motion picture industry alone—spread across the entire entertainment landscape, extinguishing the competitive fires that have brought so many benefits to consumers in recent years.

NOTES

1. The sources for our discussion of the industry's history are A. R. Fulton, *Motion Pictures: The Development of an Art from Silent Films to the Age of Television* (Norman: University of Oklahoma Press, 1960); and the

contributions by Tino Balio, Russell Merritt, Jeanne Allen Thomas, Robert Anderson, and Douglas Gomery in *The American Film Industry*, ed. Tino Balio, rev. ed. (Madison: University of Wisconsin Press, 1985).

2. *Block booking* is the forced licensing of a package of feature films, tying together high- and low-quality films and offering them on an all-or-nothing basis. *Blind selling* is the forced licensing of a film before it is made available to the theaters for commercial preview. For more details, see Balio, ed., *American Film Industry*, pt. 3.

3. For the export market, the Webb–Pomerene Act permitted the industry to establish an export cartel known as the Motion Picture Export Association (MPEA), which could establish common policies and negotiate on behalf of its members. The MPEA continues to dominate the world markets, even though the Webb–Pomerene Act was finally repealed over a decade ago.

4. The five majors were Twentieth-Century Fox, Loews, RKO, Paramount, and Warner Brothers. The three minor distributors were United Artists, Columbia, and Universal. None of the latter companies officially owned theaters, although United Artists had interlocking directorships with a major chain.

5. See Balio, ed., *American Film Industry*, "Introduction," pt. IV.

6. Barry R. Litman, "Decision Making in the Film Industry: The Influence of the TV Market," *Journal of Communication*, Summer 1982, pp. 33–52.

7. For a good overall account of the various mergers, see Michael Conant, "The Decrees Reconsidered," in *The American Film Industry*, ed. T. Balio (Madison: University of Wisconsin Press, 1985) chap. 20; Thomas Guback, "The Theatrical Film," in *Who Owns the Media? Concentration of Ownership in the Mass Communications Industry*, ed. Benjamin Compaine, 2nd ed. (White Plains, NY: Knowledge Industries, 1982), chap. 5; and end-of-year roundups in *Variety*, annual editions, January, various years.

8. *Multichannel News*, May 16, 1988, p. 1, and June 20, 1988, p. 40.

9. *Variety*, January 15, 1988.

10. According to Table 7-1, the aggregate market share of the Big Six Paramount majors and minors fell from 89 percent in 1982 to 77 percent in 1992.

11. Gorham Kindem, "Hollywood's Movie Star System: An Historical Overview," in *The American Movie Industry: The Business of Motion Pictures*, ed. Gorham Kindem (Carbondale: Southern Illinois University Press, 1982), chap. 4.

12. Thomas Simonet, "Conglomerate and Content Remakes, Sequels and Series in the New Hollywood," in *Current Research in Film: Audiences, Economics and Law*, ed. Bruce A. Austin (Norwood, NJ: Ablex, 1987), vol. 3, chap. 10.

13. Thomas Guback, "The Evolution of the Motion Picture Theater Business in the 1980s," *Journal of Communications*, Spring 1987, pp. 60–77.

14. Regrettably, the number of screens rather than box office admissions is the only available measure of assessing market shares.
15. Geraldine Fabrikant, "$23 Billion Media Acquisition Reported to Be Near Completion," *New York Times*, October 13, 1993, p. A1.
16. "Viacom–Paramount Merger in Works," *Broadcasting and Cable*, September 13, 1993, p. 10.
17. Rich Brown, "Turner Signs Paramount Titles for $30 m," *Broadcasting and Cable*, August 16, 1993, p. 12, and "Kudos for Turner's Hollywood Deals," *Broadcasting and Cable*, August 23, 1993, p. 17.
18. Jim Benson, "Fifth Web Woes: No O&Os," *Variety*, September 6, 1993, p. 15.
19. David Waterman, "Prerecorded Home Video and the Distribution of Theatrical Feature Films," in *Video Media Competition Regulation, Economics and Technology*, ed. Eli Noam (New York: Columbia University Press, 1985), chap. 7. If the movie theaters were to vary their prices according to how long the film had been at the particular location, a certain group would undoubtedly pay premium prices to see certain highly publicized movies their first night or opening week.
20. This section draws heavily on Guback, "Evolution of the Motion Picture Theater Business in the 1980s."
21. Until the judicial review is complete, the networks are still prohibited by the FCC's Financial Interest and Syndication Rules from coventuring prime-time programs with outside sources, and they are barred under consent decrees from producing themselves more than two and one-half hours of prime-time programming per week.
22. Bruce Owen, Jack H. Beebe, and Willard G. Manning, *Television Economics* (Lexington, MA: Lexington Books, 1974), chap. 2.
23. *Columbia Pictures* v. *ABC and CBS*, U.S. District Court. Southern District of New York, 1972; *U.S.* v. *CBS, NBC, and ABC*, U.S. District Court, Central District of California, 1974.
24. For more details about these decrees, see FCC Network Inquiry Special Staff, *An Analysis of Television Program Production, Acquisition and Distribution* (Washington, DC: U.S. Government Printing Office, 1980), chap. 8.
25. This section is based largely on Barry R. Litman and Suzannah Eun, "The Emerging Oligopoly of Pay TV in the USA," *Telecommunication Policy*, June 1981, pp. 121–135.
26. A corollary market of advertiser-supported cable networks, such as USA, ESPN, CNN, MTV, and several superstations, emerged and was packaged together by local cable systems and sold as "basic" cable service. The premium channels just mentioned have always been sold separately.
27. Lawrence White, "Antitrust and Video Markets: The Merger of Showtime and the Movie Channel," in *Video Media Competition*, ed. Noam, chap. 11.

28. Ibid.
29. If one defined the market narrowly as theatrical videos, their control would be approximately equal to that of the theatrical movie market. However, just as movies on television must compete with other entertainment and information programming, so must videocassette recordings.
30. This section draws heavily on Thomas Guback, "Hollywood's International Market," in *American Film Industry*, ed. Balio, chap. 17.
31. *Variety*, October 7, 1991, p. M-84.
32. Veronis, Suhler, and Associates, *Ninth Annual Communications Industry Report* (New York: Veronis, Suhler, and Associates, 1991). Another good source for financial information is Harold L. Vogel, *Entertainment Industry Economics: A Guide for Financial Analysis* (Cambridge: Cambridge University Press, 1987), especially chaps. 2–4.
33. Joseph Dominick, "Film Economics and Film Content: 1964–83," in *Current Research in Film*, ed. Austin, chap. 9; Barry R. Litman and Linda Kohl, "Predicting Financial Success of Motion Pictures: The 80s Experience," *Journal of Media Economics*, Fall 1989, pp. 35–50; and Scott Sochay, "Predicting the Performance of Motion Pictures," unpublished paper, Michigan State University, 1993.

SUGGESTED READINGS

Books
Balio, T., ed. *The American Film Industry*, 1st and 2nd eds. Madison: University of Wisconsin Press, 1976, 1985.
Compaine, B. M. *Who Owns the Media? Concentration of Ownership in the Mass Communications Industry*, 2nd ed. White Plains, NY: Knowledge Industries, 1982.
Fulton, A. R. *Motion Pictures: The Development of an Art from Silent Films to the Age of Television*. Norman: University of Oklahoma Press, 1960.
Gregory, M. *Making Films Your Business*. New York: Schocken Books, 1979.
Guback, T. *The International Film Industry: Western Europe and American Since 1945*. Bloomington: Indiana University Press, 1969.
Hampton, B. *History of the American Film Industry from Its Beginnings to 1931*. New York: Dover, 1970.
Kindem, G., ed. *The American Movie Industry: The Business of Motion Pictures*. Carbondale: Southern Illinois University Press, 1985.
Noam, E., ed. *Video Media Competition*. New York: Columbia University Press, 1985.
Vogel, H. L. *Entertainment Industry Economics: A Guide for Financial Analysis*. Cambridge: Cambridge University Press, 1992.
Wildman, S. S. *International Trade in Films and Television Programs*. Cambridge, MA: Ballinger, 1988.

Articles
Crandall, R. W. "The Post-War Performance of the Motion Picture Industry." *Antitrust Bulletin*, Spring 1975.
Gomery, D. "The Contemporary American Movie Business." In *Media Economics Theory and Practice*. Edited by A. Allison et al. Hillsdale, IL: Erlbaum, 1993, chap. 12.
Helliman, H., and M. Soramaki. "Economic Concentration of the Videocassette Industry: A Cultural Comparison." *Journal of Communication*, Summer 1985.
Litman, B. R. "Decision Making in the Film Industry: The Influence of the TV Market." *Journal of Communication*, Summer 1982.
————. "The Economics of the Television Market for Theatrical Movies." *Journal of Communication*, Autumn 1979.
Litman, B. R., and S. Eun. "The Emerging Oligopoly of Pay TV in the USA." *Telecommunication Policy*, June 1981.

Government Publications
U.S. v. *CBS, NBC, and ABC*. U.S. District Court, Central District of California (1974).
U.S. v. *Motion Picture Patents Company*. 225 Federal Register 800 (1915).
U.S. v. *Paramount Pictures*. 334 U.S. 131 (1948).
U.S. v. *Western Electric and AT&T*. Modified Final Judgment. U.S.
District Court for District of Columbia (August 24, 1982).

CASINO GAMING

Christian Marfels

I. INTRODUCTION

Casino gaming is spreading like wildfire across the United States. Ten years ago, it was legal in only two states, Nevada and New Jersey, but as of mid-1993, the number of states where it is legal has reached twenty and is rising. This rapid increase is due mainly to the recent introduction of gaming on Indian reservations and on riverboats. All told, casino gaming experienced revenues of more than $10 billion in 1992, climbing at an average annual rate of 9 percent during the past ten years.[1]

This growth reflects an acceptance by the public of gaming as a leisure-time activity, an acceptance largely attributable to the transformation of the casino industry from its former structure of predominantly owner-managed entities, with little if any accountability, to corporate controlled and -managed entities with high accountability and performance-oriented operations.[2] The innovative efforts of the gaming machine industry to provide customer-friendly slot and video machines have also proved to be an important factor in creating an atmosphere of trust and reliability. These state-of-the art machines with their microprocessor controls are a far cry from the traditional mechanical slot machines. In fact, the new generation of gaming machines has attracted players in staggering numbers and consequently has strongly contributed to the mushrooming expansion of casinos and casino revenues.

DEFINITION OF THE INDUSTRY

Casino gaming refers to the leisure-time activity of placing bets at table games (craps, poker, roulette, baccarat, and the like) and playing coin-operated slot and video machines in establishments licensed for that purpose. Casino gaming is just one specific type of gaming, albeit a very important one, in the wider realm of commercial gaming, which also includes state lotteries, pari-mutuels (track betting on horses and greyhounds), bookmaking, bingo, and so forth. Unlike their European cousins, American casinos usually include hotel and restaurant facilities on their premises. In fact, the immediate availability of hotel rooms and eateries next to the casino floor is an integral part of the "one-stop" marketing strategy of American casinos. On an even grander scale, this may include other leisure-time facilities such as health clubs, sports facilities and, most recently, theme parks for the whole family. The Standard Industrial Classification (SIC) assigns such all-inclusive operations to the four-digit industry "Hotels and Motels" (SIC 7011).

A discussion of casino gaming would be incomplete without including the gaming machine industry. Although it is dwarfed by the dimensions of casino gaming, it has nevertheless been a key element, if not the decisive one, in its rapid expansion. Furthermore, there are strong linkages between the two industries, suggesting a segmentation of the gaming industry to consist of (1) a manufacturing segment (gaming machine industry) and (2) a leisure segment (casino gaming).

The gaming machine industry is classified in the four-digit industry "Manufacturing Industries, n.e.c." (SIC 3999). Its main products are coin-operated gaming machines, which are usually categorized as "slot machines." They come in two distinct types, reel-type slot machines and video machines, in which win/loss combinations are determined by a microprocessor-controlled random number generator.[3] A variant of video machines are the video lottery terminals (VLTs) that pay wins by credit voucher rather than in coin.[4]

HISTORY: FROM DICE TO ELECTRONIC SLOT MACHINES

Games of chance in any form are perhaps as old as humankind itself. There are reports about the Chinese betting on a game of skill as early as 2300 B.C. and about the Egyptians engaging in a game resembling craps in 1573 B.C.[5] Stories about *homo ludens* abound in medieval Europe with the advent of playing cards in the twelfth century. In the nineteenth century, Europe's grand casinos emerged, with one of the oldest and still operating casinos opening its doors in 1824 in Baden-Baden, Germany, and gaming at the famous casino in Monte Carlo, Monaco, beginning in 1861.

In North America, games of chance were already popular among Native American tribes long before the *Mayflower* arrived, and in contrast with the settlers from Europe, Native Americans held skilled gamblers in high esteem. In colonial times, horse racing began as early as 1666 in New Market on Long Island, and card games were also very popular. In the nineteenth century, however, all kinds of unsupervised, if not unscrupulous, forms of gambling activities gave gaming the negative image from which it has struggled to recover. But the late nineteenth century brought also the all-important American contribution to the gaming industry, the slot machine.[6]

Gaming in the United States, and specifically casino gaming, is closely associated with the State of Nevada.[7] In fact, gaming has been legal in Nevada for most of the period since its admission to the Union in 1864, but casino gaming did not begin until 1931 when Assemblyman Tobin's "wide open gambling" bill was signed into law. The industry grew slowly for the first ten years despite some successful operations in Reno, the center of Nevada's casino industry until 1946.[8] Las Vegas was not yet on the map; it was just a railroad water stop, but this changed with the emergence of four casino hotels built in the area between 1941 and 1946. Among them was the legendary Flamingo which opened in 1946 just outside Las Vegas on Highway 15 heading toward Los Angeles, the first in a sequence of casinos on what was to become the famous Las Vegas "Strip."

Nevada enjoyed its monopoly position in gaming for another thirty years until 1976, when Resorts International began operations in Atlantic City after a referendum in New Jersey had legalized casino gaming. This event provided a welcome dose of competition, geographically speaking, inasmuch as the lopsided orientation of the gaming industry was corrected with the creation of another center of gravity on the East Coast. Atlantic City was not just another location for gaming; rather, it gave the gaming industry new dimensions, a new structure, and a new image. New dimensions could be achieved, since Atlantic City brought gaming to about 25 million people in a 150-mile radius. No wonder that buses with "day-trippers" became a common sight in Atlantic City. Gaming there also provided the final breakthrough to the corporate era in the industry.[9] Since high minimum threshold requirements for casino-hotel operations were prescribed in the New Jersey Casino Control Act, only large-scale operations could apply for a license.[10] This made access to capital markets necessary for financing, and thus casinos developed as corporate businesses. Finally, the image of the gaming industry as an industry of integrity and trust received a boost from New Jersey's strict regulatory framework.

The Atlantic City experiment of limiting gaming activities to a specific locality rather than statewide was adopted by the State of South Dakota in 1985, with the provision of limited-stake casino gaming in Deadwood and with the tax proceeds going to a redevelopment fund for the town's historic district. Similarly, the decision of the State of Colorado to allow limited-stake casino gaming in the former mining towns of Blackhawk, Central City, and Cripple Creek was designed to revitalize municipal economies. Although these two developments added a few more spots on the "U.S. gaming map," they were of minor importance.

Two other developments in the late 1980s and early 1990s, however, have created the potential for a wide proliferation of gaming. One of them was the 1987 U.S. Supreme Court decision (*California* v. *Cabazon Band of Mission Indians*) which opened the door for gaming on Indian reservations as set forth in the subsequent Indian Gaming Regulatory Act.[11] Most important among the casinos on Indian reservations are the Foxwoods Casino (Connecticut) and the Mystic Lake Casino (Minnesota). The other important development is the surge in riverboat gaming along the Mississippi. Such riverboat-casino gaming was authorized in Iowa, Illinois, Indiana, Mississippi, Missouri, and Louisiana; operations began in Iowa and Illinois in 1991 and in Mississippi in 1992.

REGULATION: THE VISIBLE HAND

Regulation has been an integral part of the gaming industry since its inception in 1931. The regulatory framework in its various facets, liberal or strict, can be credited with the creating and maintaining a more positive image of the industry. In fact, it is unlikely that casino gaming would have achieved an acceptance rate of as high as 55 percent of the American populace were it not for the efforts of the Nevada Gaming Control Board and the New Jersey Casino Control Commission to monitor and supervise gaming activities.[12]

The most important instrument of the regulatory agencies is the licensing process, which is designed to ensure that "gaming is free from criminal and corruptive elements."[13] An application for a gaming license is required to set up a new gaming establishment or to change the ownership of an existing establishment. Beyond the licensing and monitoring of gaming activities, the agencies are also responsible for auditing casinos and collecting gaming taxes and fees.

Compared with Nevada, the regulatory regimen in New Jersey is much stricter. In fact, reference has been made in this regard to a "primacy of regulation"[14] and even to "overregulation of casino operations."[15] This situation must be evaluated in the context of the objectives of the 1976 New Jersey referendum to legalize casino gaming after a similar attempt was rejected by the electorate only two years earlier. First, there was the specific goal of the urban redevelopment and revitalization of Atlantic City by limiting casino gaming to this location only.[16] There was also the fiscal aspect of raising revenues—which were earmarked to finance programs for senior citizens and disabled persons in the Garden State—without having to raise the income tax. Finally, in order to gain acceptance, the proposed Casino Control Act had to set high standards to ensure integrity and trust in Atlantic City's gaming activities. That yardstick applied not only to the process of granting a casino license, which it does in Nevada as well, but also to numerous rules for the day-to-day casino operations, which it does *not* in Nevada, or at least not to the same extent. These rules range from the mandatory thresholds for large-scale casino operations to a maximum of three casinos under one ownership, to the 50 percent rule for gaming machines,[17] to the prohibition of twenty-four-hour gaming,[18] just to name a few. On top of that, the New Jersey casino revenue tax stands at 8 percent on gross gaming revenues, almost 2 percent higher than in Nevada.[19]

THE INDUSTRY TODAY

The gaming industry is one of the fastest-growing industries in the United States. At an average annual rate of 9 percent, its growth compares favorably with that of high-tech industries. A new record of $10 billion of gaming revenues was set in 1992.[20] In the wider realm of commercial gaming, only state lotteries were ahead, with a share of 40 percent of all gaming revenues versus

35 percent for casino gaming.[21] However, the "true" dimensions of casino gaming emerge when the magnitudes of total gross wagering are considered; that is, casino gaming accounted for a staggering 77 percent of the $330 billion in gross wagers for all 1992 commercial gaming. This was eleven times higher than the total wagers on lotteries.[22]

By geographic location and excluding gaming on Indian lands, casinos were in operation in seven states in 1992[23] (see Table 9-1). Not surprisingly, the two centers of gravity are Nevada and New Jersey (Atlantic City), which took more than 93 percent of all U.S. gaming revenues. The scale of casino operations ranges from the megacasinos in Atlantic City at one end to the minicasinos in South Dakota and Colorado at the other, with the casinos in Nevada and in riverboat jurisdictions falling in between. Nevada is a multicasino market, a blend of casinos of all sizes. This makes for a low average casino size of only 11 percent of the Atlantic City average. Even the more comparable "peer group" of the fourteen large casinos on the Las Vegas Strip, which posted gaming revenues of $141.7 million on average in 1992, is only about one-half of the size of the Atlantic City average, and the Strip average is less than the gaming revenues of the Claridge, the smallest casino in Atlantic City.[24] Furthermore, in 1992 the volume of gaming activity per square foot of casino was more than three times the amount in Nevada, that is, $3,860 versus $1,250.[25]

The dynamics of gaming activity in the two premier markets during the past two decades is characterized by a volatile development (see Table 9-2). With a monopoly, gaming revenues in Nevada doubled from 1970 to 1975, and they continued to double at an even higher rate of growth during the second part of the 1970s, that is, until Nevada's monopoly gave way to the du-

TABLE 9-1 Gaming Revenues in Various Jurisdictions, 1992[a]

Jurisdiction	No. of Casinos/Riverboats	Gaming Revenues ($ Millions)	
		Total	Per Casino/Riverboat
Nevada	192[c,d]	5,584.6[c,e]	29.1
New Jersey	12	3,216.0	268.0
South Dakota[b]	80	40.3	0.5
Colorado[b]	70	180.1	2.6
Iowa[b]	5	69.5	13.9
Illinois	5	226.3	45.3
Mississippi	5	122.0	24.4

[a] Sequence in the order of commencement of Casino gaming.
[b] Limited-stake casinos/riverboats ($5 maximum bet).
[c] Nonrestricted operations with gaming revenues in excess of $1 million (FY 1992).
[d] Of which 56 in Las Vegas (Strip area and downtown).
[e] Of which $3,177.5 million (57 percent) in Las Vegas (Strip area and downtown).

Source: Ernst & Young; *Nevada Gaming Abstract 1992.*

TABLE 9-2 Gaming Revenues in Nevada and Atlantic City, 1970–1992[a]

Year	Total	Nevada	Percentage of Total	Atlantic City	Percentage of Total
1970	575	575	100	–	–
1975	1,126	1,126	100	–	–
1980	3,025	2,382	79	643	21
1985	5,453	3,314	61	2,139	39
1990	8,190	5,238	64	2,952	36
1992	9,080	5,864[b]	65	3,216	35

[a] Gaming revenues (casino revenues) is the net difference between gaming wins and losses.
[b] Nonrestricted operations with gaming revenues in excess of $1 million.

Source: Standard & Poor's Industry Surveys—Leisure Time; Ernst & Young.

opolistic market configuration. The rapid expansion of Atlantic City's casino industry to three casinos by the end of 1979 and to six casinos by the end of 1980 boosted its share of national gaming revenues to more than one-fifth. This explosive growth, coupled with the strong growth in Nevada, lifted the average annual growth of national gaming revenue to a peak of 21.8 percent, from which it has declined ever since. The next decade can best be described as a seesaw between the two markets. During the early 1980s, Atlantic City set the pace and still expanded, whereas the Nevada markets slowed down considerably and posted only a nominal growth rate at a dismal one-fourth of Atlantic City's rate. This led Atlantic City to a two-fifths share of national gaming revenue by the mid-1980s. Thereafter, the momentum switched back to Nevada, which reclaimed a share of 65 percent in 1992, up from a low of 60 percent in 1986.

Perhaps the most important catalyst for the rapid expansion of the gaming industry in the 1980s was the ever increasing popularity of slot and video machines. In Atlantic City, the number of machines almost doubled between 1982 and 1992; even more important, the gaming revenue per machine rose by 60 percent, indicating both increasing patronage and play frequency.[26] The gaming machine industry has benefited from the brisk demand for its products: Total sales rose from $250 million in 1989 to an estimated $400 million in 1992.[27]

II. STRUCTURE

CONCENTRATION

The structural setting of the gaming industry ranges from polypolistic markets to oligopolistic markets (see Table 9-1). The former configuration applies particularly to the South Dakota and Colorado jurisdictions where small-scale gaming is mandated by limiting the size of bets. In fact, the combined 1992 gaming revenues of the 150 minicasinos in both states are surpassed by those of 10 of the 12 casinos in Atlantic City. Likewise, the Nevada market resembles a many-firm market, but on a grander scale because of the statewide spread. It would appear that of all casinos in Nevada, only about 2 or 3 on the Las Vegas Strip exceeded $200 million in gaming revenues in 1992.[28] Consequently, statewide concentration levels in Nevada are low: Estimates of the 1992 concentration of gaming revenues on a company basis showed a top-firm ratio (CR1) of 9 percent (Mirage Resorts), a four-firm ratio (CR4) of 32 percent (Mirage Resorts, Circus Circus, Hilton, Promus), and a minimum estimate of the Hirschman–Herfindahl index (HHI) of 350.

In stark contrast, the $200 million threshold has been the rule in Atlantic City: It had already been exceeded by Resorts International in 1979; it became the industry average for the first time in 1986; and only two casinos were left with a lower gaming revenue in 1992.[29] The principal reason for the large size of the Atlantic City casinos is the minimum-size requirements set forth in the Casino Control Act. This framework sets the stage for high-volume casino play in order to maintain a viable operation. The number of casinos in Atlantic City rose rapidly from one in 1978 to nine in 1981 to the present twelve in 1987.[30] The transition from a one-firm market to a few-firm market caused the concentration of gaming revenues to fall to a low in 1985, from which it has subsequently risen, owing mainly to mergers and acquisitions (see Table 9-3).[31] The largest firm, as indicated by CR1, changed five times between 1978

TABLE 9-3 Concentration in the Atlantic City Gaming Industry, by Gaming Revenues, 1978–1992

Year/Measure	CR1[a]	CR[b]	HHI[c]
1978	100.0 (Resorts)	–	10,000
1979	71.6 ″	–	5,911
1980	33.2 ″	98.0	2,958
1981	17.4 (Caesars)	62.1	1,421
1982	14.4 (Resorts)	53.0	1,165
1983	14.1 (Gold. Nugget)	54.0	1,174
1984	13.2 (Resorts)	49.6	1,063
1985	14.9 (Trump Org.)	48.4	1,063
1986	19.4 ″	52.1	1,146
1987	29.0 ″	70.1	1,656
1988	19.9 ″	65.2	1,252
1989	20.3 ″	65.8	1,279
1990	27.5 ″	68.5	1,484
1991	27.1 ″	68.2	1,464
1992	28.4 ″	69.1	1,509

[a] Leading-firm ratio; identity of leading firm in parentheses.
[b] Four-firm ratio.
[c] Hirschman–Herfindahl index.

Source: Calculated from data published by the New Jersey Casino Control Commission.

and 1985. Using the gaming operations in 1981 as a benchmark, in the sense of the first year of a full complement of the present casino structure in Atlantic City, CR1 shows a substantial increase of 11 points, to the 1992 level of 28 percent held by the Trump Organization which controls the Taj Mahal, the Plaza, and the Castle. The four-firm ratio stood at 62 percent in 1981, from which it declined to 48 percent by the mid-1980s to rebound to between 65 and 70 percent in the late 1980s and early 1990s. This level can be classified as a fairly high concentration of gaming revenues. The Hirschman–Herfindahl index, in its capacity as a summary measure of concentration, provides a better structural perspective of the market. In the present context, HHI indicates levels in the middle to upper range of the moderate concentration category of the Horizontal Merger Guidelines of the Department of Justice and the Federal Trade Commission. In addition, the numbers equivalent of HHI (n-HHI) points to a few-firm market with the numbers of equal-sized firms for a given level of HHI ranging from 6.0 to 9.4 for the past 12 years.[32] It is interesting to note the transition from a structure of "equals" during the 1982–1984 period of expansion to a structure of "not so equals," if not "unequals," during the 1990–1992 period of restructuring, adjustment, and consolidation: In the former period, the gap between n, the actual number of firms, and the n-HHI was a scant 4 to 6 percent, but this jumped to between 43 and 45 percent in the latter period.

Other few-firm markets exist in the emerging riverboat-gaming markets in Iowa, Illinois, and Mississippi (see Table 9-4). The very high levels of concentration are not surprising in view of the early stage of gaming in these jurisdictions; in fact, they are reminiscent of the conditions in Atlantic City in the late 1970s.

Moving beyond jurisdictional boundaries to nationwide concentration, Table 9-5 lists the ten leading casino operators. All of them are multicasino operators, ranging from a low of two casinos (Showboat) to a high of seven

TABLE 9-4 Concentration in the Riverboat Gaming Industry, by Gaming Revenues, 1992

Jurisdiction\ Measure	CR_1^a 1992	CR_4^b 1992	HHI^c 1992
Iowa	50.2[d]	92.8	3,301
Illinois	34.8[e]	92.7	2,489
Mississippi	26.4[f]	85.4	2,083

[a] Leading firm ratio.
[b] Four-firm ratio.
[c] Hirschman–Herfindahl index.
[d] President.
[e] Joliet Empress.
[f] Two riverboats.

Source: Calculated from data from Ernst & Young.

TABLE 9-5 The Ten Leading Casino Operators, by 1992 Gaming Revenue, 1989 and 1992[a]

	Gaming Revenues ($Million)		Percentage of Total	
	1989	1992	1989	1992
Trump Organization	571	922	7.7	10.1
Caesars World	685	730	9.2	8.0
Promus Cos. (Harrah's)	679	712	9.1	7.8
Mirage Resorts	210	534	2.8	5.9
Circus Circus Ent.	329	495	4.4	5.4
Bally Mfg.	674	478	9.0	5.3
Hilton Hotels	311	439	4.2	4.9
Aztar[b]	406	432	5.4	4.8
Showboat	296	313	4.0	3.4
Boyd Group	252	291	3.4	3.2
All casinos[c]	7,454	9,080	100.0	100.0

[a] FYs of companies.
[b] TropWorld (Atlantic City), Tropicana (Las Vegas), Ramada Express (Laughlin).
[c] Nevada and New Jersey only.

Source: Standard & Poor's Industry Surveys–Leisure Time; Casino Journal; New Jersey Casino Control Commission.

TABLE 9-6 Concentration in the Casino Gaming Industry, by Gaming Revenue, 1989 and 1992

Year/Measure	CR_1[a]	CR_4[b]	CR_{10}[c]	HHI[d]
1989	9.2	35.0	59.2	409
1992	10.1	31.8	58.8	388

[a] Leading-firm ratio.
[b] Four-firm ratio.
[c] Ten-firm ratio.
[d] Hirschman–Herfindahl index.

Source: Calculations from data in Table 9-5.

casinos (Circus Circus) at the end of 1992. Multiple operations may be confined to one jurisdiction (Nevada: Circus Circus, Hilton, Mirage, and Boyd; New Jersey: Trump) or be extended to the two major markets (Aztar, Caesars, Bally, Promus, Showboat). The low concentration levels shown in Table 9-6 point to a very competitive market from a structural perspective. But be aware that the national concentration data for the gaming industry can be misleading, as there are significant differences among gaming markets that make the industry regional, if not local, in character.

MERGERS

The 1980s were a decade of mergers and acquisitions, and the gaming industry had its share of this activity. According to Table 9-7, a total of $3.3 billion was paid for the acquisition of casino-hotel assets in the seventeen larger deals since 1980. About half the volume of merger activity involved the Atlantic City market, 32 percent Las Vegas, and 19 percent Reno/Lake Tahoe.

Holiday Inns' 1980 acquisition of Harrah's casinos in Nevada from the estate of the late William Harrah set the pace for subsequent events. After a brief lull, merger activity exploded in the three-year period from 1985 to 1988 when Donald Trump and Bally went on a buying spree. In a series of strikes, Trump bought the almost finished Atlantic City Hilton when Hilton Hotels was denied a casino license, bought Resorts from the estate of the late Jim Crosby, and bought the partially constructed Taj Mahal from Merv Griffin.[33] These acquisitions made the Trump Organization the nation's leading casino operator since 1990.[34] However, the highly leveraged external growth pushed Trump almost to the brink of losing control of his casinos.

Likewise, Bally made two spectacular acquisitions in 1986 and 1987 that brought great trouble. In the Golden Nugget deal, both the price and the timing were wrong: The price was the highest ever paid for a casino hotel, and the purchase was made at a time when the growth of gaming revenues in Atlantic City had begun to slow down. In fact, gaming revenues at the renamed Bally's Grand fell for the next five years at an annual rate of 3.6 percent, when the industry still posted an average *increase* of 5.2 percent, a dismal return for a $440 million price tag. The MGM Grand in Reno did not fare much better under Bally's aegis, and it was sold to Hilton for a bargain price of $88 million, although Bally had paid approximately $230 million for the property five years earlier.[35]

All of the major mergers in Table 9-7 are horizontal mergers in which market shares add up, statistically speaking. In the competitive world of casino gaming, however, it may not be so easy to project gains in the post-merger-market share on the basis of pre-merger-market constellations, especially when a local market in the narrow confines of the Atlantic City casino industry is concerned. Here, acquisition of a competitor does not mean the end of competition for gaming patrons in the newly enlarged firm. This shift from interfirm competition to intrafirm competition can lead to divergent developments in such a multicasino operation. For example, Bally's Grand saw its

TABLE 9-7 Important Mergers and Acquisitions in the Gaming Industry, 1980–1993

Year	Acquiring Company	Acquired Company/Assets	Consideration Paid $ mill.
1980	Holiday Inns	Harrah's (Reno and Lake Tahoe)	300
1984	Elsinore	Playboy Casino and Hotel, Atlantic City[a]	51
1985	Trump Org.	Atlantic City Hilton[b]	320
1986	Trump Org.	Harrah's at Trump Plaza[c]	223
1986	Bally Mfg.	MGM Grand Hotels	573
1987	Trump Org.	Resorts Hotel and Casino, Atlantic City (73% of voting stock)	79
1987	Bally Mfg.	Golden Nugget, Atlantic City	439[d]
1988	MGM Grand[e]	Desert Inn Hotel and Casino, Sands Hotel and Casino, Las Vegas	167
1988	Griffin Co.	Resorts[f]	159
1988	Trump Org.	Taj Mahal Hotel and Casino, Atlantic City[g]	273
1989	Interface Group[h]	Sands Hotel and Casino, Las Vegas	110
1989	Trump Org.	Atlantis Hotel and Casino, Atlantic City[i]	63
1990	MGM Grand	Marina Hotel and Casino, Las Vegas	80
1991	Tracinda Corp.[h]	Desert Inn Hotel and Casino, Las Vegas	130
1992	Hilton Hotels	Bally's Casino Resort, Reno[j]	88
1993	Mirage Resorts	Dunes Hotel Casino and Country Club, Las Vegas	70
1993	ITT Sheraton	Desert Inn Hotel and Casino, Las Vegas	160

[a] Renamed Atlantis Hotel and Casino.
[b] Renamed Trump's Castle Hotel and Casino.
[c] Donald Trump buys the outstanding 50 percent from Holiday Inns and renames the property Trump Plaza Hotel and Casino.
[d] $140 million in cash plus assumption of $299 million of Nugget's debt.
[e] Incorporated in Delaware in 1986.
[f] From the Trump Organization.
[g] Merv Griffin sells the yet unfinished hotel/casino to Donald Trump.
[h] Investment holding.
[i] Donald Trump buys the former casino (no license) and operates it as a hotel named Trump Regency.
[j] Renamed Reno Hilton Resort.

Sources: Moody's Industrial Manual; Casino Chronicle; company reports; D. Johnston, Temples of Chance (New York: Doubleday, 1992).

pre-merger-market share of 11 percent in 1986 plummet to 6.2 percent in 1992; at the same time, Bally's Park Place stayed on course, increased its gaming revenues, and overtook its twin as the larger operation. This means that mergers in the casino industry are not necessarily a prescription for growth, especially when they involve combinations in the same local market.

IS BIGNESS BETTER?

With the advent of megacasinos—with a casino floor space in excess of 100,000 square feet and a hotel capacity of 4,000 rooms and more—the question arises whether and to what extent large casinos are more efficient than smaller ones. In the manufacturing economy, scale economies that may be experienced at greater plant sizes can serve as a criterion of efficient production. This concept cannot be directly applied to the gaming industry, however, because of the absence of physical capacity and output units. As a substitute indicator, albeit an imperfect one, the relationship between gaming revenues and casino size in terms of casino floor space seems best suited to trace the evidence of efficiency.

Indications from the Atlantic City industry suggest that larger casinos are not more efficient than smaller ones, other things being equal. Ratios of gaming revenues and casino-floor square footage (relative gaming revenues) have shown volatility rather than stability, which somewhat reflects the dynamics of the market in the 1980s: There was an overall increase during the years of formation (1980 to 1984), a decline in the period of mergers (1984 to 1988), and, finally, a modest recovery in the era of consolidation and maturity (1988 to 1992). The highest relative gaming revenue in the market was recorded by the Golden Nugget Casino, with a monthly average of $529 in 1983; in fact, the Golden Nugget was the leader in all of the six years from 1982 to 1987.

The relation between casino size and relative gaming revenues in 1992 is presented in Table 9-8. As an example, the casino floor space of the Taj Mahal is twice as large as that of Caesars Atlantic City, but gaming revenue was only 25 percent higher in the former; consequently, Caesars' relative gaming revenue exceeded that of the Taj Mahal. Apparently there is no consistent correlation between casino square footage and relative gaming revenues. For the whole group of twelve casinos during the thirteen years from 1980 to 1992, correlation was negative in eight years and positive in five years. Thus there was no strong tendency in either direction. Moreover, the larger casinos in Atlantic City were not more efficient in terms of generating a higher relative gaming revenue through higher play frequency, spillover effects from the variety of games, or the like.

Searching for evidence on large size and efficiency in the gaming industry on the basis of relative gaming revenues is just one aspect of the all-inclusive casino-hotel operations, albeit perhaps the most important one. When it comes to securing *sustained growth* of gaming revenues in the long run, large size in the gaming industry may nevertheless pay off, especially in an era of

TABLE 9-8 Gaming Revenue and Operating Profit of Atlantic City Casinos, by Casino Size, 1992

	Casino Square Footage	Gaming Revenue ($mill.)	Gaming Revenues Per Sq. Ft. ($)	Operating Profit Per Sq. Ft. ($)
Trump Taj Mahal	120,000	416.1	3,469	929
TropWorld	90,774	310.2	3,417	906
Bally's Park Place	64,435	280.5	4,353	1,443
Trump's Castle	62,595	240.4	3,841	599
Harrah's Atlantic City	61,183	287.5	4,699	1,287
Caesars Atlantic City	60,000	332.5	5,542	1,559
Resorts	60,000	235.5	3,925	751
Trump Plaza	60,000	264.2	4,403	1,005
Showboat	59,858	257.7	4,305	980
Sands	49,789	245.2	4,925	923
Bally's Grand	45,442	199.8	4,397	1,173
Claridge	43,579	146.4	3,359	741

Source: 1992 Annual Report of the New Jersey Casino Control Commission; Casino Chronicle.

steady proliferation of casino gaming in the 1990s and beyond. Among other things, this makes large-scale strategic marketing (including generous promotional allowances and buying power for needed inputs) the necessary instrument in the ever rising competition to attract and keep gaming patrons. Only large-scale operations have the resources to do that.

BARRIERS TO ENTRY

The most important entry barrier to casino gaming is the requirement to obtain a license from the Gaming Control Board or Commission. Beyond this artificial barrier, there are natural barriers to entry. First and foremost is the high cost of real estate in prime gaming locations like the Boardwalk in Atlantic City and the Strip in Las Vegas. On the Strip, the cost per acre in recent land acquisitions ranged from a low of $0.37 million for the 116-acre property bought by Circus Circus in 1987 for the Excalibur and Luxor casinos, to $1.05 million for the 112-acre property bought by MGM Grand in 1990 to build the world's largest casino-hotel resort.[36] On the Boardwalk, these prices would be considered bargains[37] because land prices were driven to extreme heights after the referendum to legalize casino gaming in 1976.[38] This is one of the reasons for the undue cost pressure on the Atlantic City casinos vis-à-vis their counterparts in Las Vegas.

In addition to the cost of land acquisition, there are the construction costs of modern casino-hotel complexes which can be astronomical. In 1981 it cost more than $400 million to build Ramada's Tropicana in Atlantic City, at that time the most expensive hotel construction in the world.[39] Today, we have Circus Circus's Luxor ($390 million), Mirage Resorts' Treasure Island ($475 million), and Kirk Kerkorian's MGM Grand ($1.03 billion).[40] If these amounts in excess of $100 million are regarded as a benchmark, then the natural barriers to entry to the world of gaming may be high indeed.

Gaming Machines: A Dominant-Firm Industry

There are about thirty-seven manufacturers of slot and video machines and VLTs in the United States, although most of them are small, privately owned operations.[41] More than 90 percent of the slot and video machines come from just four manufacturers: International Game Technology (Reno/Las Vegas), Bally Gaming (Las Vegas), Universal Distributing (Las Vegas), and Sigma Game (Las Vegas). The two latter companies are subsidiaries of Universal and Sigma from Japan. IGT, Bally Gaming, and Universal also manufacture VLTs, but in this special market segment Video Lottery Technologies (Bozeman) is in the lead. Another important supplier of VLTs is WMS Industries, which entered this market only recently.

To describe the gaming machine industry as highly concentrated with a CR1 of 51 percent (IGT), a CR4 of 93 percent, and a minimum estimate of the HHI of 3,347 in terms of installed machines as of 1991 is almost an understatement.[42] In specific markets, IGT's shares of installed slot and video machines are staggering: approximately 75 percent in Nevada, 50 percent in Atlantic City,[43] and more than 90 percent on riverboats.[44] What is more, IGT boasts a presence in all of the world's major gaming markets, particularly in Australia, western Europe, and, most recently, Japan.

The irony is that although IGT is today's dominant firm, it was only a distant second just seven years ago, at which time the dominant firm was Bally. With a share of 57 percent of the 1986 market in terms of installed machines, it was 30 points ahead of IGT with 27 percent; Universal and Sigma held 3 percent and 1 percent, respectively.[45] Bally had built up a reputation of experience and quality ever since its production of coin-operated amusement and slot machines began in Chicago in 1931. In the 1960s, the company formed its Bally Gaming Division and became the leading slot machine supplier to the Nevada market. Its dominance was also parlayed to the Atlantic City market, which fact persuaded the New Jersey Casino Control Commission to enact its 50 percent rule to curb an extension of Bally's dominance.[46] Subsequently, when the gaming activity accelerated, Bally's efforts shifted into low gear by either misreading the signals or paying too much attention to its casino and fitness divisions, or a combination of both. In contrast, IGT—only a $57 million company in 1985—expanded rapidly and eventually left its $1.3 billion competitor (1985) behind. In fact, IGT, the brainchild of the legendary William Redd, made the right decisions at the right time; most important, Redd had entered into a seven-year agreement with Bally in 1976 by which the partners would respect each other's territory, that is, Bally reel-type slot machines and Redd's A1 Supply Co. (the predecessor of IGT) in video-gaming machines.[47] This gave IGT an insurmountable lead in the fastest-growing segment of the industry.

No wonder that its output of gaming machines rose at an average annual rate of 35 percent during the past ten years when a total of 235,000 machines were manufactured by the company. Meanwhile, Bally took drastic measures to regain lost territory, which included the relocation of manufacturing capacities from Chicago closer to the action in Las Vegas and the spin-off of its gaming-machine division in a public offering as a separate company in 1991/1992, renamed Bally Gaming International (BGII).[48] The indications are that BGII is on a comeback trail, since the production of machines in Las Vegas more than tripled between 1990 and 1992 to reach 14,000 in 1992, which means that there is still space in the shadow of the dominant firm: space not only for BGII but also for the numerous small firms that have found lucrative niche markets, and space also for the leading foreign competitors which include the aforementioned Universal and Sigma from Japan, Aristocrat Leisure Industries from Australia, and Betstar, a joint venture of H. Betti Industries from New Jersey and Stella International from Germany. Despite the inherent dynamics of the gaming industry, the past, present, and future structure of the gaming machine industry is that of a dominant-firm industry: In the early years Mills was at the helm, then Bally, and now—and for the foreseeable future—IGT.

III. CONDUCT

PROMOTING THE PRODUCT WITH "COMPLIMENTARIES"

Other things being equal, gaming machines do not differ from casino to casino. Likewise, table games and their rules are, for the most part, the same, at least in the same location. Consequently, because American casino operators are exposed to a fiercely competitive environment, they must be world champions in the art of product differentiation, and they are indeed: Products that look so much alike are made totally different in the perception of gaming patrons.

This transformation is performed with the help of a multitude of promotional allowances and expenses designed both to attract players and to earn their loyalty. Promotional allowances or complimentary revenues[49] are unique to the gaming industry and include the provision of on-premise accommodation, food and beverages, parking, and show tickets free of charge to qualifying patrons. Qualification is based on level of play and play frequency at slot and video machines and tables in the casino as recorded on the magnetic stripe of a membership card of the casino's slot club.[50] The magnitudes of these promotional allowances are enormous: Caesars World and Aztar spent the most in relative terms, with complimentaries accounting for almost 10 percent of their total costs. On an estimated retail-value basis, these allowances exceeded $100 million annually for the past five years at Caesars.[51] Of equal importance are promotional expenses in terms of direct cash payments to qualifying players for coupon redemptions and payments on their behalf for off-premise services like travel to and from the casino. In the Atlantic City market, casinos spent a total of $799 million for promotional allowances and expenses in 1992, which represents 25 percent of their gaming revenues.[52]

DIFFERENT MARKETS—DIFFERENT STRATEGIES

Casino operators in the United States prefer to gain competitive advantage through innovative efforts to differentiate their product in "what is largely a look-alike industry."[53] The catalog of strategies and varieties of product differentiation is virtually endless, extending from building a theme-oriented casino-hotel resort complex to providing a full range of health, sports, and leisure facilities, to catering to specific target groups of players. This last strategy deserves particular attention because its implementation determines a host of other peripheral strategies.

Some casinos are clearly designated to serve the market of the upscale clientele, but the majority devote most of their attention to the so-called middle market.[54] The former strategic orientation comes at a hefty price tag of a four-star/diamond hotel, gourmet restaurants, elegant-to-luxurious casino, and leisure-time facilities. Perhaps only a handful of casinos are in this category, most prominently Caesars Palace and The Mirage on the Las Vegas Strip. Catering to the middle market requires no four-star property, but it does require additional strategic measures of differentiation because of the multitude of competitors in this segment. Great ingenuity is used to achieve such favorable perception, but competitive advantage can be gained with a single slogan such as Harrah's "Where the Better People Play," which has become an industry standard for success. Likewise, the notion that a casino can be reached by "land, air, and sea" can make a difference (Trump's Castle).

Beyond customer markets, competitive strategies differ with the geographic market and, specifically, between the Atlantic City and the Las Vegas markets. Because of its location far away from population centers, Las Vegas has never been a day-trip destination. Rather, it became a resort area early on, where casinos built nongaming facilities ranging from golf courses and RV parks to attractions such as monorails (Circus Circus, Bally's), an erupting volcano (The Mirage), and much more in order to make a stay for players *and* their families more enjoyable and longer lasting. This family-oriented approach was pioneered by Circus Circus, and it got a boost with the most recent development of casino-hotel resorts designed as theme parks (Treasure Island, Luxor, MGM Grand).[55]

Nothing of this kind exists in Atlantic City. First and foremost, space is very limited, specifically in the vicinity of the Boardwalk, and even if it were available, it would come at an almost prohibitive premium. In a cost–benefit analysis, one may also wonder whether there is a need for theme parks in Atlantic City. Unlike Las Vegas, the majority of visitors are "day-trippers" who come by chartered buses and cars.[56] Longer stays are usually confined to weekends and holidays. Most of the bus visitors are in the middle to low end of the market, which nevertheless can be lucrative because of the volume of play frequency. These differences in clientele require totally different strategies to create an identity and "product recognition," including refunds of bus fares, cash coupons, and provision of nickel and dime slot machines in addition to the regular quarter and dollar machines.

IV. PERFORMANCE

PROFITABILITY AND COMPETITIVENESS

Contrary to common belief, a casino license is not a license to print money. Gaming revenue is "produced" at a hefty price, which is ever rising in an era of proliferation of gaming destinations. Fierce competition for the players' attention and loyalty requires disproportionately more spending on complimentaries, which puts an additional burden on casino coffers. Between 1987 and 1992, expenditures on complimentaries increased by 35 percent in Atlantic City and by 60 percent in Nevada; at the same time, gaming revenues rose by only 29 percent and 52 percent, respectively.[57]

In an overall comparison, casinos in Nevada appear to be more profitable than their counterparts in Atlantic City, despite the much greater volume of gaming revenue per casino in the latter market. Costs are greater in Atlantic City mainly because of higher interest charges, higher labor costs, and a higher burden of sometimes costly regulatory requirements.[58] These higher costs take their toll on the bottom line for casinos in Atlantic City. For example, in 1981 and 1990, losses outweighed profits in the industry, and in 1992, pretax income as a percentage of net revenues stood at a scant 2.8 percent vis-à-vis 9.2 percent in Nevada.[59]

Despite this dismal record of overall profitability, there were exceptions to the rule. Harrah's Atlantic City chalked up an average annual net income of about $50 million during its thirteen years of casino operations, and Caesars Atlantic City posted an annual average of more than $16 million during its fourteen years of operations.[60] Table 9-9 presents a synoptic overview of key indicators between 1988 and 1992.[61]

CONCLUSION

The only sure bet on this industry's future is that gaming revenues will continue to grow, perhaps at even higher rates than before. Riverboat gaming is expanding rapidly with boats in Louisiana, Indiana, and Missouri joining the existing "flotilla" in 1993/1994. High gaming revenues have attracted established operators such as Harrah's, Mirage, Bally, Fitzgeralds, and Circus Circus, which have applied for licenses or have already joined the ranks of newcomers like Argosy, President Riverboats, and others.

Whither Native American gaming? At last count, there were Indian gaming sites in twenty-three states.[62] Although many of them are only small-scale

TABLE 9-9 Indicators of Competitiveness of the Leading Casino Operators, 1988–1992[a]

	Growth Rates		Operating Income	Net Income as a Percentage of		Labor Prod.[c]
	Total Revenue[b]	Casino Revenue	($mill.)	Total Revenue[b]	Stockholder's Equity	($'000)
Aztar	2.9	4.8	3.8	3.3	5.0	66.8[d]
Bally Mfg.	1.3	-8.1	107.1	-3.3	-16.5	42.4
Caesars World	2.7	3.5	138.5	6.7	16.8	89.1[e]
Circus Circus	13.3	11.5	182.1	13.6	40.0	57.3
Hilton Hotels	6.6	8.2	211.5	10.1	12.0	27.5
Mirage Resorts	47.7	47.4	106.4	0.2	3.0	76.7[f]
Promus Cos.	6.5	2.5	202.2	8.2	10.7[e]	43.5[g]
Showboat	4.7	6.5	32.2	1.6	7.1	70.8[d]

[a] Five-year averages.
[b] Excluding complimentaries.
[c] Total revenue per employee.
[d] 1991.
[e] 1990–1992.
[f] 1990 and 1992.
[g] 1989 and 1991.

Source: Company reports; Moody's Industrial Manual.

bingo operations, there also are a number of high-stakes casinos, including Foxwoods.[63] Their rapid expansion has triggered a debate between those critics who want to restore the states' rights in negotiations with Native American tribes and those who want to challenge Indian casino gaming altogether.[64] Because the management of Indian casinos is usually in the hands of non-Indian companies and because Indian gaming is outside the state regulatory control mechanism, there is also the potential for the indirect involvement of organized crime.[65]

Faced with mounting competition in domestic markets, American casino operators have begun to extend their reach to foreign jurisdictions. Heading the list are Australia and Canada, with Hilton and ITT Sheraton already operating casinos in Australia.[66] In Canada, the tense bidding process for the single casino in Windsor, Ontario, saw the consortium of Caesars World, Circus Circus, and Hilton as the successful finalist and future operator. This novel approach of joining forces in a strategic partnership may become the role model for the future when operations of single casinos in one location are at stake. In the domestic market, this has already emerged in connection with the single land-based casino in New Orleans.

Going global and forging joint ventures opens new dimensions for American casino operators, and it is yet another indicator of the spirited entrepreneurship that characterizes the firms in this industry.

NOTES

1. E. M. Christiansen, "Revenues Soar to $30 Billion," *Gaming & Wagering Business*, August 15, 1993, pp. 12–35.
2. Compare J. Rosecrance, *Gambling Without Guilt* (Pacific Grove, CA: Brooks/Cole, 1988), pp. 48–51.
3. See D. Johnston, *Temples of Chance* (New York: Doubleday, 1992), p. 57.
4. These machines are part of the realm of state lotteries and are usually found in lounges and taverns.
5. C. McQuaid, ed., *Gambler's Digest*, 2nd ed. (Northfield, MN: DBI Books, 1981), p. 6.
6. Ibid., p. 7.
7. The following account is based on Nevada Gaming Commission and State Gaming Control Board, *Gaming Nevada Style* (Carson City: Nevada Gaming Commission and State Gaming Control Board, 1989).
8. E. M. Greenless, *Casino Marketing and Financial Management* (Las Vegas: University of Nevada Press, 1988), p. 6.
9. W. R. Eadington, "The Casino Gaming Industry: A Study of Political Economy," *Annals of the Academy of Political and Social Science*, July 1984, p. 27.
10. The requirements include a hotel capacity of at least five hundred first-class rooms, minimum square footage of casino floor space, conference

and meeting room facilities, exercise and sports facilities, and nightly live entertainment.

11. For further details, see W. R. Eadington, "Recent National Trends in the Casino Gaming Industry and Their Implications for the Economy of Nevada, Reno," Institute for the Study of Gambling and Commercial Gaming, August 1992 (mimeo); and the issue of *Gaming & Wagering Business* dated March 15, 1992.

12. Another 35 percent of the respondents found casino gaming acceptable for others but not for themselves. See *The Harrah's Survey of U.S. Casino Gaming Entertainment* (Memphis, TN: Harrah's, 1993).

13. *Gaming Nevada Style*, p. 6; *1979 Annual Report of the New Jersey Casino Control Commission*, p. 8. In spite of these precautions, however, there have been instances, albeit very few, in which illegal elements were able to infiltrate the system. See Eadington, "Casino Gaming Industry," p. 30; and Johnston, *Temples of Chance*.

14. B. Ransom, "Public Policy and Gambling in New Jersey," in *Gambling and Public Policy: International Perspectives*, ed. W. R. Eadington and J. A. Cornelius (Reno: Institute for the Study of Gambling and Commercial Gaming, 1991), p. 158.

15. R. Lehne, *Casino Policy* (New Brunswick, NJ: Rutgers University Press, 1986), p. 119.

16. This is in stark contrast with the 1974 gaming bill, according to which casino gaming would be made a county option. See R. Gros, "Casino Gaming in Atlantic City—15 Years of Action," *Casino Player*, Special Souvenir Edition, Summer 1993, p. 9.

17. No more than 50 percent of the gaming machines in a casino can come from one and the same manufacturer.

18. This restriction was lifted in 1992.

19. In addition, there is also an investment alternative obligation of 1.25 percent or an investment alternative tax of 2.5 percent of gross gaming revenues. However, this tax can be offset by investment tax credits that may be obtained by purchasing bonds for housing or other development projects approved by the New Jersey Casino Reinvestment Development Authority.

20. Gaming revenue as the reference measure in the gaming industry is the casino revenue from the difference between gaming wins and losses.

21. Christiansen, "Revenues Soar to $30 Billion."

22. E. M. Christiansen, "Industry Rebounds with 8.4 Percent Handle Gain," *Gaming & Wagering Business*, July 15, 1993, pp. 12–35.

23. Except for the Foxwoods Casino in Connecticut, data on Indian gaming are not available.

24. In contrast with its New Jersey counterpart, the Nevada Gaming Control Board does not publish data on individual casinos.

25. *Standard and Poor's Industry Surveys—Leisure Time*, March 12, 1992, p. L40.

26. See *Annual Reports of the New Jersey Casino Control Commission*.

27. *New York Times*, September 16, 1990, p. F4; Company Reports.

28. *Standard and Poor's Industry Surveys—Leisure Time*, March 11, 1993, p. L37.

29. Bally's Grand missed the mark by a whisker, with $199.8 million.

30. The Atlantis Casino lost its license in 1988, but one year later, the Taj Mahal opened its doors, which brought the number back to twelve.

31. Concentration is measured in terms of casinos under common ownership (firms).

32. The numbers equivalent of HHI is the reciprocal of the index. This is a helpful instrument for the morphologic determination of market structures, inasmuch as the actual number of firms can be compared with the number of equal-sized firms corresponding to a given level of concentration.

33. The last deal was only struck after Resorts was sold to Griffin in order to avoid a potential conflict with the guidelines of the Casino Control Act, which do not permit undue concentration in the hands of a single licensee (*1987 Annual Report of the New Jersey Casino Control Commission*, p. 3).

34. The Taj Mahal opened its doors in April 1990.

35. *1992 Hilton Hotels Corporation Annual Report*, p. 4.

36. *Mirage Resorts 1992 Annual Report*, p. 42.

37. Quite apart from the fact that such huge parcels of land were never available.

38. Compare J. Rubenstein, "Casino Gambling in Atlantic City: Issues of Development and Redevelopment," *Annals of the American Academy of Political and Social Science*, July 1984, pp. 61–71.

39. Johnston, *Temples of Chance*, p. 68.

40. *Business Week*, October 18, 1993, p. 80.

41. According to the "International Supplier Directory" and the "Slot and Video Gaming Guide" of *Gaming & Wagering Business*.

42. Market-share data are from New Jersey Casino Control Commission, *In the Matter of: 50 percent Rule Hearing*, August 20, 1992, p. 107. The composition of the four-firm ratio was IGT (51 percent), Bally Gaming (21 percent), Universal (17 percent), and Sigma (4 percent). These data do not include VLTs.

43. This "low" share is due to the 50 percent rule set by the New Jersey Casino Control Commission.

44. G. Fine and A. Gros, "IGT's New World Order," *New Jersey Casino Journal*, September 1993, p. 18.

45. *New York Times*, September 16, 1990, p. F4.

46. Bally's own casinos were exempted from the rule.

47. E. Paris, "Call and Rise," *Forbes*, August 30, 1982, pp. 50–52; compare A. Fine, "The Legend Continues," *New Jersey Casino Journal*, January 1993, pp. 38–39.

48. The name is appropriate because BGII owns one of the largest manufacturers of gaming machines (wall machines) in Germany, Bally-Wulff.

49. The first term is used in the New Jersey jurisdiction, and the second one in Nevada.
50. Compare Johnston, *Temples of Chance*, pp. 57–58; F. Legato, "Crunching the Numbers," *New Jersey Casino Journal*, December 1992, pp. 36–37.
51. *Moody's Industrial Manual*.
52. *Casino Chronicle*, April 12, 1993. Among promotional allowances by category, the provision of complimentary food and beverages led with 57 percent, ahead of accommodation with 32 percent, whereas coupon redemptions accounted for 74 percent of promotional expenses. By casino operation, the ratio of total promotional costs and gaming revenue (relative promotional costs) ranged from a low of 20.5 percent for the Showboat to a high of 28.3 percent for the Castle.
53. *Standard and Poor's Industry Surveys—Leisure Time*, March 11, 1993, p. L36.
54. *Mirage Resorts 1992 Annual Report*, p. 28.
55. See *Standard and Poor's Industry Surveys—Leisure Time*, March 11, 1993, p. L36.
56. According to head counts by casinos, chartered buses bring up to 900,000 visitors per month and more, depending on the season. See *Atlantic City Action Newsletter*.
57. *Standard and Poor's Industry Surveys—Leisure Time*, March 12, 1992, p. L40.;*Casino Chronicle*, April 12, 1993; *1992 Nevada Gaming Abstract*.
58. *Standard and Poor's Industry Survey—Leisure Time*, March 11, 1993, p. L37.
59. *Casino Chronicle*, April 12, 1993; *1992 Nevada Gaming Abstract*.
60. The figure for Harrah's is an approximation because it does not fully reflect income tax provisions for the partnership. See *Casino Chronicle*, April 12, 1992.
61. Consolidated figures are not available for the Trump Organization and for the Boyd Group because of their status as private companies.
62. *Fortune*, November 1, 1993, p. 14.
63. With a casino floor space of 140,000 square feet, Foxwoods is bigger than any of the twelve casinos in Atlantic City, and its anticipated revenues of $500 million and operating profit rate of 45 percent will make it the most profitable operation nationwide. See *Business Week*, October 18, 1993, p. 82.
64. The former issue refers to the closing of loopholes in the 1988 Indian Gaming Regulatory Act, and an appropriate Senate bill was introduced in the spring of 1993. The latter issue was raised by Trump in a lawsuit filed in 1993. See *Boston Globe*, September 29, 1993, pp. 1, 24; and Eadington, "Recent National Trends in the Casino Gaming Industry." Recently, Trump reiterated his uneasy feelings regarding the Indian Gaming Regulatory Act, by observing, "Under the IGRA, the tribes lose and federal, state and city tax support suffers. The big winner is organized crime." See *USA Today*, January 6, 1994, p. 14A.

65. *U.S. News and World Report*, August 23, 1993, pp. 30–32; *Boston Globe*, September 29, 1993, p. 25.

66. Apart from the Conrad Istanbul Hotel-Casino (25 percent stake), Hilton operates the Hotel Conrad and Jupiters Casino (with Jupiters, Ltd., as a partner) in Gold Coast, Queensland, and ITT Sheraton runs the Sheraton Breakwater Casino Hotel in Townsville, Queensland. Hilton, Sheraton, and Harrah's were and are also involved in bids for other Australian casinos. For further details, see J. Beagle, "Casinos Play Monopoly," *Gaming & Wagering Business*, March 15, 1993, pp. 15–18.

SUGGESTED READINGS

Abt, V., J. F. Smith, and E. M. Christiansen. *The Business of Risk*. Lawrence: University of Kansas Press, 1983.

Eadington, W. R. "The Casino Gaming Industry: A Study in Political Economy." *Annals of the American Academy of Political and Social Science*, July 1984, pp. 23–35.

———. "Public Policy Considerations and Challenges on the Spread of Commercial Gaming." In *Gaming and Public Policy: International Perspectives*, edited by W. R. Eadington and J. A. Cornelius. Reno: Institute for the Study of Gambling and Commercial Gaming, 1991, pp. 3–12.

Greenless, E. M. *Casino Accounting and Financial Management*. Las Vegas: University of Nevada Press, 1988.

Lehne, R. *Casino Policy*. New Brunswick, NJ: Rutgers University Press, 1986.

Nevada Gaming Commission and State Gaming Control Board. *Gaming Nevada Style*. Carson City: Nevada Gaming Commission and State Gaming Control Board, 1989

Ransom, B. "Public Policy and Gambling in New Jersey." *Gaming and Public Policy: International Perspectives*, edited by W. R. Eadington and J. A. Cornelius. Reno: Institute for the Study of Gambling and Commercial Gaming, 1991, pp. 155–168.

Rosecrance, J. *Gambling Without Guilt*. Pacific Grove, CA: Brooks/Cole, 1988.

AIRLINES

William G. Shepherd and James W. Brock

Across the terrain of industrial America, the airline industry is one of the most hotly contested arenas of government regulatory policy. From 1938 to 1977, the field was dominated by a handful of carriers operating under all-encompassing regulatory oversight. Beginning in 1978, however, a dramatic reversal in public policy was launched: The industry was deregulated in order to subject it to what was hoped to be the more effective discipline of competitive market forces.

Subsequent developments have engendered a fierce controversy: Optimistic observers declare airline deregulation a success, with highly beneficial economic results. Others disagree, contending that deregulation has unleashed monopoly power and led to a deplorable degradation in the industry's performance.

Has airline deregulation been a success or a failure? Is the industry now primarily competitive, with isolated instances of monopoly? Or is it predominantly monopolistic, with only isolated pockets of competition? Is it an example of a "contestable" market to which relative firm size and industry concentration are irrelevant? Or does industry structure play a key role in shaping conduct and determining performance? What public policies should be implemented, and what lessons should be learned?

An examination of the industry's structure, behavior, and performance suggests some answers to these questions.

I. HISTORY

Airline history is colorful. The field matured from a barnstorming pack of hedge-hopping daredevils in the 1920s into a major commercial endeavor dominated by a half-dozen major firms in the 1930s. Along the way there was a considerable degree of chicanery and politicking, particularly involving government subsidies paid for carrying airmail.[1]

THE REGULATION ERA

By 1938 the largest carriers had recognized the advantages of being regulated in a way that would minimize competition while applying only light constraints. Regulation was legislated: The industry's structure was frozen; existing carriers were "grandfathered" into place; and the Civil Aeronautics Board (CAB) was established to promote the industry and to regulate its development by controlling entry, routes, mergers, and fares.

The next thirty years saw minimum CAB restraint on prices and maximum CAB restraint on competition. The preferred regulatory pattern was monopolistic: 90 percent of city-pair routes, with 60 percent of all passenger travel, were essentially monopolies. Virtually no entry was permitted by the CAB into the mainstream, scheduled part of the industry between 1938 and 1975, and consequently, the number of trunk-line carriers dwindled from sixteen to eleven. Under tight CAB controls, the movement of the airlines into others' routes was only grudgingly permitted, despite explosive growth and massive changes in air traffic patterns. By 1970 most routes had two or three airlines, but few routes had more than that; many routes featured only a single dominant carrier. The majors functioned essentially as a market-rigging cartel, agreeing on fares and then submitting them to the CAB for rubber-stamped approval and enforcement. Rather than applying ceilings to prices in order to protect consumers, the CAB seemed more concerned with placing floors under fares in order to minimize price competition.

Rivalry among the airlines was artificially channeled into nonprice areas, such as decor, personnel, meals, and frequency of flights. As a result of this exaggerated nonprice rivalry, operating costs were higher than otherwise would have been the case. The quality of service was high, but so too were prices, and so mainly only the well-to-do and business travelers on expense accounts could afford to fly. Indeed, by some estimates, regulation inflated prices and costs by as much as 50 percent.[2] The CAB had prevented the flexible, competitive price-cutting and route adjustments that would have expanded the market and brought it within the reach of the masses.

THE ADVENT OF DEREGULATION

The case for deregulating the industry began with a small corporal's guard of economists in the mid-1950s, escalated during the 1960s, and reached a critical mass by the 1970s.[3] Bolstered by an extensive body of economic studies criticizing the consequences of CAB regulation, the appeal of airline deregulation began to reach across the political spectrum: Free-market Republicans and consumer-oriented Democrats began to support deregulation and its promise to reduce airfares, improve consumer choice, enhance efficiency and productivity, and "get government out of the marketplace."

In 1978, the case for deregulation culminated in the legislative enactment of the Airline Deregulation Act, which provided for free entry by 1980 (subject to safety standards and certification) and free pricing by 1983. Following its passage, the incumbent carriers furiously adjusted their routes while a number of regional airlines (USAir, Piedmont, Delta) vigorously expanded to nationwide operations. Striking new "hub-and-spoke" patterns proliferated, with each airline routing most of its traffic through one or two hub airports. More than sixty new, maverick carriers entered the industry—including People Express, Southwest, New York Air, and World Airways—cutting fares by 20 to 40 percent and offering "few-frills" service. The slow descent of real airfares thereupon accelerated sharply (see Figure 10-1).

The age of airline deregulation had commenced, but the age of controversy was not far behind.

II. MARKET STRUCTURE

Contrary to what was anticipated, the most striking structural developments in the industry since its deregulation have been sharp increases in concentration, the advent of monopoly airport "hubs," and the erection of a number of daunting barriers to new competition.[4]

FIGURE 10-1   Airline revenue per passenger-mile adjusted for inflation (1991 dollars).

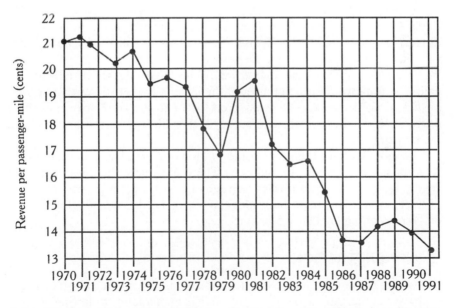

Source: Steven Morrison and Clifford Winston, *The Evolution of the Airline Industry* (Washington, DC: Brookings Institution, 1994).

DEFINING THE MARKETS

In order to assess competition, the market must first be defined. Each market is a zone of consumer choice, containing closely substitutable goods or services. Rather than sharp, bright lines, actual markets typically have shaded, ragged edges. Markets thus are defined along two dimensions: product features and geographic regions.

THE PRODUCT DIMENSION.   In most cases, airline travel is clearly distinguished from its alternatives. It is so much faster than bus, train, or automobile travel over distances of more than 150 miles that the latter are no substitute for it: That is, they are not "in the market" with airline services. For some short-range travelers, the choice may be closer; for example, driving or taking the Amtrak shuttle between New York and Washington, D.C., takes about as long as flying. A few executives may also have the option of using their company jets, but for the vast majority of travelers, there is no close (or even remote) substitute for using a scheduled airline.

THE GEOGRAPHIC DIMENSION.   Assessing this dimension is more complicated. The entire United States is not a single "market": For those needing to travel from Miami to New York, a flight from Chicago to Seattle is useless, and even a parallel flight from Atlanta to Philadelphia has little value. On the other hand, some city-pair routes overlap others quite closely, either in whole or in part. For example, the New York–Denver route is overlapped by New York–Chicago–Denver, New York–Pittsburgh–Denver, New York–St. Louis–Denver, and even New York–Dallas–Denver. Despite their physical differences, these flights may be quite close economic substitutes, both in service (getting from New York to Denver) and in price.

In general, some of the main trunk routes have a degree of substitution among alternative routes, but most of the lesser city-pair routes do not. Major hub airports may also be considered local markets of their own in some degree, because they control access along their spoke routes, into and out from the hub airport.

Economists have therefore come to have complex ideas about defining relevant markets within the industry. There are many shadings of market edges, and in some cases no clear agreement exists. Yet many city pairs are distinct markets, and many hub cities can be regarded as meaningful markets or as cores of regional air-travel markets.

CONCENTRATION

One of the most dramatic structural developments in the industry was the sharp rise in the entry and number of carriers immediately following deregulation, followed by an equally sharp decline in the number of carriers and a rapid rise in concentration thereafter.

As predicted, dozens of newcomers rushed to enter the field following deregulation. But these firms soon either failed or were taken over by the major carriers. The carnage was severe: Some fifty airlines disappeared in the first few years after deregulation.

Then, in the mid-1980s, the largest carriers embarked on a binge of mergers and acquisitions: USAir acquired Piedmont and Pacific Southwest Airlines; Frank Lorenzo combined Eastern, Continental, People Express, and Frontier; Delta purchased Western; Northwest acquired Republic; and TWA bought Ozark. Some of these mergers did not directly reduce competition, but others did. Approval of all of them by the Department of Transportation has been characterized as an "abysmal dereliction," significantly contributing to the emergence of a tight oligopoly dominating airports and communities across the country[5] (see Table 10-1).

Scores of regional feeder lines also existed before or were formed between 1978 and 1985. But they, too, have since been either absorbed or contractually tied to the largest carriers.

As these and other airlines have vanished, American, United, and Delta have emerged as the Big Three in the field in terms of size and financial strength. Northwest and USAir are less strong, and Continental and TWA have only recently emerged from bankruptcy. Eastern and Pan Am were major carriers for decades, but each has been liquidated as a result of bankruptcy.

One notable development is the recent proliferation of alliances between large U.S. carriers and major foreign airlines. Some of these joint ventures comprise cooperation in scheduling flights, transferring baggage, and issuing tickets (such as United Airline's code-sharing agreement with Lufthansa, Delta Airline's code-sharing agreement with Aeromexico, and the merger between United's "Apollo" computer reservation system and the European "Galileo" reservation system). Other alliances are more substantial, including British Air's purchase in 1993 of a $750 million ownership stake in USAir,

TABLE 10-1 Leading U.S. Airlines, by Passenger Revenues, 1992 (Domestic Operations)

	Revenue Passenger Miles ($mill.)	Revenue/Passenger-Mile (cents)	Passenger Revenues ($mill.)
American Airlines	72,234.0	12.0	8,660.9
Delta Airlines	61,141.0	13.9	8,516.9
United Airlines	59,276.0	11.8	6,988.6
USAir	32,831.0	17.2	5,633.8
Northwest	32,631.0	12.2	3,971.2
Continental	31,027.0	11.2	3,465.7
TWA	19,186.0	11.5	2,210.2
Southwest Airlines	13,788.0	11.1	1,531.8

Source: Standard & Poor.

KLM Royal Dutch Airline's co-ownership of Northwest Airlines, an exchange of 5 percent ownership stakes between Delta and Swissair, and Scandinavian Airline Systems' (SAS) purchase of a 10 percent ownership stake in the parent company of Continental Airlines. Whether these cross-national ties between some of the world's largest air carriers will serve to enhance competition, or to suppress it, remains to be seen.

These domestic developments and their structural effects are displayed in Figures 10-2 and 10-3. As Figure 10-3 shows, the merger mania that engulfed the industry dramatically raised concentration levels. The eight largest airlines now control about 90 percent of all air traffic, compared with 73 percent in 1985. Recently, a number of small carriers—Morris Air (now merged with Southwest), Reno Air, Kiwi Air (launched by pilots formerly employed by Eastern Airlines), and Southwest—have reemerged to challenge the large, high-cost carriers. The latter, in turn, have begun to pressure their workers to accept lower wages and heavier workloads, demands that have touched off a wave of severe labor confrontations. Some of the major carriers have also responded by contemplating spin-offs of their own operations into no-frills, low-cost carriers (such as Continental's establishment in 1993 of its Calite subsidiary, which operates nearly sixty routes).

Particular route markets are even more concentrated, as Table 10-2 suggests. Among city pairs, major trunk routes tend to have lower concentration, especially if parallel routings are taken into account. Lesser routes often have much higher concentration, however, and many spoke routes are near monopolies. Concentration in the industry as a whole, therefore, substantially understates the degree of market power wielded by the major airlines.

BARRIERS TO ENTRY

Between 1978 and 1983, some economists argued that airplanes were "capital with wings," that carriers could quickly dispatch their planes to any routes where a dominant line was reaping monopoly profits. They contended that high market shares (on city-pair routes or in hub cities) were irrelevant, because potential entry would be a formidable safeguard against monopolistic conduct. In other words, they asserted the applicability of the "contestable" markets theory, according to which entry is "ultrafree".[6]

But are airline markets really characterized by such ultrafree entry? The evidence suggests not. Consider entry into the industry through the creation of an entirely new airline: That kind of entry would require substantial funds and skills, months or years to build reputation and attract staff, and resources to withstand the retaliation of incumbent carriers. In fact, few of the entrants of 1978–1985 have survived the retaliatory fare wares and merger movement. Some new low-cost firms commenced operations in the 1990s, but their entry has been neither instantaneous nor easy.

Contradicting the "contestability" theory, potential competitors in airlines also confront a number of additional barriers to entry.[7]

FIGURE 10-2 Chronology of large U.S. airlines since deregulation.

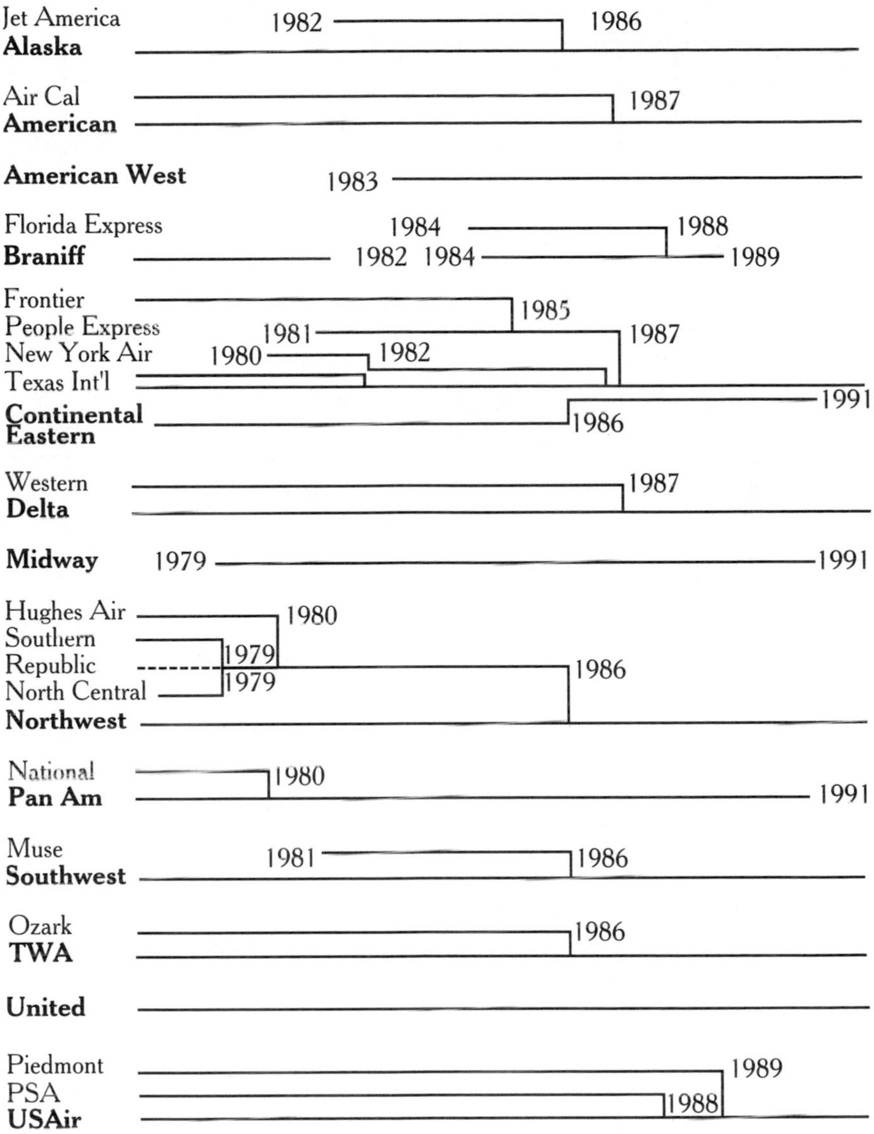

Jet America 1982 ——————————— 1986
Alaska ————————————————————————————

Air Cal ——————————————————————— 1987
American ————————————————————————————

American West 1983 ——————————————————————

Florida Express 1984 ——————————— 1988
Braniff ————————— 1982 1984 ——————————— 1989

Frontier ——————————————— 1985
People Express 1981 ——————— 1987
New York Air 1980 ———— 1982
Texas Int'l ════════════════
Continental ———————————————————————— 1991
Eastern ——————————————— 1986

Western ——————————————— 1987
Delta ————————————————————————————

Midway 1979 ——————————————————————— 1991

Hughes Air ——————— 1980
Southern ═══════
Republic -------- 1979
North Central ——— 1979 1986
Northwest ————————————————————————————

National ——————— 1980
Pan Am ———————————————————————— 1991

Muse 1981 ——————————— 1986
Southwest ————————————————————————————

Ozark ——————————————— 1986
TWA ————————————————————————————

United ————————————————————————————

Piedmont ——————————————————— 1989
PSA ——————————————————— 1988
USAir ————————————————————————————

Source: Severin Borenstein, "The Evolution of U.S. Airline Competition," *Journal of Economic Perspectives*, Spring 1992, Figure 1.

FIGURE 10-3   Percentage of domestic passenger miles controlled by the largest eight firms.

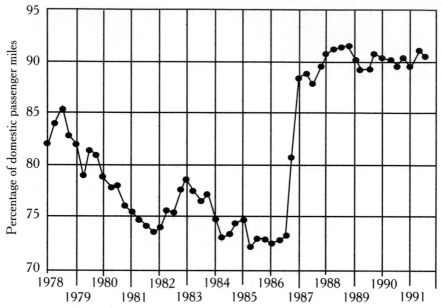

Steven Morrison and Clifford Winston, *The Evolution of the Airline Industry* (Washington, DC: Brookings Institution, 1994).

TABLE 10-2   Share of Departures of Dominant Carriers at Concentrated Airports (by Passenger Enplanements)

Airport	1979		1991	
	Share (%)	Carrier	Share (%)	Carrier
Atlanta	50.2	Delta	87.0	Delta
Charlotte	71.0	Eastern	95.5	USAir
Cincinnati	36.5	Delta	87.8	Delta
Dayton	35.3	TWA	76.0	USAir
Denver	26.3	United	46.8	United
Detroit	20.3	American	73.9	Northwest
Memphis	41.2	Delta	81.0	Northwest
Mpls.–St. Paul	40.1	Northwest	81.4	Northwest
Nashville	26.8	American	76.2	American
Pittsburgh	48.3	USAir	89.4	USAir
Releigh–Durham	17.8	Delta	81.9	American
St. Louis	43.1	TWA	71.7	TWA
Salt Lake City	42.0	Western	84.2	Delta
Syracuse	36.2	USAir	53.9	USAir

Source: Salomon Brothers.

AIRPORT GATES AND SLOTS.   Many of the nation's largest airports are extremely congested, particularly during the preferred travel times of the day. This congestion worsened during the 1980s, when air traffic doubled while airport capacity remained essentially unchanged.

Access to terminal gates and to time slots for takeoffs and landings is a key bottleneck at many airports. Those who hold the rights to use these gates and slots can fly, but others cannot. Many of these rights were retained by the dominant airlines, which obtained them under regulation. This scarcity of gates and slots has been aggravated by the proliferation of hubs and spokes. In many cases, dominant carriers effectively "own" their hub airports: They may hold leases for half or more of the terminal gates at key airports; their leases may be extremely long term in duration; they lease gates that they may not use but that they may refuse to sublease to others; and they often wield the right to veto airport construction projects that might be essential to enable new competitors to operate (USAir agreed to participate financially in the construction of a new airport for its Pittsburgh hub, but only on condition that the former airport would not be used for passenger travel).

In addition, dominant hub carriers may control critical ground-handling and baggage services and charge smaller lines prohibitively high rates for using them.

COMPUTER RESERVATION SYSTEMS.   Computerized systems for displaying flight information and issuing tickets became the norm during the 1980s. For travel agents and other users, these systems present the available menu of flights for any requested route, for any requested dates of travel.

Developed initially by American and United, which have continued to dominate them, these computer reservation systems (CRSs) have served as a significant barrier to new competition in at least two important respects: First, presentation of flight information can and has been deliberately biased in order to favor the flights of the dominant carriers over those of smaller rivals or fledgling new competitors (for example, by showing first the flights of the carrier operating the CRS).[8] This bias thus exploits dominance in one market (reservation services) to buttress market power in another market (airline transportation). Second, United and American have extracted maximum monopoly profits in pricing access to their CRSs by other carriers, in effect by subsidizing their own operations through the monopoly profits extracted from smaller rivals, which require access to computer reservation systems in order to compete. In 1986, these access fees generated revenues of $16 billion for American and United (out of total industry reservations revenue of $21 billion). Bearing little relation to cost, the CRS fees impose a substantial burden on smaller, nonintegrated rivals

MILEAGE BONUS PROGRAMS.   The larger airlines also have established mileage-plus and frequent-flyer bonus programs, which enable frequent flyers to win free flights and other benefits by piling up mileage with individual airlines. These are especially popular with business travelers, who often reap the bonuses for themselves by sticking with one airline for all their business flights. But these bonus programs are a powerful obstacle to new competition, because they wed flyers to existing carriers and make it more difficult for a new entrant to attract customers.

TRAVEL AGENT "OVERRIDES."   Some 32,000 travel agencies now account for 85 percent of all airline tickets issued, with the fees paid to agents by carriers constituting the airlines' third largest operating expense (after labor and fuel). Through commission "override" programs devised by the major carriers, travel agents have been exploited as yet an additional barrier to new competition. Override arrangements offer agents larger commissions from an airline if they sell a targeted percentage of their flights on that carrier. Because the schemes are secret, consumers are not aware of this bias in agents' efforts to steer travelers to particular carriers in order to obtain these bonus payments.

Because "[p]ractically all major airlines use overrides to increase market share in hub cities," these arrangements may strongly undermine competition and serve as yet another barrier to the entry of new, smaller carriers, which have scant resources and only tiny market-share footholds. For example, Morris Air claimed in 1993 that its low-cost flights at Delta's Salt Lake City hub were unfairly harmed by Delta's override scheme for travel agents in that city.[9]

ADVERTISING.   Finally, larger airlines have lower per unit costs of advertising because they can spread their advertising costs over larger volumes. Smaller carriers may simply be unable to fund the cost of the saturation advertising necessary in order to survive as a major competitor. Thus, advertising tends to further raise barriers to entry and to reduce competition in the industry.

THE QUESTION OF ECONOMIES OF SCALE

Why has the structure of the airline industry developed as it has? Do extensive economies of scale make it inevitable that only a few large carriers will survive? Or have mergers and other tactics been employed to artificially induce excessive concentration and high entry barriers?

ECONOMIES OF SCALE.   By 1970, the bulk of economic research revealed that economies in airline operation were exhausted at a relatively small scale: There was no significant cost advantage to extremely large firm size. Research still provides that answer, with no evidence that the largest lines have the lowest costs (see Table 10-3).[10] Indeed, the emergence of many regional "commuter" airlines since 1977 has demonstrated the great variety of economically viable organizational sizes that are feasible, as have recent reports that some large carriers are contemplating spinning off some of their operations into small, lower-cost subsidiaries.[11] In addition, the pre-1975 emphasis on ever

TABLE 10-3   Costs at Selected U.S. Airlines, 1990

Airline	Average Cost per Passenger-Mile (cents per mile)	Average Flight Distance
Southwest	11.1	376
American	14.4	776
United	14.5	809
Continental	15.0	743
Northwest	15.0	665
TWA	15.1	719
Delta	15.5	626
Pan AM	16.8	693
USAir	18.9	463

Source: Adapted from Severin Borenstein, "The Evolution of U.S. Airline Competition," *Journal of Economic Perspectives*, Spring 1992, Table 4.

larger jet aircraft has given way to smaller aircraft flying shorter routes on the hub and spoke pattern.

In short, the industry was and is inherently competitive; there is ample room for a considerable number of efficient airlines to coexist and compete effectively.

ECONOMIES OF SCOPE.   There may, however, be significant economies of scope, another term for the familiar "network effects."[12] Once an airline has established a hub airport, it may be able to add new spoke routes at a relatively low cost. These economies of network expansion may encourage the airline to expand onto those routes, frequently by cutting fares down toward the level of marginal costs in order to attract customers.

Recently, however, evidence has emerged suggesting that there are definite limits to the economies of scope obtainable from the hub-and-spoke arrangement of routes. The enormous costs required to create and maintain hub airports may have become sufficiently burdensome and uneconomic to render the traditional end-to-end route system just as efficient, if not more so. Consequently, some of the largest carriers are now openly expressing doubts about the hub-and-spoke systems they rushed to create during the 1980s and are contemplating a reconfiguration of their route patterns.[13]

III. MARKET BEHAVIOR

Within the setting of this oligopoly-and-dominance structure, the airlines have a wide range of choice in the way they behave toward one another and toward potential newcomers. They could fight fiercely and engage in sustained fare wars that would produce the main results of effective competition. Or at the other extreme, they could strive for quiet, peaceful coexistence, with a careful avoidance of fare wars and a reliance on common pricing patterns engineered through mutual cooperation.

PRICING AND DEMAND DIFFERENCES

Pricing reflects the great variety of demand conditions in the industry.

BUSINESS AND LEISURE DEMAND.   The main distinction among travelers is between business and leisure travel. Business passengers have an inelastic demand: They often need to travel on short notice, cannot stay over weekends, and are insensitive to ticket prices. Leisure travelers, on the other hand, are often price sensitive, are able to plan ahead, and are flexible in their scheduling.

Figure 10-4 illustrates this contrast. The airline maximizes its profit by setting output Q_b and Q_1 for the two groups. Q_b is the standard coach fare, and Q_1 is a supersaver or other discount. This is price discrimination, because the ratio of P_b to cost is higher than the ratio of P_1 to cost. On each plane, some full-coach-fare passengers sit side by side with others who have paid far less.

This discrimination works only when customer groups can be separated; otherwise the low-price customers would resell their cheap tickets to the high-price customers. Airlines achieve this separation by selling tickets only to specifically named customers and by tightly limiting the conditions for obtaining price discounts (for example, requiring a stay over Saturday, buying

FIGURE 10-4   How differing elasticities yield price discrimination.

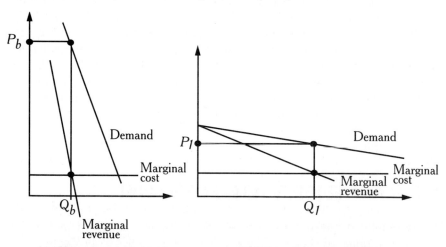

(a) When demand is highly inelastic: business.

(b) When demand is highly elastic: leisure.

the ticket in advance, imposing restrictions or penalties for changes in travel dates).

Business travelers have been the core source of airline profits. But the business share of air travel is shrinking, and the leisure customers may account for as much as 80 percent of air travel by the turn of the century, a development that, if it comes to pass, will severely squeeze the airlines' profits.

OTHER CAUSES OF PRICE DIFFERENCES. There is also great variation among city-pair routes—in their volume, seasonal travel patterns, and longer-term growth trends. For example, there are huge passenger flows between New York and Chicago or Los Angeles, compared with the trickle between, say, Nashville and Hartford. The demand for airline service also fluctuates sharply by time of day, day of week, and season, as well as during recessions and business booms. Finally, the demand for an airline's services is often shaped by the intensity of competition from other airlines along particular routes.

As a result, airline demand is enormously varied across submarkets throughout the country. This diversity strongly influences the airlines' competitive strategies, especially their pricing methods. Extreme varieties of demand invite the airlines to adopt correspondingly complex discriminatory pricing. And the airlines have indeed developed discrimination to an extraordinary degree, beyond any historical precedent in any other major industry. At times, the discounting approaches perfect price discrimination.

PRICING PATTERNS

As contrasted with the experience under regulation, the deregulation of the airlines freed routes and pricing, with both becoming highly flexible. Airlines entered one another's routes, and price-cutting became vigorous, even chaotic. By 1982 most passengers were flying at discount prices.

But in 1985, American Airlines sought to establish order. It set a basic price structure that the other carriers quickly adopted: Discounts would be confined to round-trips and advance purchases, with Saturday night stays mandatory, and little or no opportunity to revise scheduled flights. This pricing structure has been somewhat diluted at times, but its main lines have held remarkably firm for ten years.

"YIELD MANAGEMENT" TECHNIQUES. Within that structure, airlines engage in "yield management," a systematic process for extracting the maximum revenue from each of their thousands of daily flights. Months before each flight, a yield manager apportions the seats on a flight among full-fare and discount classes, on the basis of past experience. Then, as time passes, the numbers of seats in each class are adjusted regularly, in the light of subsequent ticket sales. These revisions proceed continuously as the day of the flight approaches, in order to fill seats and to make passengers pay as much as possible. The changes may even be made hourly, so that when the time for departure actually arrives, each flight has milked the maximum of revenues from the clientele on board.

Additional refinements of this technique include frequent-flyer points, preferential choice of seats, opportunity to board planes first, access to special airport lounges, and so forth—all of which have the effect of incorporating further discrimination (in service quality) even when prices do not differ.

COMPETITIVE SIGNIFICANCE OF PRICE DISCRIMINATION.   When firms set lower price–cost ratios for elastic-demand customers, they generally are cutting their prices in response to competition. Ordinarily, such price variations are normal and efficient; competition drives down prices. When implemented sporadically by firms with small market shares, such price competition can be procompetitive. In fact, discounting is often the lifeblood of effective competition, a vital means for small firms to raise their market shares and reduce the market power of their larger rivals.

But the opposite may be true when dominant firms apply systematic, sharp price cuts. Selective pinpoint price discounting may enable them to attack smaller rivals while keeping their prices up in their other market areas. Such sharpshooting may be lethal for small rivals while minimizing the dominant firm's sacrifice of revenues. Instead of enabling smaller firms to increase pressure on larger firms, price discrimination may become a technique for systematically reducing competition and maintaining market power.

Airlines are able to apply this weapon with extreme precision, using the power of the computerized reservation systems. Prices can be changed by the hour, if necessary, and focused precisely on the routes and customers where they damage small rivals the most. Of course, small rivals can use pinpoint pricing too, but their financial capacity for retaliating to attacks by industry giants is vastly more limited.

PREDATORY PRICING.   Skeptics of predatory pricing have dismissed it as highly unlikely, on the grounds that the costs of engaging in predation are borne primarily by the perpetrator, because a large firm must suffer lower prices and profits on its correspondingly larger sales. This view, however, neglects pinpoint predation that is targeted to those areas where a troublesome smaller rival operates. In these cases, price discrimination may be "predatory"—unfair and specifically destructive to competition.

In 1993, for instance, Reno Air, a maverick new entrant, found that Delta Air Lines was matching Reno's low fares on the Minneapolis–Salt Lake City route. Because Delta's service quality was higher (aircraft, flight frequency), Delta was effectively undercutting Reno's fares as part of a strategy to eliminate Reno altogether. When Clinton administration officials warned Delta that its actions would be prosecuted as anticompetitive, Delta backed off.

Similarly, in 1984 People Express entered the Newark–Minneapolis route, with fares of $99 on weekdays and $79 on evenings and weekends (compared with conventional fares of $149 to $263 for that route). Northwest, the dominant carrier, promptly slashed its fares to $75 to $95. Desperately, People Express reduced its fares further, to $79 and $59; Northwest matched those cuts too. People Express eventually succumbed to this and other attacks by the dominant airlines and disappeared. Northwest acquired Republic Airlines (its only other significant rival at the Minneapolis hub), obtained control of more than 80 percent of the Minneapolis traffic, and sharply raised its fares thereafter.

Some experts now consider predatory pricing to constitute one of the most important barriers to competition in the airline industry, especially "the certainty that any such direct competitive challenge would be met immediately, selectively, and hard."[14]

COLLUSION AND FARE CODES. During the 1980s, the airlines adopted a complex system of price signaling in which they used computerized reservation systems to communicate with their rivals, attaching special codes to various prices to indicate their precise intentions. They also attached beginning and ending dates to their planned fare reductions, communicating to rivals exactly what to expect, when, and for how long. These were methods for discussing prices, which facilitated collusion and reduced competition. They also enabled the carriers to communicate their displeasure swiftly and forcefully should one carrier reduce prices to a degree that the others might consider "excessive" or threatening. As described in one account,

> The most common—and perhaps most questionable—"discussion" between airlines is played out like this: Carrier A—often a smaller operator such as Midway Airlines or America West—attempts to boost its business by lowering ticket prices. It enters lower fares in the industry's computer system. In response, Carrier B—the dominant carrier at the affected airport—not only matches the new fares, but lowers them in other markets that are served by Carrier A.
>
> Carrier B may also attach special codes to its new fares to get its message across. Pricing executives say some carriers have been known to prefix new fares with the letters "FU" to indicate an indelicate expletive. The end result is that Carrier B often cancels its reduction, depriving consumers of a lower fare.[15]

These practices were challenged by antitrust suits, and during the 1992/1993 period, most of the airlines abandoned them. The pricing-dates matter was complex; although the Justice Department's Antitrust Division emphasized their anticompetitive effects, some consumer groups contended that such advance notification of the effective dates for fares benefited travelers by enabling them to plan their trips better.

IV. PERFORMANCE

For a few years, deregulation replaced a government-enforced cartel with active competition, and it yielded benefits to passengers estimated at some $6 billion per year. Prices fell more rapidly, the volume of air travel rose, and air travel became accessible to more segments of the population. Airline load factors went up, from 56 percent in 1977 to 62 percent in the mid-1980s. Airline profits were fairly normal during the 1980–1988 period, suggesting that the carriers were not extracting large monopoly gains from their customers. Also, the rate of airplane crashes has declined since 1977, so that today flying appears to be safer.

But closer inspection suggests that the net benefits may have been marginal or even negative and that the recent rise of market power may be eroding the initial gains achieved under deregulation.

PRICE AND SERVICE QUALITY TRENDS

There are estimates that ticket prices (in cents per passenger mile) fell by about 30 percent after deregulation, compared with the levels that otherwise might have prevailed.[16] Fuel prices rose rapidly between 1978 and 1981, reflecting the rise in oil prices from $14 to nearly $30 per barrel. Other input prices rose, too, including average wages and the cost of new aircraft. When these rises are accounted for statistically, relative fares appear to show a significant fall.

Yet there are a number of contrary considerations: First, fares already were declining in real terms before deregulation, as Figure 10-1 shows. Second, the reduction in competition after 1986 has probably reversed some of the earlier competitive effects, leading to higher fares. Indeed, specific fares went up quickly from 1986 to 1989, following the mergers that created dominance on a number of routes and hubs.

Third, although prices may have fallen relatively, so too has the quality of service, perhaps even more noticeably. Service quality includes such elements as the time interval taken by the flight (including time spent at airport check-in counters as well as transfers at hubs), the risk of missing flights (the frequency of cancellations of flights for revenue reasons has increased), the risk of lateness or loss of baggage, the crowding in lines and in the planes themselves, and the overall quality of the travel experience. Moreover, most discounts can be obtained only by conforming to tight restrictions. The traveler's freedom of choice has thus been curtailed, as many passengers are required to lock in their plans weeks or months ahead, with virtually no chance of refunds should conditions change. Although these hidden economic costs to passengers are hard to estimate, they are real.

AIRLINE PROFITS

Despite the turbulence of competitive changes, the airlines' rates of profit on total invested capital were actually as high between 1978 and 1987 (6.4 percent) as they were over the decade before deregulation in 1978. The years between 1989 and 1993, however, brought large financial losses to the airlines: $1.9 billion in the aggregate in 1991 and more than $4 billion in 1992.

The cause and significance of these losses are the subject of spirited debate: Some point out that the airlines traditionally have been highly sensitive to business cycles, even under regulation, and that the recessionary macroeconomic conditions of the 1989–1993 period are primarily to blame. Others construe recent multibillion-dollar airline losses as proof that competition in the field is "ruinous." Some point to the excessive debt loads taken on by corporate raiders in their takeovers of Northwest, Continental, and TWA as an important part of the problem. Others argue that the airlines' losses over this period have been exacerbated by American Airlines' all-out effort to establish pricing discipline in the industry, by slashing fares in order to force other carriers into compliance with its price preferences, and to drive financially weak carriers into bankruptcy. Some (most prominently, Robert Crandall, chairman of American Airlines), contend that deficiencies in the country's bankruptcy laws enable insolvent carriers to continue to operate and to drive down prices to ruinously low levels. Still others suggest that large carriers are to an important extent the victims of the high-cost hub-and-spoke systems they created; they note that Southwest, one of the smallest of the major carriers, has been robustly profitable through the 1989–1993 years, in important part due to its lower cost structure and avoidance of the hub-and-spoke routing pattern.

A national airline study commission advocated in 1993 a number of proposals for addressing the airlines' financial problems, ranging from the creation of a private-sector financial advisory committee to modifications of the nation's bankruptcy laws as they apply to air carriers.

EFFICIENCY

A major anticipated benefit of deregulation was to remove the incentives under regulation to drive up costs. By shifting from nonprice competition, deregulation offered the possibility of removing excess airline costs that appeared to make fares as much as 50 percent higher than necessary.

Various efficiency indicators, including load factors and passenger miles per worker, reflect gains from deregulation. Yet the largest airlines continue to operate with unnecessarily high costs (as Table 10-3 shows). They have sought savings in part by limiting wages, but this is merely a transfer of funds out of workers' pockets rather than a real reduction in resource use.

The large U.S. carriers have also struggled with the difficulty of combining trunk-route and peripheral-route operations, as jumbo jets and standard staffing are too costly for low traffic routes to small cities. As mentioned before, some large airlines are moving to create low-cost, no-frills, smaller-plane systems to parallel their regular operations. This would enable them partly to escape labor unions, but it could also improve efficiency and variety while intensifying competition on some routes. The major airlines thus can choose between linking with commuter lines or creating their own subsidiaries. The evolving mix of these alternatives presumably will reflect relative technical and market-power considerations.

INNOVATION

Innovation can occur as product changes (new products and services) or as process innovations (new methods of organizing production that reduce the cost of producing the same goods).

Before 1978, airline innovations of both types came reasonably rapidly. New aircraft types were adopted without delay, and nonprice competition encouraged a variety of service amenities (meals, decor, and so on).

Deregulation brought a burst of innovation in services, particularly with the introduction of the "no-frills" service at very low prices. People Express, World Airways, and Air Florida led this innovation. At the upper end, some new airlines (most recently, Ultrair) have offered luxury service at extremely high fares, and at least one entrepreneur has recently begun cross-country air service for smokers only! Sophisticated computer reservation systems were improved rapidly. New fare discount programs proliferated, from extremely low "peanuts" fares to off-peak discounts, and most of these were initially offered without restrictions.

New aircraft types were adopted about as fast as before, with one exception. Small commuter-type aircraft were quickly introduced on many small-city routes, to fill the gap left as large aircraft were withdrawn. Recently, however, aircraft producers have begun to produce small, comfortable jets which some of the larger commuter airlines (such as Comair) are finding economical and attractive to use on small-volume, small-city routes.

The pace of innovation was greatest in the early deregulation years, when competition and new entry were strongest. But the subsequent rise in market power and the diminution of competitive entry may have served to slow the rate of innovation in the field.

SMALL-AIRPORT SERVICE

Deregulation raised the specter that small cities would lose vital service (as they previously lost train and bus service and post offices). Hence, there were strong fears that airline deregulation would eliminate flights to small airports. In fact, full-size jet aircraft were withdrawn from many smaller airports. But in most cases, new, small-craft commuter airlines emerged to provide service. Although some cities have suffered severe service cutbacks, smaller airports have generally gained in flight frequency since 1978. Fares to small towns have risen faster, however, and small aircraft are noisier and less comfortable—considerations that offset the gains from deregulation.[17]

SAFETY

Safety is an extremely sensitive dimension of airline performance, particularly in the wake of deregulation and concerns that competitive pressures might tempt carriers to compromise safety by skimping on maintenance.

Statistically speaking, airline travel remains the safest mode of travel. In fact, the safety risks are far greater in driving a car, riding a bike, cooking a meal, or crossing a moderately busy street. Between 1978 and 1988, the fatality rates per passenger miles flown fell by one-third or more. Since 1978, however, there has been an unsettling number of reports of near-misses and errors in flight guidance systems. Moreover, a number of airlines have been fined record amounts for falsifying aircraft repair records or failing to perform required repairs.

Continued safety regulation thus is clearly required and, in fact, must be strengthened to ensure that the potential gains from deregulation are not jeopardized by safety concerns.

V. PUBLIC POLICY

Deregulation was appropriate and effective, but it elevated antitrust enforcement to critical importance as the guardian of the competition that was its objective to induce. Instead, antitrust protection was fumbled and withdrawn, permitting dominance and market control to develop in much of the industry. The experience thus offers a number of important lessons about the relationship between deregulation and antitrust policy.

WHAT WORKED IN DEREGULATION

DEREGULATION WAS NOT PREMATURE. Deregulation worked (briefly) because it removed all of the CAB's control only after competition was already effective. One great danger is premature deregulation, which frees a dominant firm from regulatory restraints before competition has become effective. In that case, the dominant firm can use its advantages and resources to crush rivals and fledgling new entrants. In the airlines, deregulation unleashed an initial flood of competition as incumbent carriers shifted routes, smaller regional lines expanded nationwide, and newly formed airlines entered the field. Because the largest carriers were initially unable to control the markets, route entry and fare cutting were turbulent and powerful.

A SERIES OF CAREFUL STEPS. In addition, CAB controls were removed in a series of steps, with continuous attention to the results. The process could have been stopped or modified at any point if a monopoly were emerging. Care was taken—at least in the beginning—to ensure that competition was effective as regulations were relaxed.

WHAT FAILED IN DEREGULATION

CONTESTABILITY. Relying on ultrafree entry (or "contestability") has failed to check the market power of the dominant carriers at their monopolized hubs and on their routes. High market shares have bred high fares on many routes, despite the early promises that the fear of instantaneous entry would prevent monopoly behavior. Moreover, the entry of entirely new airline companies has been substantially blocked by the existing airlines' strategies and their control over gates and time slots, computer reservation systems, and mileage bonus programs, as well as by their predatory pricing tactics.

PERMITTING MERGERS. The Transportation Department's "abysmal dereliction" permitted a sharp—and unnecessary—consolidation of the industry. The result has been an increasingly tight oligopoly, with large elements of dominance. Maverick low-cost carriers have been eliminated, taken over, or subordinated as affiliates of the major airlines.

CONTROL OF GATES AND SLOTS. The scarcity of gates and slots at crowded airports and their control by the largest airlines have raised important barriers to competition and generated increased monopoly power.

PREDATORY PRICE DISCRIMINATION. Contrary to what the skeptics claim, systematic price discrimination has both reflected and intensified monopoly elements in the industry. Severe price discrimination has helped drive independent airlines from the market, an outcome that could have been prevented by vigilant antitrust action. As in the case of mergers, however, the Reagan administration's laxity in antitrust enforcement took effect just when the transfer to vigorous antitrust supervision was most important.

SCARCITY OF AIR CONTROLLERS. The firing of striking air controllers in 1981 (plus the apparent lifetime prohibition on rehiring any of them) further reduced the system's capacity and strengthened the impact of the anticompetitive practices in which the leading carriers have engaged.

FUTURE PUBLIC POLICY STEPS

The airline industry's competitiveness and performance could be improved with a number of reforms.

GATES AND SLOTS. Expanding the capacity of airports to handle flights is widely favored. But it may not be easy to achieve, because it requires large capital investments, which take many years to complete. (Denver and Pittsburgh have the only new airports to have been constructed over the past two decades.) There also is neighborhood resistance to the consequent increase in noise levels.

Some contend that if local airports were privatized, they would function more efficiently and effectively than they have. In a related vein, the National Commission to Ensure a Strong Competitive Airline Industry proposed in 1993 that the air traffic control functions of the Federal Aviation Administration be assigned to an independent government entity with greater autonomy in performing and financing these functions.

Others suggest that the implementation of more creative airport pricing and allocation steps would alleviate some of these capacity pressures. For example, much private small-plane activity could be moved to smaller airfields. Currently, small executive and pleasure aircraft pay only a few dollars for landings that delay hundreds of airline passengers. Raising these fees would more closely reflect true social costs and discourage the wasteful use of scarce airport space.

Gate capacity is largely controlled by the large airlines, especially at their hub airports, thereby enabling them to block entry. Alternative gate allocation systems—such as "use it or lose it" or perhaps more frequent auctioning of gates—might remedy the problem.

PREVENTING ANTICOMPETITIVE PRICING. The Clinton administration's action in 1993 to prevent predatory behavior against Reno Air sets a sound precedent, which should be continued. Vigilance in prosecuting tacit collusion through computer pricing systems and codes must also be maintained.

FREQUENT FLYER PROGRAMS. Discounts for frequent flyers might be prohibited. They are solely a marketing device for tying flyers to the large airlines, and they only make more difficult the survival of smaller airlines or new entrants.

COMPUTER RESERVATION SYSTEMS. Perhaps the major airlines should be required to divest their computer reservation systems, as Alfred Kahn and others have suggested. These systems, combined with commission "overrides," create an artificial barrier by tilting the choices made by travel agents toward the largest carriers.

THE QUESTION OF FOREIGN COMPETITION. Finally, there is the issue of foreign competition in air transportation.

Some point to the potential entry into U.S. air service by large foreign carriers as the ultimate solution to the domestic industry's monopoly problem. If these foreign carriers were permitted to fly between American cities, it is argued, they could dissipate the incumbent firms' market control. According to this view, government restrictions preventing foreign carriers from flying domestic routes ("cabotage"), as well as government regulations preventing foreign carriers from owning U.S. lines, should be abolished. The bulk of the U.S. airline industry is vehemently opposed to such steps, however, without a corresponding reduction in regulatory barriers abroad, which they claim prevent them from reciprocally competing in foreign markets and territories.

Seen in this light, the recent proliferation of cooperative agreements, joint ventures, and cross-ownership positions between U.S. carriers and major foreign lines pose some nettlesome questions: Do these arrangements mark the beginning of a more competitive global marketplace in air service that should be encouraged? Or instead, do they have an anticompetitive effect in linking as partners the dominant domestic carriers with their only remaining potential source of competition? If the latter, then perhaps these joint ventures and cooperative arrangements should be prosecuted under the antitrust laws.

VI. CONCLUSION

Every weekend the security fences surrounding airports are lined with small children, their parents, and assorted onlookers of all ages. They are mesmerized by the deafening roar of planes landing and taking off. They marvel at the miracle of how such monstrous gleaming machines, weighing thousands of tons, are able to rise and land so effortlessly. With varying degrees of success (and patience!), parents attempt to instruct their children in the rudimentary principles of aerodynamics—how the shape of the airfoil rather than the flapping of the wings produces lift.

At first blush, the dynamics of an industry confront the casual observer with a similar mystery. Once armed with facts and the tools of economic analysis, the student of industrial organization can penetrate the mystery of market forces and public policy issues buffeting the field.

NOTES

1. For good reviews of the industry's history, see Richard E. Caves, *Air Transport and Its Regulation* (Cambridge, MA: Harvard University Press, 1962); William A. Jordan, *Airline Regulation in America: Effects and Imperfections* (Baltimore: Johns Hopkins University Press, 1970); and Steven A. Morrison and Clifford Winston, *The Evolution of the Airline Industry* (Washington, DC: Brookings Institution, 1994).
2. Two kinds of research suggested that costs were raised by 30 to 50 percent on many routes as an effect of regulation. First, costs on *intrastate* routes in Texas and California, which were not regulated by the CAB, were compared with costs on *interstate* routes, which the CAB did regulate. The costs of flights on intrastate routes were lower by roughly 30 percent on average. See especially Jordan, *Air Regulation in America*.

    Second, statistical analysis was used to factor out the sources of cost levels, including regulatory effects. Regulation's net cost-raising effect was found to be about 30 to 50 percent on typical routes. See Theodore E. Keeler, "Airline Regulation and Market Performance," *Bell Journal of Economics*, Autumn 1972, pp. 399–424.
3. Before 1978, economists (especially Caves, Jordan, Keeler, and Kahn) did voluminous research, most of which showed that airline regulation

was harmful. After 1978, extensive research showed that deregulation was successful but that substantial monopoly remained.

The industry generates a great volume of facts (about traffic patterns, pricing, safety, and the like), and the vested interests in the research results have been intense. The Suggested Readings at the end of this chapter note a few of the leading studies.

4. See especially the summary discussion of these and other surprises in Alfred E. Kahn, "Surprises of Airline Deregulation," *American Economic Review*, May 1988, pp. 316–322. As chairman of the CAB in the late 1970s, Kahn played a leading role in promoting deregulation of the industry.
5. The characterization is Alfred E. Kahn's, in "Airline Deregulation—A Mixed Bag, but a Clear Success Nevertheless," *Transportation Law Journal* 16 (1988): 229–252.
6. See Elizabeth E. Bailey, "Contestability and the Design of Regulatory and Antitrust Policy," *American Economic Review*, May 1981, pp. 178–183; Elizabeth E. Bailey and John C. Panzar, "The Contestability of Airline Markets During the Transition to Deregulation," *Law and Contemporary Problems*, Winter 1981, pp. 125–145; and William J. Baumol, John C. Panzar, and Robert D. Willig, *Contestable Markets and the Theory of Industry Structure* (San Diego: Harcourt Brace Jovanovich, 1982).
7. For an extensive documentation of these barriers to competition, see U.S. General Accounting Office, *Airline Competition: Industry Operating and Marketing Practices Limit Market Entry*, report GAO/RCED-90-147 (Washington, DC: U.S. General Accounting Office, August 1990).
8. See the CAB's analysis of this problem in its *Report to Congress on Computer Reservation Systems*, Washington, D.C., 1983. As John R. Meyer and Clinton V. Oster, Jr., note, "A bias that diverts only one passenger per month with a $200 round-trip flight for each SABRE computer terminal would net American Airlines, and cost American's competitors, $120 million per year." See *Deregulation and the Future of Intercity Passenger Travel* (Cambridge, MA: MIT Press, 1987), p. 132. Since marginal costs are low, the diverted revenues are virtually all profit.
9. James S. Hirsch, "Delta's Bonuses to Travel Agents Spur Inquiry on Anticompetitiveness Question," *Wall Street Journal*, October 11, 1993, p. C11.
10. See Morrison and Winston, *Evolution of the Airline Industry*.
11. For example, see Carl Quintanilla, "United Airlines May Spin off Short Routes," *Wall Street Journal*, June 25, 1993, p. A3.
12. See Baumol, Panzar, and Willig, *Contestable Markets*, among others.
13. See James S. Hirsch, "Big Airlines Scale Back Hub Airport System to Curb Rising Costs," *Wall Street Journal*, January 12, 1993, p. 1; and Wilton Woods, "Goodbye Hub and Spoke?" *Fortune*, December 13, 1993, p. 160.
14. Alfred E. Kahn, "Thinking About Predation—A Personal Diary," *Review of Industrial Organization* 6 (1991): 137–146.

15. Asra Q. Nomani, "Airlines May Be Using a Price–Data Network to Lessen Competition," *Wall Street Journal*, June 28, 1990, pp. A1, A6. For additional examples and discussions, see Nomani, "Fare Warning: How Airlines Trade Price Plans," *Wall Street Journal*, October 9, 1990, p. B1, and "Dispatches from the Air-Fare Front," *Wall Street Journal*, July 11, 1989, p. B1; Agis Salpukas, "American: Disciplinarian of the Skies," *New York Times*, May 29, 1992, p. C1, and "A Higher Rock Bottom for Many New Air Fares," *New York Times*, April 22, 1992, p. C1.

16. See Morrison and Winston, *Evolution of the Airline Industry*; and Meyer and Oster, *Deregulation*.

17. See Meyer and Oster, *Deregulation and the Future of Intercity Passenger Travel*, and Morrison and Winston, *Evolution of the Airline Industry*, for varying views on these particular issues.

SUGGESTED READINGS

Books

Bailey, Elizabeth E., David R. Graham, and Daniel P. Kaplan. *Deregulating the Airlines*. Cambridge, MA: MIT Press, 1985.

Baumol, William J., John C.Panzar, and Robert D. Willig. *Contestable Markets and the Theory of Industry Structure*. San Diego: Harcourt Brace Jovanovich, 1982.

Caves, Richard E. *Air Transport and Its Regulators: An Industry Study*. Cambridge, MA: Harvard University Press, 1962.

Dempsey, Paul S., and Andrew R. Goetz. *Airline Deregulation and Laissez-Faire Mythology*. Westport, CT: Quorum Books, 1992.

Jordan, William A. *Airline Regulation in America*. Baltimore: Johns Hopkins University Press, 1970.

Meyer, John R., and Clinton V. Oster, Jr. *Deregulation and the Future of Intercity Passenger Travel*. Cambridge, MA: MIT Press, 1987.

Morrison, Steven, and Clifford Winston. *The Evolution of the Airline Industry*. Washington, DC: Brookings Institution, 1994.

Sampson, Anthony. *Empires of the Sky: The Politics, Contests and Cartels of World Airlines*. New York: Random House, 1984.

Transportation Research Board. *Winds of Change: Domestic Air Transport Since Deregulation*. Washington, DC: National Research Council, 1991.

Articles

Adrangi, Bahrom, Garland Chow, and Richard Gritta. "Market Structure, Market Share, and Profits in the Airline Industry." *Atlantic Economic Journal*, March 1991, pp. 98–107.

Bailey, Elizabeth E., and William J. Baumol, "Deregulation and the Theory of Contestable Markets." *Yale Journal on Regulation* 1 (1984): 111–137.

Borenstein, Severin. "The Evolution of U.S. Airline Competition." *Journal of Economic Perspectives*, Spring 1992, pp. 45–73.

Buetel, Phillip A., and Mark E. McBride. "Market Power and the Northwest–Republic Airline Merger: A Residual Demand Approach." *Southern Economic Journal*, January 1992, pp. 709–720.

Evans, William N., and Ioannis Kessides. "Structure, Conduct, and Performance in the Deregulated Airline Industry." *Southern Economic Journal*, January 1993, pp. 450–467.

Keeler, Theodore E. "Airline Regulation and Market Performance." *Bell Journal of Economics*, Autumn 1972, pp. 399–424.

Levine, Michael E. "Airline Competition in Deregulated Markets: Theory, Firm Strategy, and Public Policy." *Yale Journal on Regulation*, Spring 1987, pp. 393–494.

Meyer, John R., and John S. Strong. "From Closed Set to Open Set Deregulation: An Assessment of the U.S. Airline Industry." *Logistics and Transportation Review*, March 1992, pp. 1–21.

Moore, Thomas G. "U.S. Airline Deregulation: Its Effects on Passengers, Capital, and Labor." *Journal of Law & Economics*, April 1986, pp. 1–28.

Pustay, Michael W. "Toward a Global Airline Industry: Prospects and Impediments." *Logistics and Transportation Review*, March 1992, pp. 103–128.

Review of Industrial Organization, August 1993 (issue devoted to articles assessing airline deregulation).

Saunders, Lisa M., and William G. Shepherd. "Airlines: Setting Constraints on Hub Dominance." *Logistics and Transportation Review*, September 1993, pp. 201–220.

Vietor, Richard H. K. "Contrived Competition: Airline Regulation and Deregulation, 1925–1988." *Business History Review*, Spring 1990, pp. 61–108.

Government and Other Reports

Air Transport Association. "Air Transport: Annual Report of the U.S. Scheduled Airline Industry."

Aviation Safety Commission. "Final Report and Recommendations." Washington, DC, April 1988.

Maldutis, Julius. "Airline Competition at the 50 Largest U.S. Airports." Annual: Salomon Brothers.

National Commission to Ensure a Strong Competitive Airline Industry. "Final Report and Recommendations." Washington, DC, 1993.

U.S. Congress. House. Subcommittee on Aviation. *Hearing: Airline Computer Reservation Systems*. 100th Cong., 2nd sess., 1988.

U.S. Congress. House. Subcommittee on Aviation. *Hearing: Leveraged Buyouts and Foreign Ownership of United States Airlines*. 101st Cong., 1st sess., 1990.

U.S. Congress. Senate. Subcommittee on Aviation. *Hearing: Airline Concentration*. 101st Cong., 1st sess., 1989.

U.S. Congressional Budget Office. *Policies for the Deregulated Airline Industry*. Washington, DC, July 1988.

U.S. Department of Transportation. *Report of the Secretary's Task Force on Competition in the U.S. Domestic Airline Industry*. Washington, DC, February 1990.

U.S. General Accounting Office. "Airline Competition: Fare and Service Changes at St. Louis Since the TWA–Ozark Merger." Washington, DC, Report GAO/RCED-88-217, September 1988.

U.S. General Accounting Office. "Airline Competition: Higher Fares and Less Competition Continue at Concentrated Airports." Washington, DC, Report GAO/RCED-93-171, July 1993.

U.S. General Accounting Office. "Airline Competition: Industry Operating and Marketing Practices Limit Market Entry." Washington, DC, Report GAO/RCED-90-147, August 1990.

11

COLLEGE SPORTS

Cecil Mackey

I. INTRODUCTION

The *New York Times* ran a lead editorial on Sunday, August 29, 1993, under the headline "Cleaning up College Football."[1] Noting that the college football season was "starting with a familiar flurry of resignations and probations," the editorial stated flatly that "big-time college football has a corruption at its core that can and must be cured." It expressed concern about graduation rates for athletes, money generated by big-time college athletics, the fact that the college athletes are not allowed to share in what they help generate, and why corruption and abuse exist and why they continue to be tolerated.

Sports in the United States is a huge enterprise; from high school through the professional ranks it is big entertainment, big business, and big money. College sports, especially the so-called revenue sports at the "big-time" National Collegiate Athletic Association (NCAA) Division I–A schools, are all these things—entertainment, business and money, and an important part of a complex and interrelated picture.

More than 250,000 student-athletes participate in the various intercollegiate varsity sports at NCAA schools, but football and men's basketball are the principal revenue sports, especially at the (approximately 105) NCAA Division I–A schools. In some regions, hockey or baseball may also produce significant revenue. These major revenue sports generate the great majority of media interest, fan support, television coverage, and revenue. They provide the channel for the most talented athletes into the high-paying, high-visibility ranks of the pros. They provide major benefits, both tangible and intangible, to their universities. They represent the dominant and controlling force in the primary governing body of intercollegiate athletics, the NCAA. They are, in addition, the sports in which the great majority of what is characterized as corruption, abuse, infractions, and rule violations occur.

273

II. HISTORY AND STRUCTURE

The United States is alone among nations in the relationship it has created between sports and its institutions of higher education, a unique arrangement that has evolved over time rather than having been specifically planned. It is unclear, however, whether the majority of individuals who have responsibility for such programs, particularly the college and university administrators, would choose the present arrangement if a choice were possible today. But the fundamental link between competitive athletics and the country's colleges and universities that produces what we know as intercollegiate athletics may be too well established to sever now, even if that were a goal desired by most decision makers.

THE NATIONAL COLLEGIATE ATHLETIC ASSOCIATION

The National Collegiate Athletic Association is a private membership organization composed primarily of more than eight hundred four-year colleges and universities and two-year upper-level institutions. It also has as members athletic conferences (for example, the Pacific Ten [Pac Ten] and the Atlantic Coast Conference) and groups related to intercollegiate athletics (for example, the Basketball Coaches Association). The current institutional membership is approximately one thousand. The NCAA states that its basic purpose is "to maintain intercollegiate athletics as an integral part of the educational program and the athlete as an integral part of the student body and, by so doing, retain a clear line of demarcation between intercollegiate athletics and professional sports." Its ostensible purpose is "to initiate, stimulate and improve intercollegiate athletics programs for student-athletes and to promote and develop educational leadership, physical fitness, athletic excellence and athletics participation as a recreational pursuit" and "to study all phases of competitive athletics and establish standards whereby the colleges and universities of the United States can maintain their athletics programs on a high level."[2]

The NCAA's predecessor organization, the Intercollegiate Athletic Association of the United States, was formed in 1906, largely as a response to public reaction to the violence that had become a part of college football. Although it may come as a surprise to some, football was a more violent game at the turn of the century than it is in the 1990s. There were no standardized rules and no governing or supervising organization to oversee or regulate play. Nor was there the protective equipment that is available to players today, much of which is now required. The extent of serious injuries and deaths reported among players aroused concern among educators, the press, and even President Theodore Roosevelt. Serious questions were raised as to whether the game should be abandoned or whether reform was possible, so that the sport could be retained. It would be fair to say, therefore, that the NCAA (which became the official name in 1910) was itself a creature of the first reform movement in college sports.

During the 1920s, college sports were popular; there was a building boom in football stadiums; and there were sports scandals for the public to read about. In 1929, the year of the Great Depression, the Carnegie Foundation for the Advancement of Teaching produced a major report on American college athletics.[3] Neither the issues nor the abuses in college sports were much different after the formation of the NCAA from what they had been before. Although pervasive and flagrant abuse was decried, basic reform did not result, a pattern that was to become all too familiar.

The years after World War II witnessed a massive influx of college students, another expansion in college sports (see Table 11-1), and a new era in regulation when the NCAA adopted what was called the Sanity Code. Again there was hope. Some saw this action as foretelling an end to widespread problems in recruiting and financial aid, as well as an improvement in academic standards within intercollegiate athletics. But the new plan failed to live up to either expectation. Sanity, it seems, was too much to expect.

In 1980, women joined the NCAA when it made women's intercollegiate athletic programs an offer they could not refuse. Its offer of membership plus championships from an organization that had far more financial strength and public recognition than could be offered by the existing women's association, the Association for Intercollegiate Athletics for Women (AIAW), caused its membership to move to the NCAA and effectively ended the role of the AIAW.

In 1991 the Knight Commission, a twenty-two-member group made up mainly of university representatives, published its report.[4] It addressed most of the usual problems, proposed no major or radical reform, and accordingly has met with essentially the same fate as its predecessors.

When there has been an occasional setback for the NCAA, as, for example, in the courts in 1984 concerning the control of TV rights, or an internal problem with dissatisfied influential members, as with the major football schools that formed the College Football Association, there has been compromise, accommodation, and regrouping. Essentially, the NCAA's philoso-

TABLE 11-1 U.S. College Football Attendance, 1948–1988

Year	Attendance
1948	19,100,000
1958	19,300,000
1968	27,000,000
1978	34,300,000
1988	35,581,790

Source: Arthur A. Fleisher, Brian L. Goff, and Robert D. Tollison, *The National Collegiate Athletic Association* (Chicago: University of Chicago Press, 1992), p. 54.

phy has been to yield when it is necessary and to control when it is possible. This philosophy has allowed the NCAA bureaucracy and the athletic establishment of the most powerful member institutions to retain control of the organization and its regulatory apparatus, to weather the periodic storms of controversy and intense scrutiny, and, in the process, to resist all efforts at fundamental reform.

As a private organization, the NCAA includes both public and private colleges and universities; its recent annual revenues are shown in Table 11-2. Its membership is divided into three divisions for purposes of athletic competition and for legislative purposes within the NCAA's governing structure: Division I, Division II, and Division III. For purposes of football only, Division I is divided into Divisions I–A and I–AA. Division II is made up of relatively smaller schools with smaller student enrollments that offer less financial aid to student-athletes than do Division I schools. Division III schools do not offer financial aid to athletes as such; that is, they do not have what are termed *athletic scholarships*.

There is, in addition to the NCAA, the National Association of Intercollegiate Athletics (NAIA). Originally formed as a basketball-based organization, its scope now extends to all sports. Its approximately five hundred members are smaller schools that do not have big-time sports programs. NAIA athletics seem to generate little money, limited media interest, and virtually no scandal or corruption.

An important distinction should be kept in mind. The NCAA is a private organization governed by rules and regulations that have been adopted by representatives of its member institutions. These rules and regulations are not statutes or laws enacted by a governmental agency. Even though discussion of NCAA violations in the media and even within the academic community is frequently cast in terms of legality or illegality, the conduct being considered

TABLE 11-2   NCAA Revenues, 1991

Source	Amount (\$mill.)
Television	118.5
Championships	20.8
Royalties	5.3
General	4.6
Publishing	1.4
Grants	1.0
Communications	.6
Visitors' Center	.3
Total	\$ 152.5

Source: U.S. General Accounting Office, "Intercollegiate Athletics: Revenues and Expenses, Gender and Minority Profiles, and Compensation in Athletic Departments," report GAO/T-HRD-92-25, April 9, 1992, p. 3.

is normally not appropriately measured by legal standards. Often the acts that constitute NCAA violations and are dealt with as opprobrious in the context of college sports would be treated as normal, rational, and economically sound and might well be applauded in any other setting. It is to the advantage of the NCAA and its members to have violations of its rules viewed as equivalent to illegal acts, even though such a characterization may have devastating consequences for the individuals involved.[5] Having been found to have violated NCAA rules does not, however, seem to present an obstacle to a talented student-athlete in landing a position with the pros, nor does it appear to prevent college coaches from moving into the professional coaching ranks or to the media when their services have economic value. The market apparently understands reality.

INTERCOLLEGIATE SPORTS AS AN INDUSTRY

College sports have clearly become an "intercollegiate athletics industry" when (1) the NCAA has a $1 billion, seven-year television contract with CBS Sports to televise the NCAA Division I championship basketball tournament, (2) high-profile football and men's basketball coaches are collecting as much as a $1 million a year in total compensation, (3) a Rose Bowl appearance pays each team $6.5 million, and (4) Chris Webber, at the end of his second season as star of the University of Michigan's "Fab Five" basketball team, decides to forgo further education for an NBA contract reported at $74.4 million for fifteen years.[6] (Selected financial data for Division 1-A colleges are presented in Tables 11-3 through 11-5.)

If intercollegiate sports is an industry, what is the NCAA? A cartel has been defined as "a formal agreement among independent businesses or governments of countries to control production, sales, and/or purchases for the benefit of the members. A cartel is used to regulate markets and fix prices as a monopolist would."[7] That is just what many people would label the NCAA—a cartel—even though its members are neither businesses nor governments of countries but, rather, not-for-profit entities, mainly colleges and universities. Those who dislike the way that the NCAA conducts its business or who believe that they (or others they care about) are being exploited by the NCAA

TABLE 11-3 Financial and Operating Measures; Division I-A College Athletic Programs, 1989

Average total revenues	$ 9,685 thousand
Average total expenses	$ 9,646 thousand
Average number of sports	18
Average number of athletes	468

Source: Mitchell H. Raiborn, *Revenues and Expenses of Intercollegiate Athletics Programs: Analysis of Financial Trends and Relationships, 1985–1989* (Overland Park, KS: National Collegiate Athletic Association, 1990), p. 9.

TABLE 11-4 Average Revenue Sources, Division I-A College Athletic Programs, 1989

	Amount ($ thousands)	Percentage of Total
Ticket sales	3,399	35
Student fees	1,840	7
Guarantees and options	792	8
Contributions from alumni and others	1,546	15
Bowl games, tournaments, and television revenues	1,470	14
Direct government	1,363	5
Other	1,558	16

Source: Mitchell H. Raiborn, *Revenues and Expenses of Intercollegiate Athletics Programs: Analysis of Financial Trends and Relationships, 1985–1989* (Overland Park, KS: National Collegiate Athletic Association, 1990), p. 17.

are likely to characterize it as a cartel. Those who support the NCAA, who believe in what the NCAA asserts as its fundamental principles, or who have a personal stake in the organization as it is structured and operates are more likely to characterize the NCAA as essential to the integrity and functioning of college sports. The final conclusion may, in fact, simply be a matter over which reasonable persons differ.

Division I–A schools are the principal focus here. In addition to being the schools with the country's major athletic programs, they are, for the most part, large, complex institutions that offer graduate and professional as well as undergraduate education. Most have substantial research programs. Their annual budgets range from roughly $100 million to more than $1 billion, with funding support from many sources. The budgets for the athletic departments of Division I–A schools vary widely but typically range from $3 million to $25 million. Although athletic department budgets may seem large in some respects, they are a very small percentage of the annual budget of the Division

TABLE 11-5 Average Revenues by Sport: Division I-A Men's College Athletic Programs, 1989 ($ 000's)

Football		Basketball		Other Sports	
Amount	Per cent	Amount	Per cent	Amount	Per cent
$4,340	47%	$1,640	18%	$3,290	35%

Source: Mitchell H. Raiborn, *Revenues and Expenses of Intercollegiate Athletics Programs: Analysis of Financial Trends and Relationships, 1985–1989* (Overland Park, KS: National Collegiate Athletic Association, 1990), p. 18.

I–A universities. For example, if a university had an annual budget of $500 million (below average for the Big Ten) and an athletic department budget of $12 million (about average for the Big Ten), athletics would constitute only 2.4 percent of the school's budget. There is no simple answer to what this kind of relationship means. Clearly it does not mean, nor should anyone even suggest, that the importance of intercollegiate athletics at a particular institution can be measured by the size of the athletic department's budget relative to the school's total operating budget. Yet more than a few university presidents would argue that no other entity in a university is likely to require so much time, cause so much grief, or take on such importance while accounting for such a small share of the overall budget.

III. CONDUCT AND PERFORMANCE

THE REACH OF NCAA REGULATIONS

Cartels tend to be more effective when membership is restricted to a few. Yet the NCAA is a large organization by cartel standards. Membership is generally open to accredited colleges and universities that agree to abide by NCAA rules and regulations, but acceptance into the elite status of Division I–A is considerably more restricted. Conference membership is even more limited and results in smaller groupings, usually about ten, of more homogeneous member institutions. The divisional and conference membership requirements are viewed by some as barriers to entry, imposed to limit cartel membership and to facilitate discipline and compliance within a smaller group. Membership eligibility for Division I–A, for example, is stated in terms of such criteria as stadium capacity, average attendance, and the number of varsity sports offered. Nevertheless, divisional and conference groupings also facilitate cooperation in academic and administrative areas not related to athletics.[8]

The NCAA has developed an extensive set of rules and regulations, rules that cover in minute detail virtually every aspect of the operation of an athletic program. More important, the rules apply to essentially everyone who works or has worked for or with a school's program, paid or unpaid, and to anyone who has ever had any association with any part of a school's athletic program. The NCAA effectively extends its jurisdiction by making a member institution responsible for the actions of what the NCAA terms the school's "representatives of athletics interests." An individual becomes a "representative" of athletics interests if she or he has at any time

- been a member of a sports booster club.
- made a donation to any of the university's men's or women's athletic programs.

- arranged for or provided summer employment for enrolled student-athletes.
- contacted (by letter, telephone, or in person) a high school student, grades nine to twelve, for the purpose of encouraging the student to participate in the university's athletic program.
- assisted in providing any benefit to enrolled student-athletes or their families.
- been involved in any way with the university's athletic program.
- held season tickets in any sport.

The NCAA stipulates that once a person has become an "athletic representative," he or she retains that status *forever*, even if the individual no longer contributes to or supports the athletics program.[9]

For a better appreciation of the detail of NCAA regulations, consider the following items in the *NCAA Manual*: (1) A prospective student-athlete (PSA) is a person who has begun classes for the ninth grade. It is possible, however, for younger student-athletes to be considered as prospects for the purposes of NCAA rules. (2) Recruiting is any solicitation of a prospective student-athlete or the PSA's family or guardian by a university staff member or by a representative of the university's athletics interests for the purpose of securing the PSA's enrollment at the university and/or participation in its intercollegiate athletics program. (3) The extra benefit rule provides that university athletic representatives cannot provide an extra benefit or special arrangement to a prospective or enrolled student-athlete. Activities that are prohibited include, but are not limited to, the following:

- Providing gifts or free or reduced-cost services.
- Providing a loan or arranging or cosigning for a loan.
- Employing relatives or friends of a PSA as an inducement for the enrollment of a prospect.
- Providing use of an automobile.
- Providing rent-free or reduced-rent housing.
- Providing tickets to an athletic, institutional, or community event.
- Entertaining or contacting a PSA or a PSA's family on or off campus.
- Promising to provide any of the above.

A person *may* provide benefits or services that are demonstrated to be generally available on the same basis to the *entire* student body.[10]

It is interesting that universities, through the NCAA, set such a limitation on their quest for excellence in athletics without imposing similar limits or adopting a similar philosophy for students other than athletes. Universities have neither ideological objections to nor rules against generous stipends for National Merit Scholars, students who march in the band, sing in the glee club, or act in campus plays, and not for students who play the violin, write poetry, debate, edit campus publications, or design computer software. Universities pay students to come to their institutions, calling the payments scholarships, fellowships, grants, gifts, or awards, and they compete vigorously for the best of the lot.

If a dean successfully solicits from a donor a large gift to support a scholarship that allows the school to enroll the nation's outstanding violinist among the year's high school graduates, the reaction would be immediate: The dean would be applauded for vision and skill, the donor honored for philanthropy, and the student praised and put on public display. A similar scenario in athletics (for example, bringing to a school the nation's premier running back) would make the coach a crook and the donor a bum and would bring the student ridicule, disgrace, and ineligibility.

The NCAA's technique of holding a member institution responsible for the actions of individuals over whom the NCAA has no jurisdiction is both creative and effective. The NCAA assumes no responsibility for the way in which schools are to supervise and regulate the actions of their alumni, donors, friends, supporters, or season ticket holders, but the consequences of any actions that constitute NCAA violations by any such individuals or groups may have the most serious consequences for the university involved. The university may be placed on probation, denied the right to appear on television or play in postseason bowls, have limits placed on the number of athletic grants-in-aid it can offer, or even suffer the so-called death penalty (be prohibited from playing a sport at all), as was the case with Southern Methodist University (a seven-time repeat violator), a few years ago. Beyond that, the NCAA may use the activities of alumni, booster groups, or others as a basis for finding a school guilty of the violation called "lack of institutional control," a highly subjective violation but one considered very serious. With regard to individuals who are found to have violated NCAA rules, the NCAA may require that the university "disassociate" a person from any future contact with any aspect of the institution's athletic programs, either for a specified term or indefinitely if the infraction is sufficiently serious.

One need consider only momentarily the very large number of people involved—coaches, players, potential student-athletes, recruits, friends, alumni (the list goes on and on)—and the potential opportunities for contact of one kind or another, to understand that there is absolutely no way to supervise the universe of interactions or detect any significant portion of what knowledgeable people are confident is a huge and ongoing parade of violations. In this arcane world of unenforceable regulations, the NCAA goes one step further and imposes on those involved in an institution's athletic programs the obligation to report any suspected violations by the member or its representative. Failure to report is itself a violation, and each year university officials from the president through athletic department staff must sign affidavits indicating that they are not aware of any unreported violations.

Amateurism as Defined by the NCAA

The concept of amateurism is at the center of the NCAA's philosophy and purpose, and the NCAA's definition of *amateur* is at the heart of a controversy that continues to engulf intercollegiate athletics. The NCAA constitution states:

> Student-athletes shall be amateurs in an intercollegiate sport, and their participation should be motivated primarily by education and by the physical, mental and social benefits to be derived. Student participation in intercollegiate athletics is an avocation, and student-athletes should be protected from exploitation by professional and commercial enterprises.[11]

It is this view of amateurism, dating back to its origin, that provides the basis for the NCAA's claim to special status and privilege before the courts and supports the kind of regulation by the NCAA of student-athletes that many view as exploitative and at odds with basic economic understanding.

The NCAA's position is supported by most college presidents and coaches. It is their contention that the student-athlete is given a unique opportunity and what amounts to favored treatment that can in no way be equated with exploitation. They insist that amateurism is a worthy ideal, well matched to the ideal of the college student in pursuit of learning and truth. Moreover, they point out, even the most successful athletic programs of Division I–A schools—the few that are said to show a profit—could not survive if it were necessary to share with the players the revenues generated by the program. They see no incongruity in a situation in which everyone else associated with the athletic program derives direct financial benefit from the sport while the players are excluded. Nor do they seem troubled by the fact that the amateur student-athletes function in an environment increasingly commercialized in almost every way at the hands of the not-for-profit institutions on whose behalf the student-athletes compete.

In a recent U.S. Court of Appeals decision, the NCAA got a significant boost for its view of amateurism. A former Notre Dame fullback and top-rated draft prospect, Braxton Banks, challenged the rules on eligibility as a restraint of trade. He had been injured, tried out for the NFL, was not drafted, and wanted to return to college and play out a remaining year of eligibility. The court upheld the NCAA and its rule prohibiting college play after entering the NFL draft. The majority held that NCAA member schools could not be "purchasers of labor" because "the operation of the NCAA eligibility and recruiting requirements prohibits member colleges from engaging in price competition for players."[12] The court went further, stating that "NCAA regulations preserve the bright line of demarcation between college and 'pay for play' football."[13] A strong dissent took the contrary view and would have found the NCAA regulations involved to have an anticompetitive effect in the market for college football players.

To sense the full incongruity of the NCAA's position, one need only consider the case of University of Michigan sophomore Chris Webber, whose skills as a basketball player were valued on one day, by NCAA standards, at tuition and fees, room, board, and books, and the next day were valued, by NBA standards, at $74.4 million as a rookie starting forward for the Golden State Warriors. Even allowing for the fact that Webber's contract is for fifteen years and, as he told the *Detroit Free Press*, Golden State "doesn't have more than $1.6 million to give me for the first year" (limited by the NBA salary cap), the discrepancy is striking.[14] Clearly, there was enormous value to Webber's talent as a college player that the NCAA's system did not allow him to realize.

Why, then, has the NCAA been able to enforce its curious definition of amateurism and safeguard its position as a monopolistic price setter for athletic talent? Why is it powerful enough to impose its strict and far-reaching regulations governing recruiting, financial aid, employment, participation in nonsanctioned athletic events, relationships with sports agents, tryouts for professional teams, eligibility, academic load, progress toward a degree, ability to transfer, and receipt of anything of value from a college or its athletic representatives? The answer is simple: College sports are the avenue to a professional career in football and men's basketball. The NCAA serves not only as the gatekeeper for the NFL and NBA, but it is also the virtually monopoly purchaser of premier football and basketball talent coming out of high school.

Unfortunately, NCAA "amateurism" is far less pure than many of its advocates suggest or perhaps would like to believe. The "athletic scholarship" (an intriguing term in itself) is clearly a means of compensating a student for coming to an institution to play a sport. So it is not that universities do not compensate the athletes. They do. It is how much the universities pay them and how they go about doing it that is at issue. The NCAA's rules may be a convenient mechanism for the schools to determine, collectively, what the maximum cost of acquiring the student-athletes' services will be and then to enforce that limit on the student-athletes and eliminate price competition in recruiting.

VIOLATIONS OF NCAA RULES

Most coaches of major football and men's basketball programs know what goes on in the recruiting process in their relevant market and are therefore one of the best and most likely sources of information concerning violations of NCAA rules. They know who among their colleagues cheat and who do not. They know the types of NCAA violations that are likely to be committed or that have been committed. Some can tell you the details of recruiting visits and even the terms of the offers made to individual athletes in violation of NCAA rules.

Representatives of one school may or may not choose to report violations by other schools. The reasons for reporting or not reporting are as varied as the personalities of individual coaches, athletic directors, or university presidents and as varied as the particular circumstances of the violation. One coach is less likely to report another coach if the two are personal friends, if their programs have strong historical relationships, if both coaches have shoe contracts or consulting arrangements with the same athletic equipment firm, if they work together in sports camps or clinics, or if they do not often compete head-to-head, recruiting for the blue chip athletes. Other reasons exist for reluctance to report violations. If a school is on probation and cannot appear on television, then a scheduled (innocent) opponent may lose an opportunity for television exposure and revenue. If a team in a conference is on probation and not allowed to appear in bowl games, then other schools in the conference may lose money that would have been shared among conference members had the team on probation gone to a bowl. In a broader sense, it is simply not good for a conference or other member schools to have a conference member under NCAA sanctions. It is bad for credibility, and it hurts financially.

There are, on the other hand, circumstances that increase the likelihood of one coach or school reporting another for known or suspected violations. Sometimes a school or coach feels a special sense of damage, insult, or resentment as the result of particular violations. Special dislike or disrespect for those committing the violations, either the coach or the school involved, may lead to reporting an infraction. This is more likely to be a factor when a transgressor has a history or pattern of known or suspected violations. Furthermore, some coaches believe that it is always an advantage in recruiting to have a competitor on probation.

It would be difficult to overemphasize the subjective and *ad hominem* aspects of the decision-making process regarding reporting a fellow coach or school.

NCAA CERTIFICATION OF ATHLETIC PROGRAMS

The 1993 NCAA Convention established a new program of athletics certification for Division I institutions.[15] The certification program represents a significant extension of the NCAA's ability to establish criteria for institutional conduct and strengthens immensely its ability to enforce its policies and its values on individual schools and to punish those who stray from its behavioral norms. Each Division I member institution is now required to complete, at least once every five years, an institutional self-study that must be verified and evaluated through external peer review. The peer review and evaluation is to be conducted according to NCAA guidelines and procedures.

The certification process itself is based on an examination of the member institution's operations and compliance in four basic areas: institutional mission, academic integrity, fiscal integrity, and commitment to equity. In the area of academic integrity, the self-study and peer review must address such matters as the admission and graduation of athletes, academic authority, and responsibility for academic decision making relative to matters affecting student-athletes. Concerning fiscal integrity, the guidelines call for an examination of the control and oversight of all funds from all sources for athletics, the establishment of audit procedures, and a requirement that all expenditures for athletics be handled in accordance with NCAA rules. Equity includes issues of gender, minorities, and matters of general student-athlete welfare, particularly fair treatment of student-athletes in their academic role as students.

The sanctions made available through the certification process, when combined with its now frequently used violation of "lack of institutional control," give the NCAA a wide range of powers that can be exercised subjectively to enforce its rules and regulations against offenders. In November 1993, twenty-five universities were under NCAA sanctions, nine having been cited for violations in men's basketball, eight in football, four in wrestling, two each in men's and women's track, and one each in women's gymnastics and men's cross-country. More significantly, fourteen of the twenty-five schools had been cited for lack of institutional control.[16] If Fleisher, Goff, and Tollison are correct in their contention that historically the NCAA's enforcement process has been used to protect the revenues and the athletic programs of traditional football powers from relative newcomers who aspire to prominence in their developing programs, then the establishment of the NCAA certification program will be bad news for the rising aspirants and the mavericks, and a powerful new tool will be in the hands of the establishment.[17]

TELEVISION

Television changed college sports in many ways, but none was more important than the access to revenue from the sale of television rights (see Table 11-2). The NCAA was quick to recognize both the earning potential of televised college sports (initially football) and the advantages of centralized control of the market for televised college sports. Any monopolist worth its salt understands the desirability of being able to control supply so as to be in a position to maximize profit. The product was amateur sports or, more specifically, televised NCAA football games, a product that the NCAA has worked hard to differentiate from competitive products in the form of professional sports. The seller was the NCAA, which, as early as 1951, moved quickly to impose restrictions on televising football games. As the sole supplier of college football for television, the NCAA could negotiate the best exclusive contract possible with the broadcasters and deal with the problem of distributing revenue as an internal matter—or, as some would say, within the cartel. Complex institution that the NCAA is, its membership was not of one mind as to which teams should be on television, how often they should appear, or how the television revenues should be divided.

It was in this context that some old differences as well as a few new ones came into sharp focus. The NCAA was forced to address such questions as the following: Should top-ranked teams have their appearances limited, held below what the viewing public and the broadcasters would prefer, in order that more teams might have television exposure? On what basis should funds be allocated, according to national ranking, television appearances, winning records, or a more egalitarian basis? Should revenues be allocated to all member institutions or only those who helped generate them? How much of the revenue should be allocated to members, and how much of the revenue should go to the NCAA itself, and for what purposes? What new functions, if any, should be undertaken by the NCAA in its own name, using money from this new source? To what extent should advertising and promotion showcase the NCAA as an organization, building its identity and name recognition through its role as the supplier of this product it now controlled? The same questions concerning the division of revenues exist today with regard to income from televising the NCAA Division I basketball tournament.

Clearly, the advent of television brought with it a period of increased affluence and popularity for big-time college sports, primarily football and men's basketball. It allowed the NCAA to acquire importance, influence, and power that would not have been possible otherwise.

THE ROAD TO THE FINAL FOUR

When a once unprofitable activity begins to make money, and especially if it seems to be making a great deal of money, it comes to be viewed quite differently by many who have, or would like to have, "a piece of the action." This is what happened in intercollegiate basketball. Unlike football, with its postseason conglomeration of bowl games (a mix that seemed to change, at least at the margin, almost every year), NCAA basketball had a national championship tournament. Each year, after regular season play and conference play-offs, sixty-four teams embarked on the highly publicized "road to the final four." Basketball, as it turned out, might well have been a sport designed for television, as it televises extremely well. The play tends to be fast and exciting; viewers easily identify with the players; and the basic structure of the game is easy for fans to grasp, all of which made for a very marketable product. Furthermore, the NCAA tournament, with teams from all parts of the country, provides good demographics for a national market, and the timing is such that other sports calendars do not offer major competition. Not only that, the tournament schedule can be arranged in such a way that it will not be a major intrusion on the academic life of the players. The result has been that the NCAA Division I Men's Basketball Tournament has become both popular and profitable. It has become, among other things, the NCAA's cash cow.

The fact that the tournament is an NCAA national championship event means that its conduct and control belong to the NCAA. In football, by contrast, the NCAA sanctions the bowls and has rules that govern certain aspects of them, but it does not operate the bowl games, negotiate for them, or share directly in the revenues from them. The bowls, the conferences, and/or the independent schools involved are the responsible parties. These differences are not lost on those who discuss the pros and cons of an NCAA national championship football play-off.

THE COLLEGE FOOTBALL BOWL COALITION

The College Football Bowl Coalition was officially established on January 23, 1992. It is a contractual relationship among six bowls (Orange, Fiesta, Cotton, Sugar, Gator, and John Hancock), six conferences (Atlantic Coast, Big East, Big Eight, Pac Ten, Southeastern, Southwest), and one independent university (Notre Dame). The coalition describes its main purpose as being "to create and perpetuate a fair and mathematically formulated selection process" for filling positions in the coalition's four New Year's Day bowls, plus the Gator and John Hancock bowls, which are played on days other than New Year's Day. Four of the six bowls had prior contractual agreements with individual conferences specifying that the conference champion would be the host team for that bowl. This arrangement means that after the end of the regular season, the coalition's selection process on "Bowl Day USA" fills eight berths for the New Year's Day bowls and three of the four berths for the two remaining bowls. The selections are determined on the basis of final regular season rankings in the Bowl Coalition Poll, which is a combination of the Associated Press sports writers' poll and the *USA Today*/CNN coaches' poll. To be considered for a bowl invitation, a team must have at minimum the NCAA-mandated six wins against Division 1–A opponents.

The coalition speaks of its desire to "facilitate" a matchup of the first- and second-ranked teams when that is possible but insists that "the Coalition was never meant to *create* a national championship nor to be a forerunner for the creation of a national championship."[18]

Those who formed the coalition and those who favor it cite what is conceded to have been a long and often unsatisfactory history of bowl selection. When prior contractual arrangements did not identify the participants, the process tended to be dominated by a few of the more influential coaches of perennial powerhouse football programs. Favoritism was suspected and openly discussed, and the influence of the dominant persons and schools was resented. Competition increased as the number of bowls and the amounts of money involved grew. Commitments were commonly sought and given by bowls and universities with little regard for rules, regulations, or established dates for agreements, a situation that might be described as an actively competitive market.

The coalition promises to bring order and stability to the process, an argument frequently made by those who wish to restrict a market. The restrictive nature of the coalition's agreement among its members—the bowls, conferences, and Notre Dame—has caused unhappiness not only among outsiders but even among some universities that are members but nonetheless feel that their treatment by the coalition has brought them less prestigious bowl berths than their records warrant. There is the possibility that the members of the coalition might choose not to continue the arrangement in its present form. There is, in addition, the possibility of legal action against the coalition under the antitrust laws as an unlawful restraint of trade. It does seem likely, however, that some system designed to maintain a relatively stable and predictable bowl environment will be attractive to the major bowls and the traditional football powers unless and until some form of play-off for the national championship can be arranged.

From the NCAA's standpoint, there could be significant advantages to a Division I–A play-off monopolistically operated as an NCAA national championship event. One can see what is at stake from the list of bowls and their payouts in Table 11-6.

COACHES AND COMPENSATION

Television has also made possible the creation of the celebrity college coach. College sports have never been without coaching heroes—legendary figures like Knute Rockne, football coach at Notre Dame, and Adolph Rupp,

TABLE 11-6   Division I-A 1993–1994 Bowl Games

Bowl	Per Team
Aloha	$ 750,000
Alamo	700,000
Carquest	1,000,000
Citrus	2,500,000
Copper	700,000
Cotton	3,100,000
Fiesta	3,000,000
Freedom	700,000
Gator	1,500,000
Hall of Fame	1,000,000
Holiday	1,700,000
Independence	700,000
John Hancock	1,100,000
Las Vegas	228,000
Liberty	1,000,000
Orange	4,200,000
Peach	1,125,000
Rose	6,500,000
Sugar	4,150,000

basketball coach at Kentucky. The earlier heroes were covered by newspapers and later radio and even newsreels shown at movie theaters, but nothing approached the coverage that became possible with television. Celebrity status for a college coach means money, recognition, notoriety, influence, marketability, and a power base outside the university. It also means great pressure to win. Television coverage of a school's games can bring the school favorable publicity that might influence enrollments, private donations, and even state appropriations. It also makes the coach, especially one who wins consistently, a very valuable and often a highly prized asset. Examples are Lou Holtz, football coach at Notre Dame, Bob Knight, basketball coach at Indiana University, and Mike Krzyzewski, basketball coach at Duke.

A major consequence of this status is the compensation packages that such coaches are able to command. The compensation packages for winning college coaches at the highest competitive level vary in detail among universities, but the general level of total compensation is now so high that it raises serious questions within the university community and in the media. In addition, some of the sources of coaches' compensation raise questions of propriety and possible conflict of interest. The total annual compensation for a successful head coach of football or men's basketball in a Division I-A school might range from a low of $200,000 to $250,000 to as much as $1 million or more. It is not unusual for such a head coach to be paid a higher university salary than the university's president, though in recent years there has been a tendency to keep coaching salaries, as shown on the university's payroll, more in line with the salaries of high-level university administrators and distinguished faculty. However, a coach's total compensation package usually far exceeds his official salary because of supplementary perquisites such as deferred compensation and annuity plans, supplemental insurance coverage, television and radio shows, sports camps, clinics, speaking engagements, personal appearances, corporate board memberships, product endorsements, low-interest loans, housing, and, for the more fortunate, shoe and clothing contracts and consulting arrangements.

The shoe and clothing contracts, in particular, have generated allegations of abuse and conflict of interest. As the amounts involved rose—in some cases reportedly as high as $1 million for individual coaches—faculty members and university administrators, among others, began asking more seriously, Why should it be the coach, rather than the university, who reaps the financial benefit from deciding what brand of shoe the team will wear? Such pressure led to an amendment of the NCAA's rules requiring Division I and Division II coaches to have prior approval in writing each year for athletically related income from nonuniversity sources and for use of the school's name or logo in connection with any endorsements for personal gain. Some schools have gone further. The University of North Carolina, for example, recently announced a four-year programwide contract with Nike that will give the university $4.7 million.

IV. CURRENT ISSUES

ANTITRUST AND THE COLLEGE FOOTBALL ASSOCIATION

Through the years, discontent over television policy grew within the NCAA. Many of the major football powers, dissatisfied with their television exposure and their share of television revenues, demanded more, if not complete, control over their own destiny.

The College Football Association (CFA) was an outgrowth of this dissatisfaction in the 1970s. The CFA had as its members most of the Division I–A schools (that is, most of the big-time football programs) except for two conferences, the Big Ten and the Pac Ten. There was some discussion among CFA members about leaving the NCAA, but with the Big Ten and the Pac Ten unwilling to join the movement, the issue was never brought to closure. Instead of leaving the NCAA, the CFA decided to challenge the NCAA directly by seeking a television contract separate from the NCAA's exclusive agreement. In 1981 the CFA and NBC signed an agreement that would address both concerns of the CFA schools: It would mean more television appearances and more money. The NCAA responded by threatening disciplinary action, indicating that sanctions would apply to football and other sports as well. The issue was drawn later that same year when the University of Oklahoma and the University of Georgia filed suit against the NCAA, alleging that the NCAA television plan was an unlawful restraint of trade under the Sherman Antitrust Act. In 1984 the U.S. Supreme Court decided the case in favor of the universities and against the NCAA, holding that the NCAA television plan was an unreasonable restraint of trade and therefore invalid. Without resolving the uncertainty and ambiguity that continue to surround intercollegiate athletics, the Court stated:

> The NCAA plays a critical role in the maintenance of a revered tradition of amateurism in college sports. There can be no question but that it needs ample latitude to play that role, or that the preservation of the student-athlete in higher education adds richness and diversity to intercollegiate athletics and is entirely consistent with the goals of the Sherman Act. But consistent with the Sherman Act, the role of the NCAA must be to preserve a tradition that might otherwise die; rules that restrict output are hardly consistent with this role. Today we hold only that the record supports the district Court's conclusion that by curtailing output and blunting the ability of member institutions to respond to consumer preference, the NCAA has restricted rather than enhanced the place of intercollegiate athletics in the Nation's life.[19]

This decision broke the NCAA's monopoly over the televising of college football and cleared the way for the CFA, the conferences, and individual schools to seek their own television contracts. The result has been far more extensive telecasting of college football games and a broader choice for the viewing public. Ironically, less than ten years after the Supreme Court decision to open the market, the Federal Trade Commission is now examining the CFA's television contract for possible antitrust violations!

GENDER EQUITY

Gender equity deals with a form of discrimination that is one of the most difficult and potentially divisive issues in college sports. To some it seems a simple and straightforward matter of complying with the unambiguous requirements of the law and fundamental notions of fairness. To others it is a concept that, though desirable in theory, is unrealistic and threatens the viability of football and may well mean the end of men's nonrevenue sports.

The basic legislation at issue is Title IX of the Education Amendments of 1972, which prohibits discrimination on the basis of sex in any educational program or activity receiving federal financial assistance. When the Department of Health, Education and Welfare (HEW) (now the Department of Health and Human Services) began developing regulations to implement Title IX and to achieve equity for women in college sports, the NCAA and many of the universities with major football programs urged that the revenue-producing sports, especially football, be excluded from coverage. The concern was that it would be impossible to fund comparable numbers of men and women student-athletes in various sports without either eliminating many or all of the men's nonrevenue sports or bankrupting the football and other men's revenue sports, especially men's basketball. It was argued that there were no sources of additional funds to support both men's and women's programs. Assertions were also made that not enough female athletes were graduating from high school with sufficient athletic skills to justify financial aid for intercollegiate competition to fill all the women's programs that would have to be established to achieve equity or comparability between men and women.

HEW was not convinced, and Congress was unwilling to create a statutory exception. The result was that the Title IX regulations issued by HEW and approved by President Gerald Ford in 1975 extended to all athletic programs. That was not the end of the matter, however. A subsequent court ruling held that Title IX applied only to those programs or areas within an institution that received federal financial aid, with the result that intercollegiate athletic programs, which typically receive no direct federal financial aid, were excluded from coverage. The respite was only temporary, for in 1988 Congress again addressed the matter and enacted legislation establishing that Title IX applies to all programs of an institution if *any* part of the institution receives federal financial assistance. Thus the universities and their athletic departments, having successfully avoided the issue for more than a decade, were face to face with a reality that was not going to go away.

In March 1992, the NCAA released a study that lent a quantitative dimension to the gender equity issue. The study revealed a fundamental imbalance in the allocation of resources in intercollegiate athletics based on gender: Although the total enrollment of men and women in colleges and universities was approximately equal, men's teams received 70 percent of grant-in-aid funds, 77 percent of operating funds allocated to athletics, and 83 percent of funds available for recruiting.[20]

The U.S. Office of Civil Rights, the federal agency responsible for enforcing Title IX, has indicated that intercollegiate athletic programs are expected to (1) provide participation opportunities for male and female students in numbers substantially proportionate to their respective enrollments, (2) show a continuing practice of program expansion for women (the underrepresented gender), and (3) demonstrate that the interests and abilities of women are fully and effectively accommodated by the program, to the same degree as those of men, including levels of competition and the range of sports offered.

To say that the gender equity issue has raised a storm of controversy would be an understatement. It comes to the fore at a time when higher education is confronting harsh budget problems and cutbacks. Cost containment has become a major thrust of the NCAA, and the media are filled with stories indicating that most intercollegiate athletic programs, including many of the big-time winners, are losing money rather than making a profit. Cutting costs and providing equity for women are operating at cross-purposes. In its starkest terms, the issue boils down to one side saying "Equity now!" while the other insists that gradualism is necessary to preserve the system. The NCAA is still searching for a compromise.

ADMISSION STANDARDS AND RACE

Admission standards for student-athletes have been an issue for decades. Admitting to a college those athletes who fail to meet its academic admission standards is a practice that dramatizes what so many people decry as an overemphasis on athletics and on winning, to the detriment of academic values. Over the years, the NCAA has dealt with the question of academic standards for athletes, but never with any real success. One might think that this would be a subject that the universities could easily agree on and that they could be expected to enforce the rules enthusiastically themselves. Nothing could be further from the truth. An unending succession of violations and abuses have been perpetrated by school after school of virtually every academic quality and standing. Apparently, the desire to win, the pressure to win, and the benefits of winning are too powerful for university officials to resist.

Despite the continuing calls for reform, however, the NCAA's more recent efforts to raise admission standards for athletes have not been greeted with support from all quarters. Prominent black basketball coaches, such as John Thompson of Georgetown and John Chaney of Temple, and other members of the Black Coaches Association, voice strong opposition to higher admission standards, which they believe reduce opportunities for black athletes to get "athletic scholarships." Proposition 48, originally adopted in 1983, established minimum admission requirements of a combined 700 on the SAT (or a composite 15 on the ACT) and a 2.0 high school grade point average (GPA) in an eleven-course core curriculum of high school academic courses. It was strongly opposed as discriminatory and was asserted to have a disproportionate impact on minorities. As originally adopted by the NCAA, Proposition 48 allowed what was termed a "partial qualifier" (a student-athlete with a 2.0 high school GPA but not meeting the other academic requirements) to receive financial aid in the freshman year but not be eligible for varsity competition or practice. In addition, the partial qualifier lost a year of eligibility. Proposition 42, adopted subsequently, revised Proposition 48, eliminating the category of partial qualifier and with it the opportunity for first-year financial aid for any student-athlete not meeting all the requirements of Proposition 48. It also promulgated a new sliding scale, requiring a student-athlete to have a 2.5 high school GPA in thirteen core academic courses and a 700 SAT or 17 ACT score. The scale is adjustable to a 2.0 high school GPA with a 900 SAT or 21 ACT.

The opposition to these admission standards, which are still considered too low by some, has nevertheless grown stronger and louder. Black coaches have threatened protest actions and have combined their strenuous objections to Proposition 48's standards with other grievances against the NCAA and its member institutions. These include (1) the absence of black officials on the NCAA staff and in the athletic departments of universities, (2) what they view as the small number of black coaches, and (3) too few black faculty members in the universities. The black coaches also protest the NCAA's cost-containment measures that reduce the number of scholarships in basketball (from fifteen to thirteen) and reduce the size of coaching staffs. They contend that these actions, although taken in the name of budget cutting, have the effect of severely limiting opportunities for minority student-athletes and coaches. They also are apprehensive about the impact of the move toward gender equity on themselves, their programs, and their student-athletes. They fear that minorities, especially black males, will bear the brunt of budget reallocations, cutbacks, and other actions taken to help achieve gender equity.

On the issue of admission requirements, the NCAA points to early and still limited data indicating that Proposition 48's standards and the response to them in the high schools have already had a favorable effect on graduation rates for student-athletes. The NCAA leadership still insists on the necessity of raising academic standards, including admission requirements, and improving the academic performance of student-athletes. It is not clear how this dramatic conflict in values will be resolved.

V. CONCLUSION

One of the long-standing criticisms of intercollegiate athletics has been that college presidents assert relatively little leadership or influence on NCAA policy. The charge has been that presidents as a group have somehow defaulted, leaving actual decision making within the NCAA and on campus to their athletic bureaucracy—the athletic director, an influential coach, the faculty athletic representative, or some group of athletic functionaries. The argument has been that this absence of presidential involvement has contributed to the problems of college athletics at both the national and the campus levels.

The conclusion seems inescapable that fundamental change or basic reform in college sports and the NCAA is highly unlikely without the emergence of some external force, not yet on the horizon, capable of introducing a new sense of reality and overcoming the deeply entrenched special interests that have long dominated the world of intercollegiate athletics.

NOTES

1. "Cleaning up College Football," editorial, *New York Times*, August 29, 1993, p. E14.
2. National Collegiate Athletic Association, *1993–94 NCAA Manual* (Mission, KS: National Collegiate Athletic Association, 1993), p. 1.
3. Howard J. Savage et al., *American College Athletics* (New York: Carnegie Foundation for the Advancement of Teaching, 1929).
4. *Keeping Faith with the Student-Athlete: A New Model for Intercollegiate Athletics*, Report of the Knight Foundation Commission on Intercollegiate Athletics, March 1991.
5. Recent legal actions involving former University of Nevada at Las Vegas (UNLV) head basketball coach Jerry Tarkanian pose some interesting questions of law, NCAA policy, and university ethics. Tarkanian was a successful, winning coach at UNLV, but he had a long history of problems with the NCAA over charges of UNLV's recruiting violations and improper benefits for players. After years of investigations, hearings, and lawsuits, Tarkanian ended up in the U.S. Supreme Court. He alleged, among other things, that he, an employee of the State of Nevada, was entitled to certain protection under the U.S. Constitution, including the right to due process of law, before personnel action could be taken suspending him from his job. The Supreme Court rejected Tarkanian's plea, however, and held that the NCAA is a private organization and therefore does not have to afford Tarkanian due process before requiring UNLV to suspend him as its head basketball coach. UNLV, on its own initiative, would not have been able to suspend Tarkanian without according him due process, because the university is a public institution and would be engaged in a "state action" and thus would be obligated to provide constitutional safeguards. The Supreme Court stated, therefore, that when UNLV is in its role as an NCAA member and acting at the direction of the NCAA enforcement mechanism, its head basketball coach loses his rights under the U.S. Constitution.

The State of Nevada, following the Supreme Court decision, enacted a statute specifying the legal rights of Nevada state employees accused of wrongdoing by any national college sports association. The statute was intended to ensure employees of the State of Nevada basic constitutional rights, such as the right to a hearing before an impartial arbiter, the right to be represented by legal counsel, the right to confront accusers, and the right to appeal to a court of law. The statute also set certain restrictions on the admissibility of evidence. Some, but not all, of these rights are provided under existing NCAA rules.

In reviewing the Nevada statute, the U.S. Court of Appeals for the Ninth Circuit found it to be unconstitutional, concluding that it was an invalid attempt to regulate interstate commerce. The NCAA reportedly views this latest decision as sending a strong message to those states that want to regulate the NCAA's enforcement procedures. Three states— Florida, Illinois, and Nebraska—have enacted legislation in this area.

Tarkanian's lawyer had a different interpretation: "[T]he NCAA will be able to continue to act as police department, prosecution, judge, jury, the Court of Appeals and the Supreme Court without providing due process to those that they're prosecuting."

The NCAA now seems to have an enforcement mechanism, protected from judicial review, that would be the envy of any cartel anywhere in the world.

6. *Lansing State Journal* (Lansing, MI), November 25, 1993, p. C3.
7. A. Knopf Kenyon, *A Lexicon of Economics* (San Diego: Academic Press, 1991), p. 37.
8. For example, the presidents of the Big Ten Conference, an athletic conference, regularly discuss academic and budget matters of common interest to the member universities at Council of Ten meetings. In addition, the academic vice-presidents of the Big Ten schools (plus the University of Chicago) meet regularly as the Committee on Institutional Cooperation (CIC). One might note, by the way of analogy, that the six sheikhdoms of Saudi Arabia, Kuwait, Bahrain, Qatar, the United Arab Emirates, and Oman, a six-nation subgroup of the Organization of Petroleum Exporting Countries (OPEC), the world's oil cartel, meet regularly as the Gulf Cooperation Council to discuss national security and other matters of common concern in the Arab Gulf.
9. This example is based on information contained in a publication of the University of Illinois, 1st ed. (Champaign: Division of Intercollegiate Athletics, 1992), intended for its alumni and friends, to help them avoid violations that would be the responsibility of the university. It is an illustration of the ways in which universities must try to protect themselves and attempt to demonstrate their institutional control of athletics to the NCAA's satisfaction.
10. Ibid.
11. NCAA, *1993–94 NCAA Manual*, p. 4.

12. *Banks* v. *NCAA*, United States Court of Appeals, Seventh Circuit, 977 F. 2d 1081, 1091 (1992).
13. Ibid., p. 1090.
14. *Detroit Free Press*, October 18, 1993, p. A1.
15. NCAA, *1993–94 NCAA Manual*, pp. 435–439.
16. *Chronicle of Higher Education*, November 17, 1993, p. A44. The schools under sanction were Ashland University, Auburn University, Clemson University, Howard University, Jackson State University, Lamar University, Lock Haven University of Pennsylvania, Middle Tennessee State University, Mississippi College, Oklahoma State University, St. Bonaventure University, Simpson College, Southeastern Louisiana University, Syracuse University, University of the District of Columbia, University of New Mexico, University of Nevada at Las Vegas, University of Texas at El Paso, University of Texas at Pan American, University of the South, University of Tulsa, University of Virginia, Upsala College, Virginia Polytechnic Institutes and State University, and Winston-Salem State University.
17. Arthur A. Fleisher, Brian L. Goff, and Robert D. Tollison, *The National Collegiate Athletic Association: A Study in Cartel Behavior* (Chicago: University of Chicago Press, 1992), esp. chap. 5.
18. *College Football Bowl Coalition Media Handbook* (New York: College Football Coalition, 1993–1994), p. 5. In recent years, financial support for women's intercollegiate athletics has increased markedly, and interest and participation have grown, in large measure as a result of federal mandate. Although a few women's sports, most notably basketball and volleyball, have begun to attract significant fan support and television coverage of women's competition has increased, none has yet reached the point of being a revenue sport. Nor is it possible to determine when a women's sport may become a true revenue sport or even to know with certainty if this will happen. In regard to violations, corruption, and abuse, the men have had what could almost, but not quite, be called a monopoly. Whether gender equity will bring equality of corruption will have to be determined in the future.
19. *NCAA* v. *Board of Regents of the University of Oklahoma and University of Georgia Athletic Association*, 468 US 85, 120 (1984).
20. "Gender Equity," *Athletics Administration*, April 1993, p. 22.

SUGGESTED READINGS

Albom, Mitch. *Fab Five: Basketball, Trash Talk, the American Dream*. New York: Warner Books, 1993.
Bailey, Wilford S., and Taylor D. Littleton. *Athletics and Academe: An Anatomy of Abuses and a Prescription for Reform*. New York: American Council on Education and Macmillan, 1991.

Bosworth, Brian, with Rick Reilly. *The Boz: Confessions of a Modern Anti-hero*. New York: Doubleday, 1988.

Fleisher, Arthur A., Brian L. Goff, and Robert D. Tollison. *The National Collegiate Athletic Association: A Study in Cartel Behavior*. Chicago: University of Chicago Press, 1992.

Guttmann, Allen. *A Whole New Ball Game: An Interpretation of American Sports*. Chapel Hill: University of North Carolina Press, 1988.

Michener, James A. *Sports in America*. Greenwich, CT: Fawcett, 1976.

Sperber, Murray. *College Sports Inc.: The Athletic Department vs the University*. New York: Henry Holt, 1990.

Weiler, Paul C., and Gary R. Roberts. *Sports and the Law: Cases, Materials and Problems*. St. Paul: West, 1993.

Weistart, John C., and Cym H. Lowell. *The Law of Sports*. Charlottesville, VA: Michie, 1979.

PUBLIC POLICY IN A FREE-ENTERPRISE ECONOMY

Walter Adams and James W. Brock

Controlling power in a free society and guarding against its abuse are the core of the American political–economic experience. Indeed, the American nation was forged from the colonists' protest against the arbitrary power of the British Crown.

Once liberated, the nation's founders understood that in creating a governance structure for a free society, they must provide for a government strong enough to prevent individuals from infringing on the liberties of one another. At the same time, they also understood that additional safeguards were required in order to prevent government itself from being transformed into an instrument for oppressing the citizens. Throughout their deliberations, the founders displayed a deeply rooted, dyspeptic view of human nature; they recognized (as Thomas Burke put it in 1777) that "power of all kinds has an irresistible propensity to increase a desire for itself"; that "power will sometime or other be abused unless men are well watched, and checked by something they cannot remove when they please"; and that the "root of the evil is deep in human nature."

Their solution for resolving these dilemmas was incorporated in the Constitution and predicated on two transcending principles: first, that it is the *structure* of government, not the personal preferences and predilections of those who govern, that is of utmost importance; and, second, in Jefferson's words, that "it is not by the consolidation or concentration of powers, but by their distribution, that good government is effected." The master plan was to construct a system of checks and balances, a Newtonian mechanism of countervailing powers, operating harmoniously in mutual frustration. The foremost goal was to prevent what the founders considered the ultimate evil: the concentration of power and the abuses that flow from it.

In *Federalist Paper 51*, James Madison summed up the challenge as follows:

It may be a reflection on human nature, that such devices should be necessary to control the abuses of government. But what is government itself, but the greatest of all reflections on human nature? If men were angels, no government would be necessary. If angels were to govern men, neither external nor internal controls on government would be necessary. In framing a government which is to be administered by men over men, the great difficulty lies in this: you must first enable the government to control the governed; and in the next place oblige it to control itself. A dependence on the people is, no doubt, the primary control on the government; but experience has taught mankind the necessity of auxiliary precautions.[1]

I. THE AMERICAN ANTITRUST TRADITION

Subsequent events, however, demonstrated that in a free society the power problem is not confined to the political arena alone. A century after the Constitution was ratified, during the post–Civil War era, a proliferation of pools, trusts, cartels, and monopolies revealed the need for controlling economic as well as political power. In order to guard against excessive private, as well as governmental, concentrations of power, Americans perceived the need to prevent the "autocrats of trade" from enslaving the people in a new kind of feudalism. "If we will not endure a king as a political power," Ohio's Republican Senator John Sherman warned, "we should not endure a king over the production, transportation, and sale of any of the necessaries of life." Unless Congress addressed the private economic power problem, he stated, there soon would "be a trust for every production and a master to fix the price for every necessity of life."[2]

In theory, Adam Smith had demonstrated how the competitive marketplace would regulate and neutralize economic power—how it would disperse economic decision-making power in the hands of a multitude of rivals, how it would compel each to perform well in the public interest, and how it would thus transform the private vice of self-interest into the public virtue of good economic performance. In theory, Adam Smith had demonstrated how the competitive market system would channel private economic decision making into socially beneficial outlets—how it would compel innovation, technological advance, productivity, and efficiency in allocating resources in accordance with society's collective preferences. In theory, he had showed how the competitive market system would maintain economic freedom and opportunity for all while rendering private economic decision making accountable to the public.

In reality, however, the corporate combination and trust movement of late-nineteenth-century America demonstrated that as a system of governance, the competitive market is neither a self-perpetuating nor an immutable artifact of nature. It demonstrated that without strictly enforced rules of the game, the competitive market can be eroded and subverted, through agreements not to compete as well as through monopolization by dominant firms.

Maintaining the competitive market system as the prime regulator of America's economic affairs is the central objective of the antitrust laws. Like the Constitution, the American antitrust laws provide for a structure of governance—a social blueprint for organizing decision making and for guarding against its abuse. Like the Constitution, the antitrust laws aim to disperse power into many hands rather than tolerating its concentration in the hands of a few. And just as the purpose of the Constitution is to prevent any political "faction" from monopolizing the coercive power of the state, so the basic objective of the antitrust laws is to prevent private firms from monopolizing economic decision making in a free society.

The Sherman Antitrust Act, the nation's first antitrust law, enacted in 1890, outlaws two major types of interference with competitive free enterprise: collusion and monopolization. Section 1 of the Sherman Act, dealing with collusion, states: "Every contract, combination. . .or conspiracy, in restraint of trade or commerce among the several States, or with foreign nations, is hereby declared illegal." As interpreted by the courts, this renders it unlawful for businesses to engage in such collusive action as agreeing to fix prices, agreeing to restrict output or productive capacity, agreeing to divide markets or allocate customers, and agreeing to exclude competitors by systematic resort to oppressive tactics and discriminatory policies—in short, all joint actions by competitors aimed at controlling the market and short-circuiting its regulatory discipline (see Figure 12-1).

Section 2 of the Sherman Act, which deals with structural concentrations of power, provides that "every person who shall monopolize or attempt to monopolize, or combine or conspire to monopolize any part of the trade or commerce among the several States, or with foreign nations, shall be deemed guilty. . .and. . .punished." Section 2 thus makes it unlawful for firms to obtain a stranglehold on the market, by either forcing rivals out of business or absorbing them. It forbids a single firm (or group of firms acting jointly) from dominating an industry or market area. Positively stated, Section 2 encourages a decentralized industry structure in which there are a sufficient number of independent rivals to ensure effective market competition.

The Sherman Act's provisions are general, perhaps even vague, and essentially negative. Directed primarily against *existing* dominant firms and *existing* trade restraints, the act proved incapable of addressing specific practices that could be employed to realize the proscribed results. Armed with the power to dissolve existing monopolies, the enforcement authorities could not, under the Sherman Act, attack the growth of monopoly in advance and prior to its realization. For this reason, Congress passed supplementary legislation in 1914 in order "to arrest the creation of trusts, conspiracies and monopolies in their incipiency and before consummation." In the Federal Trade Commission Act of 1914, Congress established an independent regulatory commission to police the industrial field against "all unfair methods of competition," as well as to undertake expert studies of conditions of competition and monopoly in the American economy. In the Clayton Act of the same

FIGURE 12-1   Free competition versus price-fixing.

Economics of a Free Market

Producers   Consumer

Economics of Security

Producers   Consumer

Source: Thurman W. Arnold, *Cartels or Free Enterprise?* Public Affairs Pamphlet 103, 1945. Reproduced by courtesy of Public Affairs Commission, Inc.

year, Congress targeted four specific practices that experience had shown to be favorite means for creating monopoly positions: (1) price discrimination (that is, cutthroat competition and targeted, localized price-cutting), (2) tying contracts and exclusive dealership agreements, (3) mergers and acquisitions, and (4) the formation of interlocking directorships among competing corporations. These practices were declared unlawful, not in and of themselves, but only when their effect may be to substantially lessen competition or tend to create a monopoly. Thus, price discrimination, for example, would be illegal only if used as a systematic device for destroying competition, as it was in the hands of the Standard Oil and American Tobacco trusts. Similarly, the Clayton Act's merger provisions (as amended in 1950 by the Celler–Kefauver Act) prohibit any corporate mergers and acquisitions whose effect may be to substantially lessen competition or tend to create a monopoly. The emphasis in the Clayton Act, then, is on preventing anticompetitive problems from occurring, rather than struggling to correct them once they already are in place.

The general purpose of American antitrust policy was perhaps most cogently articulated by Judge Charles Wyzanski:

> Concentrations of power, no matter how beneficently they appear to have acted, nor what advantages they seem to possess, are inherently dangerous. Their good behavior in the past may not be continued; and if their strength were hereafter grasped by presumptuous hands, there would be no automatic check and balance from equal forces in the industrial market. And in the absence of this protective mechanism, the demand for public regulation, public ownership, or other drastic measures would become irresistible in time of crisis. Dispersal of private economic power is thus one of the ways to preserve the system of private enterprise. . . .[Moreover,] well as a monopoly may have behaved in the moral sense, its economic performance is inevitably suspect. The very absence of strong competitors implies that there cannot be an objective measuring rod of the monopolist's excellence. . . .What appears to the outsider to be a sensible, prudent, nay even a progressive policy of the monopolist, may in fact reflect a lower scale of adventurousness and less intelligent risk-taking than would be the case if the enterprise were forced to respond to a stronger industrial challenge. . . .[Economic progress] may indeed be in inverse proportion to economic power; for creativity in business as in other areas, is best nourished by multiple centers of activity, each following its unique pattern and developing its own esprit de corps to respond to the challenge of competition.[3]

II. ANTITRUST POLICY UNDER FIRE

In recent years, however, these precepts have been attacked from both ends of the political–economic spectrum.

THE NEOLIBERAL CHALLENGE: "COALITION CAPITALISM"

Coalition capitalism, grounded in "industrial policy," is what the neoliberal left advocates as a substitute for America's traditional antitrust principles.

Lester C. Thurow, economist and erstwhile dean of the MIT Sloan School of Management, is a leading proponent of this view. Interest in industrial policy, he noted, "springs from a simple four letter word—fear. American industry is being beat up by the international competition, and business and labor are both afraid that American industry is going down for the count. . . .What used to work won't work. For the world has changed." As Thurow sees it, the central challenge for public policy "is not 'the market' versus 'planning,' but how planning can be used to improve one's market performance."[4]

According to this view, the only viable course is a consciously constructed national "industrial policy," comprising tripartite planning, negotiation, bargaining, and compromise among representatives of management, labor, and government. Such industrial policies, their advocates claim,

> are both an expression of and a vehicle for bringing about a strategic consensus among government, industry, and labor as to the basic directions in which the economy might be moving. Such a strategic consensus allows each of the individual decision makers to undertake actions that will jointly increase the likelihood that all the economy's economic actors will be successful in reaching their desired collective and individual objectives. . . .Business might, for example, promise new investment and labor changes in work rules if government were willing to help finance additional research-and-development expenditures in a particular industry. Or government might promise changes in the tax code which would make it easier to finance new start-up ventures if labor and business would agree to restrain wage and price increases.

Allied steps could include creating a ministry of technology; implementing a financial infrastructure of public and private banks able to facilitate resource flows and to aid in restructuring "sick" industries; and establishing industrial restructuring boards made up of representatives of labor, management, and government.[5]

According to industrial policy advocates, the overall purpose would be to nurture a "cooperative bubble-up relationship where government, labor, and industry can work together to create world-class competitive American industries"—an associational relationship in which each constituency "can learn about the problems of the others and how they can mutually interact to solve their joint problems." They point to Japan's postwar economic miracle as evidence of what such consciously cultivated cooperation can achieve.[6]

Advocates of coalition capitalism dismiss traditional antitrust concerns about economic power as irrelevant, outmoded, and counterproductive to the challenges posed by the world economy of the twenty-first century. They

> don't care whether General Motors is the only car manufacturing company. It's still in a competitive fight for its life with the Japanese and the Germans. And it doesn't make sense to hamstring General Motors or anybody else with antitrust laws since they must operate in an international competitive environment, whether or not there are other domestic producers.[7]

Instead of concerning ourselves with giant American firms that dominate their markets, they contend, the more important question is, "How can we encourage an intelligent combination of the second rank and smaller companies so that we might have a second successful giant?"[8] They consider it "far better that small American companies be crushed by big American companies than that they be crushed by big foreign companies."[9]

As a guide for national economic policy, however, this vision suffers from at least four fundamental shortcomings.

1. The first drawback is the inevitable tendency toward the coalescence of power among organized groups that are predisposed to protect their pri-

vate interests rather than to promote the public interest. Coalition capitalism naively assumes that powerful private groups will act in ways, and toward ends, that promote better economic performance. But such power blocs may instead recognize their mutual interest in preserving the status quo, opting instead to aggrandize their own power and influence. Such collaboration may immunize them from competition, global and domestic alike. It may enable them to inflate prices and wages while slumbering in a quiet life of inefficiency and technological stagnation. Government participation may serve to legitimize and cement the counterproductive schemes collaboratively contrived by these coalescing power blocs.

This concern is not idle theoretical conjecture but the product of past experience. For example, the National Recovery Administration (NRA), established during the Great Depression, was at the outset intended to achieve the kind of comprehensive management–labor–government cooperation advocated by today's coalition capitalists. It featured management–labor planning, replete with "codes of fair competition" and with overall supervision and enforcement by the government. But the results were scarcely what was hoped for. The NRA enabled business groups to cartelize industries and labor groups to raise wages while government rubber-stamped their self-serving decisions. In 1934, at the conclusion of congressional hearings on the NRA, Senator Gerald P. Nye (Republican, North Dakota) was

> forced to the conclusion that the power of monopoly has been greatly increased during the stay of the NRA; that invitation to monopoly in the United States is greater than ever before. In view of what amounts to suspension of the antitrust laws, the small independent producers, the small business man generally, whether buyer from, competitor of, or seller to large monopolized industries, and the great mass of ultimate consumers, are seemingly without protection other than that given by the NRA. And the NRA is not giving this protection. On the contrary, it has strengthened, not weakened, the power of monopoly.[10]

Advocates of coalition capitalism are strangely silent about this fiasco in American economic statecraft.

The nuclear electric power debacle provides a more recent illustration of this problem.[11] Here is a field where government and industry cooperated for decades to promote what both considered to be a promising "sunrise" technology. Tens of billions of dollars were expended over four decades on state subsidies for research and production of nuclear power plants, equipment, and fuels. Nuclear power provided cost-plus revenues for the monopoly utility firms that commissioned the plants. It provided a seemingly endless gravy train for the labor unions and construction trades that initially built the plants, dismantled them when they constantly required redesigning, and reconstructed them over and over again, drawing their pay all the while. But the consequences for the country have been less than inspiring: astronomical cost overruns, billions of dollars of abandoned nuclear power plants, multimillion-dollar rate hikes, deception of the public regarding risks and health threats, and a lethal, steadily mounting harvest of highly radioactive wastes for which no acceptable method of disposal has yet been devised.

Defense weapons procurement affords another example of coalition capitalism in action.[12] Here, too, there has been close cooperation among government, industry, and labor—so close, in fact, that the same people regularly exchange jobs through a well-lubricated "revolving door" of contractor–government–contractor employment. Here, too, government and industry have operated in close collaboration, as a steady stream of indictments and criminal convictions attest. But with what consequences for the public who pays the bills? Mind-boggling cost overruns; weapons that do not perform to contractual specifications; and ineffective and, at times, nonexistent "testing": a coalition carnival of $600 hammers and $7,000 coffeepots!

In airlines and trucking, management and organized labor cooperated for decades to cartelize the fields under the guise of government regulation "in the public interest." Aided by government, they erected insuperable barriers to new entry. Insulated from new competition, they parceled out routes, divided territories, and fixed prices. And what were the results of such cooperative capitalism for the public? Inefficiency, waste, highly inflated prices and wages, and an almost complete denial of freedom of economic opportunity.

2. Advocates of coalition capitalism also display a misplaced faith in megamergers, corporate giantism, and the consolidation of American industry as prerequisites for world-class economic performance.

Is it not fatuous to suggest that the problems of the American automobile industry are somehow due to the fact that General Motors is too small, when the firm ranks as the world's very largest industrial concern, when it is bigger than its two largest Japanese rivals combined, and when its annual sales exceed the gross national products of all but nineteen of the world's nations? Is it not more realistic to conclude (with *Business Week*) that the "basic problem nagging this biggest, most diverse, and most integrated of car companies is [that] it is just too big to compete in today's fast-changing car market"?[13]

Similarly, can the shortcomings of the American steel industry be attributed to the fact that America's steel giants are too small or that they have been prevented from merging and consolidating, when merger-induced giantism in the industry has proceeded virtually unmolested from the formation of the United States Steel Corporation in 1901 to the consolidation (and subsequent financial collapse) of LTV and Republic Steel in the 1980s? Today, it is the sleek, small "minimills," not Big Steel, that are revolutionizing steel production. It is these technologically sophisticated minimills that not only are profitable but also are capturing one market after another from such allegedly "invulnerable" foreign producers as "Japan Inc." and South Korea.

Nor has elephantine size been an asset in the computer industry. In spite of its massive size, IBM has staggered from one crisis to another during the latter 1980s and 1990s, announcing layoffs of some 140,000 employees, and suffering a catastrophic decline of $75 billion in the stock market value of its securities.[14] "If bigger were better in research and development," says the British *Economist*,

> IBM would be the best. It spends $6 billion a year on R&D. Its researchers have won two Nobel prizes in the 1980s, and thickened IBM's portfolio of patents to over 33,000. Yet, despite all that, IBM's new technologies have time and again been beaten to market by those of smaller, nimbler firms.[15]

Ironically, IBM now seems to be doing what the Justice Department originally sought to achieve in its 1969 monopolization suit against the firm: dismantling its bloated bureaucratic structure.

Abroad, two decades of merger-induced corporate giantism failed to provide salvation for the west European nations (especially Britain and France), which became imbued in the late 1950s with the bigness complex. The putative "national champions" that were merged to dominate their national industries (including British Leyland in automobiles; British Steel; ICI; and Bull in computers) became the lame ducks of the 1970s and 1980s: They lost sales, market share, and employment, and after suffering billions of dollars in losses, they are now dependent on financial life support from their home governments in order to forestall outright collapse.[16]

3. Proponents of coalition capitalism may also have misdiagnosed the source of Japan's postwar economic "miracle." Rather than tightly knit government–industry collaboration or excessive industrial concentration, Japan may have benefited from an intensely competitive industrial milieu: a competitive domestic market structure put into place by the American occupational authorities during the years immediately following the end of World War II.[17]

Under the leadership of General Douglas MacArthur, a massive deconcentration program was implemented in postwar Japan: Sixteen of the country's largest holding companies were dissolved outright; twenty-six were dissolved and reorganized; eleven were reorganized with dissolution; and another nineteen firms with "excessive concentration" were dismembered. The two largest Japanese holding companies, Mitsui and Mitsubishi, were divided into some two hundred successor firms, and an accompanying divestiture program forced the sale of possibly half the 1945 paid-in value of all Japanese corporate securities.

During the 1950s and 1960s, there was a sharp divergence of concentration trends in Japan and the United States, drifting downward in Japan while rising in the United States (even though the U.S. economy is 40 percent larger than Japan's). At the same time, Japanese industrialists stubbornly resisted government efforts to promote oligopoloid giantism, by limiting major fields to one or a few "national champions." In automobiles, for example, Mr. Honda and other entrepreneurs ignored government efforts to restrict the field to just two firms. In machine tools, the president of one successful firm recounts that Japan's Ministry of International Trade and Industry (MITI) "told us to form into larger companies. We told them 'the hell with that' and refused."[18] More generally, as George Gilder pointed out,

> in every one of the industries in which the Japanese did prevail, they generated *more* companies and *more* intense domestic competition than the United States. . . .The Japanese created three times as many shipyards, four times as many steel firms, five times as many motorcycle manufacturers, four times as many automobile firms, three times as many makers of consumer electronics, and six times as many robotics companies as the United States.

Gilder observed that it was established American oligopolies in consumer electronics, such as RCA, Zenith, and General Electric, that gave up the color television industry to the more entrepreneurial Japanese. It was vertically integrated American automobile firms that succumbed to the more fragmented and entrepreneurial Japanese automobile industry.[19]

Perhaps, then, the Japanese succeeded in the old-fashioned way—not by elaborate collaborationist schemes or megamergers, but by building better products, of higher quality, in modern state-of-the-art facilities, and by competing intensely at home to produce them more efficiently. Perhaps the most significant policy lesson of the Japanese "miracle" is that the best way to promote competitiveness abroad is to maintain a competitive industry structure at home.[20]

4. In the final analysis, the Achilles heel of coalition capitalism, "industrial policy," and other modern-day reincarnations of the economic syndicalist schemes of the 1930s is the specter of interest-group politics. As Henry C. Simons incisively asserted, "Bargaining organizations will contest over the division of the swag, but we commonly overlook the fact that they have large common interests as against the community and that every increase of monopoly power on one side serves to strengthen and implement it on the other." To facilitate such coalescing power, Simons warned, would be to "drift rapidly into political organization along functional, occupational lines—into a miscellany of specialized collectivisms, organized to take income away from one another and incapable of acting in their own common interest or in a manner compatible with general prosperity."[21]

THE NEOCONSERVATIVE CHALLENGE: "ECONOMIC DARWINISM"

From the right of the political spectrum, neo-Darwinists deprecate America's traditional antitrust philosophy and urge its replacement with a policy of untrammeled laissez-faire. This, they claim, will bolster the nation's economic performance by encouraging the natural selection of the economically fittest.

Professor Robert H. Bork, perhaps the most sophisticated advocate of this view, sees a striking analogy between a free-market system and the Darwinian theory of physical evolution:

The familiarity of that parallel, and the overbroad inferences sometimes drawn from it, should not blind us to its important truths. The environment to which the business firms must adapt is defined, ultimately, by social wants and the social costs of meeting them. The firm that adapts to the environment better than its rivals tends to expand. The less successful firm tends to contract—perhaps, eventually, to become extinct. The Stanley Steamer and the celluloid collar have gone the way of the pterodactyl and the great ground sloth, basically for the same reasons. Since coping successfully with the economic environment also forwards consumer welfare (except in those cases that are the legitimate concern of antitrust), economic and natural selection has normative implications that physical natural selection does not have. At least there seems to be more reasons for enthusiasm about the efficient firm than about the most successful physical organisms, the rat and the cockroach, though this view is, no doubt, parochial.[22]

There is little justification, Bork argues, to interfere with the "natural" operation of a free-market system. Laissez-faire, he believes, can be trusted to produce optimal results:

> It is a common observation of biologists that whenever the physical environment provides a niche capable of sustaining life, an organism will evolve or adapt to occupy the place. The same is true of economic organisms, hence the fantastic proliferation of forms of business organization, products, and services in our society. . . .To expand, or even to survive, every firm requires a constant flow of capital for employees' wages, raw material, capital investment, repairs, advertising, and the like. When the firm is relatively inefficient over a significant time period, it represents a poorer investment and greater credit risk than innumerable alternatives. If the firm is dependent upon outside capital, the firm must shrink, and, if no revival in its fortunes occurs, die.

Monopoly or market power, according to Bork, are of little social concern because neither is endowed with any significant durability: "A market position that creates output restriction and higher prices will always be eroded if it is not based upon superior efficiency."[23] Of course, if it is based on superior efficiency, Bork would say that it serves the best interest of society and should therefore be immune from antitrust attack.

In his view, then, a firm achieves market power or giant size because of superior efficiency, and it would be wrongheaded for public policy to punish such a firm for its success. The winner of the race deserves the prize: Punishing industrial size, the economic Darwinists contend, would result in a diminution of consumer welfare while destroying economic incentives to excel.

This doctrine, which was enshrined as official U.S. government policy during the Reagan years (1980–1988), suffers from a number of serious defects.

1. It is based on a *post hoc ergo propter hoc* fallacy: The mere existence of a monopolist, oligopolist, or conglomerate giant proves that it must have achieved its market position exclusively or predominantly because of superior

performance. This is no more than an assertion—devoid, more often than not, of any empirical substantiation.

2. Although economic Darwinism makes superior economic performance the centerpiece of its policy position, its advocates concede that such economic performance is difficult, if not impossible, to measure scientifically. Professor Bork, for example, admits that

> the real objection to performance tests and efficiency defenses in antitrust law is that they are spurious. They cannot measure the factors relevant to consumer welfare, so that after the economic extravaganza was completed we should know no more than before it began. In saying this I am taking issue with some highly qualified authorities. Carl Kaysen and Donald Turner proposed that "an unreasonable degree of market power as such must be made illegal," and they suggested that all the relevant dimensions of performance be studied. Their idea, essentially, is that a court or agency determine, through a litigation process, whether there exists in a particular industry a persistent divergence between price and marginal cost; the approximate size of the divergence; whether breaking up, say, eight firms into sixteen would reduce or eliminate the divergence; and whether any significant efficiencies would be destroyed by the dissolution.[24]

Bork seems to despair about the possibility of measuring performance, even though he posits superior performance as the ultimate goal of economic policy.

3. The new economic Darwinism is concerned primarily with static, managerial efficiency rather than with dynamic social efficiency. It thus falls victim to the sin of suboptimization. Perhaps the relevant policy question is not whether General Motors produces gas-guzzling automobiles at the lowest average cost but whether it should be producing such automobiles at all. Perhaps the relevant policy question is whether the tight oligopoly that dominates the U.S. automobile industry is more likely than a more competitively structured automobile industry to "reinvent the automobile," for example, to come up with a safer, more fuel-efficient, more pollution-free prototype.

4. The new economic Darwinism assumes that any firm that no longer delivers superior performance will automatically be displaced by newcomers. This neglects the ability of powerful established firms to erect private storm shelters—or to lobby government to build public storm shelters for them—in order to shield themselves from the Schumpeterian gales of creative destruction. It ignores the difference between the legal freedom of entry and the economic realities deterring the entry of potential newcomers into concentrated industries. The American airline industry provides a graphic case in point: Having deregulated the field in order to enable competition to function, the Reagan administration (true to its laissez-faire principles) refrained from challenging the anticompetitive merger mania that consumed the industry and transformed it into an oligopoly of giants—carriers that now dominate hub cities across the country and whose deteriorating performance is precipitating cries to reregulate the field.

5. Proponents of the new laissez-faire also overestimate the disciplining effect of international competition in the "new" global market. They underestimate the capacity of powerful domestic firms, once dominant, to join with their foreign rivals to neutralize the threat of international competition from abroad.[25] In automobiles, for example, an exploding web of "joint venture" agreements has organizationally conjoined the Big Three U.S. producers with virtually every one of their major foreign rivals. Similarly in airlines, merger-induced domestic concentration and monopolistic hub control have recently been buttressed by a spate of joint operating agreements between the dominant American airline giants and their very largest potential foreign rivals. As Sir Alfred Mond, organizer of the giant British chemical combine Imperial Chemical Industries, pointed out in the 1920s, "The old idea of the heads of great businesses meeting each other with scowls and shaking each other's fists in each other's faces : nd. . .trying to destroy each other's business may be very good on the films, but it does not accord with any given facts."[26]

6. Economic Darwinists decry counterproductive and anticompetitive government policies that, they say, are the prime evil to be avoided. But this ignores the fact that government does not operate in a vacuum and that the anticompetitive government policies that they decry are the result of lobbying by powerful economic groups capable of capturing the power of the state. It ignores the fact that in a representative democracy, disproportionate economic size implies disproportionate influence in the political arena and that corporate giants can mobilize the vast political resources at their command—executives and labor unions, suppliers and subcontractors, governors and mayors, senators and representatives, Republicans and Democrats—in their efforts to pervert public policy to antisocial ends.[27]

Corporate bigness complexes, for example, can lobby the state to immunize them from foreign competition at exorbitant expense to the economy (as the steel industry has done for three decades). They can obtain hundreds of millions of dollars in tax favors, tax loopholes, free land, and interest-free government loans by whipsawing states and communities against one another in bidding wars for the location of plants (as United Airlines and Northwest Airlines have recently done in Indiana and Minnesota, respectively). And if they are big enough and incompetent enough, corporate giants can enjoy the ultimate perversion of private enterprise—government bailouts—because they are believed to be too big to be allowed to fail. As the Chrysler, Lockheed, and Continental Illinois episodes illustrate, giant firms can survive, not because they are better, but because they are bigger—not because they are fitter, but because they are fatter.

Because they are willing to tolerate private economic power while ignoring its damaging political ramifications, today's economic Darwinists are like Henry David Thoreau's neighbors who, he observed, "invite the devil in at every angle and then prate about the garden of Eden and the fall of man."

7. Economic Darwinism fails to make the critical distinction between individual freedom and a free economic *system*. As Jeremy Bentham pointed out over two hundred years ago, it is not enough to shout laissez-faire and op-

pose all government intervention: "To say that a law is contrary to natural liberty is simply to say that it is a law: for every law is established at the expense of liberty—the liberty of Peter at the expense of the liberty of Paul."[28] If individual rights were absolute and unlimited, they would, as Thomas Hobbes pointed out long ago, mean license to commit the grossest abuses against society, including destroying the freedom of others. As Lord Robbins suggested, public policy must "distinguish between (government) interventions that destroy the need for intervention and interventions that tend to perpetuate it." Viewed in this light, the Darwinist admonition not to penalize the winner of the race is irrelevant to public policy purposes. Rather, the relevant problem is how to reward the winner without including in its trophy the right to impose disabling handicaps on putative competitors, or the power to determine the rules by which future races shall be run, or the discretion to terminate the institution of racing altogether.

8. Finally, the new economic Darwinism fails to appreciate the linkage between industrial structure and market behavior and the ultimate consequences for economic performance. As John Bates Clark warned more than a half-century ago,

> In our worship of the survival of the fit under free natural selection we are sometimes in danger of forgetting that the conditions of the struggle fix the kind of fitness that shall come out of it; that survival in the prize ring means fitness for pugilism, not for bricklaying nor philanthropy; that survival in predatory competition is likely to mean something else than fitness for good and efficient production; and that only from strife with the right kind of rules can the right kind of fitness emerge. Competition is a game played under rules fixed by the state to the end that, so far as possible, the prize of victory shall be earned, not by trickery or mere self-seeking adroitness, but by value rendered. It is not the mere play of unrestrained self-interest; it is a method of harnessing the wild beast of self-interest to serve a common good—a thing of ideals and not of sordidness. It is not a natural state, but like any other form of liberty, it is a social achievement, and eternal vigilance is the price of it.[29]

III. CONCLUSION

The overarching purpose of American antitrust policy is to sustain the structural preconditions for the maintenance of effective competition. The purpose is to substitute the decentralized decision-making system of the competitive market for central planning, whether by the state or, alternatively, by private monopolists, oligopolists, or cartels. The purpose is to substitute regulatory control by the invisible hand of the competitive market for the visible fist of corporate power blocs operating as private governments—free from competitive checks and balances, free from accountability, and operating with no compulsion or assurance that their decisions will promote the public interest.

History is replete with ironies, and American antitrust policy is no exception. One can only speculate on how different the U.S. industrial landscape would be today if the antitrust authorities had not been frustrated in major antimonopoly suits against such dominant firms as U.S. Steel, American Can, United Shoe Machinery, Eastman Kodak, International Harvester, and IBM.[30]

What if the triopoly in American automobiles had been restructured pursuant to an antitrust suit that was rumored in the early 1960s but that was never filed? Would a more competitively structured domestic automobile industry have been so vulnerable to foreign competition as it was during the 1970s and 1980s? Would it have suffered such catastrophic financial losses, plant closings, and worker layoffs? Would it have become so technologically moribund? Or instead, would it have been more innovative, more efficient, more productive—in a word, more competitive at home and abroad?

What if the United States Steel Corporation had been reorganized as the Justice Department sought in its antitrust suit eighty years ago? Would not the American steel industry have been rendered far more competitive in structure? Might not a vigorously competitive domestic structure have rendered American steel firms stronger and more capable of resisting the onslaught of foreign competition? Might not jobs have been gained rather than lost? Might not the industry have been more efficient at home and more competitive abroad?

The contrasting outcomes in two recent Section 2 cases underscore this point: In 1982 IBM was widely considered to have "won" when the Reagan Justice Department dropped its monopolization suit against the firm, whereas AT&T was considered to have "lost" its monopolization suit when it agreed in the same year to a consent decree mandating the divestiture of its local operating companies. Yet IBM's subsequent performance has been abysmal, manifested by a loss in its stock market value equivalent to the gross domestic product of Sweden.[31] AT&T and the divested "Baby Bells," on the other hand, have performed superbly following their antitrust "loss" and dismemberment, recording stock market gains of 272 percent in value over the 1982–1992 period.[32]

Similarly, what if the antitrust agencies had enforced Section 7 of the Clayton Act to halt the merger–takeover–buyout mania of the 1980s? Instead of a trillion dollars being devoted during the decade to shuffling paper securities, might not that trillion dollars have instead been invested directly into new research and development, bringing new products to market, building new state-of-the-art production facilities, and creating new jobs? Might this not have enhanced U.S. global competitiveness, rather than submerging corporate America in a sea of debt and bankruptcies while its fiercest foreign rivals continued to forge ahead?[33] (see Figure 12-2).

The evidence is clear: A decentralized economic power structure—the root principle of antitrust policy—does not have to be sacrificed in order to obtain efficiency in production or innovation in technology. One can hardly disagree with the central finding by Michael Porter—based on his exhaustive analysis of the competitive advantage of nations, conducted country by coun-

FIGURE 12-2 Public policy alternatives: the road ahead.

| Security Through Government Control | Free Enterprise | Security Through Monopoly |

Source: Thurman W. Arnold, *Cartels or Free Enterprise?* Public Affairs Pamphlet 103, 1945. Reproduced by courtesy of Public Affairs Commission, Inc.

try and industry by industry—that the single most important source of international competitiveness is intense competition in the home market. States Porter, "What is needed today in American industry is not less competition but more. Instead of relaxing antitrust enforcement, we should be tightening it." The lesson of American industrial history, he concludes, affirms the wisdom of recommitting ourselves to one of America's unique national values—competition enforced by law. As uncomfortable as it may be, it "remains the only way that American industry will ever truly prosper."[34]

NOTES

1. *The Federalist* (Middletown, CN: Wesleyan University Press, 1961).
2. Quoted in Hans B. Thorelli, *The Federal Antitrust Policy: Origination of an American Tradition* (Baltimore: Johns Hopkins University Press, 1955), p. 180. For additional, excellent surveys of the problem from a va-

riety of perspectives, see James May, "Antitrust in the Formative Era: Political and Economic Theory in Constitutional and Antitrust Analysis, 1880–1918," *Ohio State Law Journal* 50 (1989): 257–395; and David Millon, "The Sherman Act and the Balance of Power," *Southern California Law Review* 61 (1988): 1219–1292.

3. *United States* v. *United Shoe Machinery Corp.*, 110 F. Supp. 295, (D. Mass. 1953).

4. House Committee on Banking, Finance, and Urban Affairs, Subcommittee on Economic Stabilization, *Hearings: Industrial Policy, Part 1*, 98th Cong., 1st sess., 1983, pp. 173–178.

5. Lester C. Thurow, *The Zero-Sum Solution* (New York: Simon & Schuster, 1985), pp. 263–266.

6. Ibid., pp. 264ff.

7. Lester C. Thurow, "Abolish the Antitrust Laws," *Dun's Review*, February 1981, p. 72.

8. Joseph L. Bower, "The Case for Building More IBM's," *New York Times*, February 16, 1986, sec. 3, p. 2.

9. Thurow, *Zero-Sum Solution*, p. 182.

10. Quoted in Leverett S. Lyon et al., *The National Recovery Administration: An Analysis and Appraisal* (Washington, DC: Brookings Institution, 1935), p. 709. For an equally critical appraisal, see Clair Wilcox, *Public Policies Toward Business* (Homewood, IL: Irwin, 1971), p. 680.

11. See Walter Adams and James W. Brock, *The Bigness Complex* (New York: Pantheon Books, 1986), chap. 21.

12. Ibid., chap. 24.

13. "GM: What Went Wrong?" *Business Week*, March 16, 1987, p. 110.

14. Generally, see Paul Carroll, *Big Blues: The Unmaking of IBM* (New York: Crown, 1993).

15. *Economist*, November 30, 1991, p. 81.

16. Walter Adams and James W. Brock, "The Bigness Mystique and the Merger Policy Debate: An International Perspective," *Northwestern Journal of International Law & Business*, Spring 1988, p. 1.

17. Ibid, pp. 36–43.

18. David Friedman, *The Misunderstood Miracle: Industrial Development and Political Change in Japan* (Ithaca, NY: Cornell University Press, 1988), p. 100.

19. George Gilder, *Microcosm* (New York: Simon & Schuster, 1989), p. 341 (italics in original).

20. For extensive documentation of this point, see Michael E. Porter, *The Competitive Advantage of Nations* (New York: Free Press, 1990). See also Karl Zinsmeister, "MITI Mouse: Japan's Industrial Policy Doesn't Work," *Policy Review*, Spring 1993, pp. 28–35.

21. Henry C. Simons, *Economic Policy for a Free Society* (Chicago: University of Chicago Press, 1948), pp. 119, 219.

22. Robert H. Bork, *The Antitrust Paradox* (New York: Basic Books, 1978), p. 118.

23. Ibid., pp. 119, 133.

24. Ibid., pp. 124–125.

25. See Walter Adams and James W. Brock, "Joint Ventures, Antitrust, and Transnational Cartelization," *Northwestern Journal of International Law & Business*, Winter 1991, pp. 433–483.

26. Quoted in George Stocking and Myron Watkins, *Cartels in Action* (New York: Twentieth Century Fund, 1946), p. 429.

27. For an extended debate on this point, see Walter Adams and James W. Brock, *Antitrust Economics on Trial: A Dialogue on the New Laissez-Faire* (Princeton, NJ: Princeton University Press, 1991), pp. 117–127.

28. J. Bowring, ed., *The Works of Jeremy Bentham* (Edinburgh: W. Tait, 1843), vol. 3, p. 185.

29. John B. Clark, *The Control of Trusts* (New York: Macmillan, 1912), pp. 200–201.

30. For detailed explorations, see William Comanor and F. M. Scherer, "Rewriting History: The Early Sherman Act Monopolization Cases," and William G. Shepherd, "The Lawyers Win, the Company Suffers: IBM, General Motors, and Section 2 Since the 1960s," both papers presented at the annual meetings of the Industrial Organization Society, New Orleans, January 1993.

31. Carroll, *Big Blues*, pp. 6, 265.

32. Randall Smith, "The Biggest Stocks Aren't Always a Safe Bet," *Wall Street Journal*, November 19, 1992, pp. C1, C14; James B. Stewart, "Whales and Sharks: The Unexpected Fates of IBM and A.T.&T. May Offer a Lesson to the Clinton Justice Department," *New Yorker*, February 15, 1993, pp. 37–43.

33. Walter Adams and James W. Brock, *Dangerous Pursuits: Mergers and Acquisitions in the Age of Wall Street* (New York: Pantheon Books, 1989).

34. Michael E. Porter, "Japan Isn't Playing by Different Rules," *New York Times*, July 22, 1990, p. F13. See also Michael E. Porter, *The Competitive Advantage of Nations* (New York: Free Press, 1990).

SUGGESTED READINGS

Adams, Walter, and James W. Brock. *The Bigness Complex: Industry, Labor and Government in the American Economy.* New York: Pantheon Books, 1986.

———. *Antitrust Economics on Trial: A Dialogue on the New Laissez-Faire.* Princeton, NJ: Princeton University Press, 1991.

Adams, William J., and Christian Stoffaes, eds. *French Industrial Policy.* Washington, DC: Brookings Institution, 1986.

Bork, Robert H. *The Antitrust Paradox.* New York: Basic Books, 1978.

Brandeis, Louis D. *The Curse of Bigness.* New York: Viking, 1934.

Commission of the European Communities. *Report on Competition Policy.* Annual.

De Jong, H. W., ed. *The Structure of European Industry.* 3rd rev. ed. Boston: Kluwer Academic Publishers, 1993.

Dirlam, Joel B., and Alfred E. Kahn. *Fair Competition: The Law and Economics of Antitrust Policy.* Ithaca, NY: Cornell University Press, 1954.

Edwards, Corwin D. *Maintaining Competition.* New York: McGraw-Hill, 1949.

Fetter, Frank A. *The Masquerade of Monopoly.* New York: Harcourt Brace, 1931.

Fox, Eleanor M., and Lawrence A. Sullivan. *Cases and Materials on Antitrust.* St. Paul: West Publishing, 1989.

Friedman, David. *The Misunderstood Miracle: Industrial Development and Political Change in Japan.* Ithaca, NY: Cornell University Press, 1988.

Galbraith, John Kenneth. *The New Industrial State.* Boston: Houghton Mifflin, 1971.

Graham, Otis L., Jr. *Losing Time: The Industrial Policy Debate.* Cambridge, MA: Harvard University Press, 1992.

Greer, Douglas F. *Business, Government, and Society.* 3rd ed. New York: Macmillan, 1993.

Hadley, Eleanor M. *Antitrust in Japan.* Princeton, NJ: Princeton University Press, 1970.

Hayek, F. A. *The Road to Serfdom.* Chicago: University of Chicago Press, 1944.

Heilbroner, Robert. *21st Century Capitalism.* New York: Norton, 1993.

Hofstadter, Richard. *Social Darwinism in American Thought.* Boston: Beacon Press, 1944.

Jacquemin, Alexis, et al. *Merger & Competition Policy in the European Community.* Oxford: Basil Blackwell, 1990.

Kovaleff, Theodore P. *The Antitrust Impulse.* 2 vols. Armonk, NY: M. E. Sharpe, 1993.

Lippmann, Walter. *The Good Society.* Boston: Little, Brown, 1937.

Machlup, Fritz. *The Political Economy of Monopoly.* Baltimore: Johns Hopkins University Press, 1952.

Peterson, Wallace C., ed. *Market Power and the Economy.* Boston: Kluwer Academic Publishers, 1988.

Porter, Michael E. *The Competitive Advantage of Nations.* New York: Free Press, 1990.

Reich, Robert B. *The Next American Frontier.* New York: Times Books, 1983.

Samuels, Warren J., ed. *Fundamentals of the Economic Role of Government.* Westport, CT: Greenwood Press, 1989.

Scherer, F. M., and David Ross. *Industrial Market Structure and Economic Performance.* 3rd ed. Boston: Houghton Mifflin, 1990.

Shepherd, William G. *The Economics of Industrial Organization.* 3rd ed. Englewood Cliffs, NJ: Prentice Hall, 1990.

Simons, Henry C. *Economic Policy for a Free Society.* Chicago: University of Chicago Press, 1948.

Smith, Adam. *The Wealth of Nations.* New York: Modern Library, 1937.

Stocking, George W., and Myron W. Watkins. *Monopoly and Free Enterprise.* New York: Twentieth Century Fund, 1951.

Thorelli, Hans B. *The Federal Antitrust Policy: Origination of an American Tra-dition.* Baltimore: Johns Hopkins University Press, 1954.

Thurow, Lester C. *The Zero-Sum Solution: Building a World-Class American Economy.* New York: Simon & Schuster, 1985.

Wills, Robert L., Julie A. Caswell, and John D. Culbertson, eds. *Issues After a Century of Federal Competition Policy.* Lexington, MA: Lexington Books, 1987.

NAME INDEX

SUBJECT INDEX